ECOLOGICAL ECONOMICS OF SUSTAINABLE WATERSHED MANAGEMENT

ADVANCES IN THE ECONOMICS OF ENVIRONMENTAL RESOURCES

Series Editor: Richard Howarth

Volume 5: *Perspectives on Climate Change: Science, Economics, Politics, Ethics*
Edited by Walter Sinnott-Armstrong & Richard B. Howarth

Volume 6: *Living On The Edge: Economic, Institutional and Management Perspectives on Wildfire Hazard in the Urban Interface*
Edited by Austin Troy & Roger G. Kennedy

ADVANCES IN THE ECONOMICS OF ENVIRONMENTAL RESOURCES VOLUME 7

ECOLOGICAL ECONOMICS OF SUSTAINABLE WATERSHED MANAGEMENT

EDITED BY

JON D. ERICKSON
Rubenstein School of Environment and Natural Resources, University of Vermont, USA

FRANK MESSNER
Department of Economics, Helmholtz Centre for Environmental Research - UFZ, Leipzig, Germany

IRENE RING
Department of Economics, Helmholtz Centre for Environmental Research - UFZ, Leipzig, Germany

Emerald

United Kingdom – North America – Japan
India – Malaysia – China

Emerald Group Publishing Limited
Howard House, Wagon Lane, Bingley BD16 1WA, UK

Reprints and permission service
Contact: booksandseries@emeraldinsight.com

British Library Cataloguing in Publication Data
A catalogue record for this book is available from the British Library

ISBN: 978-0-7623-1448-5

CONTENTS

LIST OF CONTRIBUTORS

James W. Boyd	Resources for the Future, Washington, DC, USA
Ines Dombrowsky	Department of Economics, Helmholtz Centre for Environmental Research – UFZ, Leipzig, Germany
Jon D. Erickson	Rubenstein School of Environment and Natural Resources, University of Vermont, Burlington, VT, USA
Horst Gömann	Institute of Rural Studies, FAL Federal Agricultural Research Centre, Braunschweig, Germany
John Gowdy	Department of Economics, Rensselaer Polytechnic Institute, Troy, NY, USA
Myrna H. Hall	State University of New York, College of Environmental Science and Forestry, Syracuse, NY, USA
Bernd Hansjürgens	Department of Economics, Helmholtz Centre for Environmental Research – UFZ, Leipzig, Germany
Caroline M. Hermans	United States Geological Survey, Menlo Park, CA, USA
Sylvia Herrmann	Institut für Landwirtschaftliche Betriebslehre (410a), Universität Hohenheim, Schloss Osthof Süd, Stuttgart, Germany
Bongghi Hong	State University of New York, College of Environmental Science & Forestry, Syracuse, NY, USA

Michael Kaltofen	WASY Gesellschaft für wasserwirtschaftliche Planung und Systemforschung mbH, NL Dresden, Dresden, Germany
Melinda Kane	1166 N. Country Club Drive, Niskayuna, NY, USA
Hagen Koch	Brandenburgische Technische Universität Cottbus, Lehrstuhl für Hydrologie und Wasserwirtschaft, Cottbus, Germany
Peter Kreins	Institute of Rural Studies, FAL Federal Agricultural Research Centre, Braunschweig, Germany
Ralf Kunkel	Systems Research and Technological Development (FZJ-STE), Research Centre Jülich, Jülich, Germany
Karin E. Limburg	State University of New York, College of Environmental Science & Forestry, Syracuse, NY, USA
Frank Messner	Department of Economics, Helmholtz Centre for Environmental Research – UFZ, Leipzig, Germany
Audra A. Nowosielski	Private Economist, Albany, NY, USA
Daniel Petry	Global Water System Project, International Project Office, Bonn, Germany
John M. Polimeni	Albany College of Pharmacy, Albany, NY, USA
Irene Ring	Department of Economics, Helmholtz Centre for Environmental Research – UFZ, Leipzig, Germany
Thomas G. Schmidt	Institute of Rural Studies, FAL Federal Agricultural Research Centre, Braunschweig, Germany
Leonard A. Shabman	Resources for the Future, Washington, DC, USA

Kurt Stephenson	Department of Agricultural and Applied Economics, Virginia Polytechnic Institute and State University, Blacksburg, VA, USA
Austin Troy	Rubenstein School of Environment and Natural Resources, University of Vermont, Burlington, VT, USA
Herwig Unnerstall	Department of Environmental and Planning Law, Helmholtz Centre for Environmental Research – UFZ, Leipzig, Germany
Frank Wendland	Systems Research and Technological Development (FZJ-STE), Research Centre Jülich, Jülich, Germany

PREFACE

The term sustainability has evolved rapidly over the past two decades beyond the general definition of the 1987 report *Our Common Future* (the Brundtland report) which defined "sustainable development" as that which "meets the needs of the present without compromising the ability of future generations to meet their own needs." Although the Brundtland definition provided a common rallying point for all those concerned with the environmental and social consequences of global economic development, it quickly became apparent that there were deep divisions among the advocates of sustainability.

Two broad outcomes of the sustainability debate are: (1) a large and growing literature on sustainability theory, including weak versus strong sustainability, models of intergenerational welfare, well-being versus per capita income measures; and (2) a large number of case studies involving more practical on-the-ground analyses of sustainability. The papers in this volume are representative of the second vision of sustainability research. The diverse articles in this volume illustrate the difficulty of "rules for sustainability." Sustainability is an on-going process more than a fixed set of objectives to be achieved. Sustainability calls for continued adjustment and reassessment as economic, environmental, social, and technological conditions change. A simple and universal indicator of sustainability is an unachievable goal.

The papers in this volume illustrate the power of a scientific approach to ecological economics. Good science is a careful blend of theory and empirical testing. Theory without empirical grounding is of no practical value and random case studies without a theoretical context are not generalizable. The back-and-forth interplay between theory and evidence is apparent in the modeling exercises, evaluation studies, and policy design described in this book.

Watershed management has been chosen as a concrete focus to illuminate the facets of sustainability. It requires both a deep understanding of the natural processes in watersheds and of the societal processes which strongly depend on the natural watershed services. Furthermore, country-specific governance structures need to be considered to fine-tune the design of sustainable watershed policies in order to approach an interaction of society and nature, which ensures a long-term use of water resources without adverse effects on society and the environment. This book has accepted the challenge to tackle the complex scientific underpinning of sustainable watershed analysis and management and will reveal basic ecological economic knowledge and methodological approaches in this prominent field of research.

John Gowdy and Bernd Hansjürgens
Troy, New York, USA and Leipzig, Germany
October 2006

ACKNOWLEDGMENTS

This book originated from a collaboration between the Helmholtz Centre for Environmental Research (UFZ, Leipzig, Germany) and the Ecological Economics graduate programs at Rensselaer Polytechnic Institute (RPI, Troy, NY, USA) and the University of Vermont (UVM, Burlington, VT, USA). We thank the German Academic Exchange Council (DAAD) and the U.S. National Science Foundation (NSF) for their generous support through the Project Based Personnel Exchange Program, allowing for mutual visits of project partners in the U.S. and Germany. We also thank the Hudson River Foundation whose support initiated and sustained much of the U.S. research. In its early days, Helga Horsch (UFZ) and especially Sabine U. O'Hara (formerly RPI, now Roanoke College, Salem, USA) strongly supported the development of close links between ecological economists at RPI and UFZ. It soon became clear that sustainable watershed management and resource conservation were foci in both research groups and a sound basis for benefiting from comparative research.

Special thanks go to Darwin Hall who initiated and constructively encouraged our book project with Elsevier Science as part of the *Advances in the Economics of Environmental Resources* Series, and Richard Howarth who continued the job and completed this project with us. We also greatly acknowledge the support and patience of our partners at Elsevier Science itself, including Joy Ideler, Mark Newson, Lisa Muscolini, and Monique Wilburs. Abigail Hood and Sandra Schneider provided valuable support in the course of manuscript preparation. Last, but not least, we gratefully acknowledge the cooperation from all contributors, both authors and reviewers of articles.

Irene Ring, Jon D. Erickson, and Frank Messner
Leipzig, Germany and Burlington, Vermont, USA
October 2006

PART I:
INTRODUCTION

ECOLOGICAL ECONOMICS AT THE WATERSHED SCALE: COMPARING AND CONTRASTING THE UNITED STATES AND GERMAN EXPERIENCES AND APPROACHES

Jon D. Erickson, Frank Messner and Irene Ring

Over the past three decades ecological economics has emerged as a coherent transdisciplinary approach to environmental problem solving. However, its evolution has been quite dissimilar in different parts of the world. In the US and UK, ecological economics evolved as a critique of and alternative to a comparatively strict application of economic theory to environmental decision making. In particular, the narrow application of benefit–cost analysis often reduced environmental decisions to one metric within a single value system (the market economy). The attractiveness of these traditional economic approaches to environmental policy has always been their "one size fits all" approach. No matter what the problem faced, the same methods were applied with a primary goal of cost effectiveness. But it has become increasingly clear that the ease of application of a strict economic approach is outweighed by its failure to capture the social and environmental contexts and realities of specific environmental problems. In contrast, ecological economics has been more problem-oriented, incorporating multiple stakeholder and disciplinary perspectives in specific contexts to shape the

Ecological Economics of Sustainable Watershed Management
Advances in the Economics of Environmental Resources, Volume 7, 3–7

ISSN: 1569-3740/doi:10.1016/S1569-3740(07)07001-0

methods that define policy choices. Furthermore, ecological economic approaches involve multiple metrics, multiple points of view, and evolutionary and flexible policy recommendations.

While the picture looks totally different in continental Europe, the outcome with regard to the increasing application of ecological economics is quite similar. European environmental policy making in the past mostly relied on natural science and engineering knowledge to support decision processes. Environmental economics had a much more restricted influence, e.g., related to the design of economic instruments in the course of introducing the German Waste Water Levy Act in 1976 and following amendments. However, during the recent decade there developed the recognition that socio-economic expertise must be included as well at a much more general scale – as a complement to the technically oriented existing supporting tools and concepts. Contrary to the traditional economic school of thought, ecological economics approaches are based on the premise that the economy is a dependent part of the natural world and cannot survive without receiving continuous ecosystem services – hence its general orientation is much more open to develop integrated approaches together with scientists of other disciplines. As a result of this advantage, ecological economic concepts gained in importance in practical application in continental Europe during the last decade.

At the watershed scale, conflicts over water and land resources are inherently multi-attribute, multi-stakeholder, and multi-discipline decision problems. Watershed systems – from those with many small tributaries to large-scale lake systems and river basins – provide direct inputs to economic processes, serve as waste sinks for economic output, and provide ecosystem services that make life possible. For instance, a single watershed may provide water for consumption, transport, and recreational use, a depository for treated sewage and industrial pollutants, flood control services, as well as a multitude of aquatic and terrestrial habitats that form the source of life and sanctuary for diverse species. A watershed perspective is more holistic than standard analysis of use or exchange value, explicitly recognizing emergent properties of the system, feedback loops between natural and societal system components, and conflicts among competing uses. Conflicts arise over the use and allocation of these resources from diverse actors in watershed economies. Characterizing the degree of trade-offs or incommensurability between the conflicting goals that often arise requires expertise drawn from diverse disciplinary perspectives.

The challenge to ecological economics at the watershed scale is thus to provide an analytical framework for decision support that is both normative and positive. The description of the system (positive analysis) should be open to multiple metrics and points of view, while advice to decision makers (normative analysis) must be grounded in sound methods and an accepted valuation framework. Since ecological economics takes a problem-oriented approach, a further challenge is to investigate the transferability of integrated methods across diverse environmental and social contexts.

The objective of this book is to present new developments in and new approaches to watershed management grounded in principles of ecological economics. Methods are developed and applied to case studies that emerged from a collaboration between the Helmholtz Centre for Environmental Research (UFZ, Leipzig, Germany) and the Ecological Economics graduate programs at Rensselaer Polytechnic Institute (RPI, Troy, NY, USA) and the University of Vermont (UVM, Burlington, VT, USA). Positive analysis is reflected in chapters on scenario analysis and integrated modeling, which offer approaches to understand and simulate complex watershed landscapes, economies, and environmental conflicts. Conflicts include US and German cases arranged around diverse goals of an integrated management of watersheds and river basins, including economic output, water quality, surface and ground water availability, and land and nature conservation. Normative analysis then builds from these detailed studies to evaluate management alternatives, participatory processes, and ultimate policy design. The book would not be complete without a suitable introduction to the history of watershed and river basin management in the United States and Germany, setting a context to compare different governance structures, policy strategies, and policy instruments.

Despite the distinct differences in context there are many opportunities for transferability of methods and general ecological economic conclusions – provided that basic data, modeling tools, and comparable institutional conditions are available. For instance, in Part III of the book, scenario analytic tools and modeling techniques are described and applied in context-specific case studies. General conclusions can be drawn on how to link different disciplinary modeling approaches, how to construct multiple future scenarios with long time horizons, or how to incorporate context-specific factors in modeling approaches. Part IV of the book addresses evaluation tools and participatory approaches to support decision making. The methodological contributions include participatory integrated assessment approaches for complex decision-making conditions, linking economic evaluation to large-scale water management modeling, and applying multi-criteria approaches as a substitute for or supplement to benefit–cost analysis. Each approach is in principle transferable to other locations and to other fields of environmental policy. Finally, Part V of the book offers US and German examples for innovative watershed policies and instruments. Beyond all doubt, the transferability of the features in this part of the book must be considered limited, because the success of specific innovative policies and instruments highly depend on the institutional and governance conditions of a country. However, the examples show how compensation schemes for ecological services, water quality emission rights, or water pricing instruments work under specific circumstances and what kind of bottlenecks they entail. This might at least stimulate innovative thinking about new policy solutions under different institutional conditions.

While the following 15 contributions in this book might be insufficient to deal with all socio-economic aspects of watershed and river basin management, there are five principal lessons learned we identify from the perspective of ecological economics.

First, watersheds have large spatial scales and are highly complex systems that require for its positive analysis an intelligent interdisciplinary analytical and modeling framework. Such a framework is not to be understood as the pure addition of many disciplinary approaches, but requires a common understanding and a common approach to simulate the complexity of the system in a proper way.

Second, context specificity is extremely important in watershed and river basin management. What is true or appropriate for one context could be totally false or inappropriate for another. Therefore, translating context conditions into the analytical and normative tools of watershed management is an often underestimated but vital prerequisite for successful decision support and management.

Third, dealing with multiple criteria in the decision support of watershed and river basin management is inevitable, because there is no unique measure that is able to reflect all aspects of economic efficiency, social equity, cultural value, and ecological sustainability – and all these aspects play a crucial role in watershed management.

Fourth, the concept of uncertainty analysis should become a central part of any scientific approach to support decision making in watershed and river basin management. Since the future development paths of the world, input data for models, as well as the values and preferences of systems of actors over time are all highly uncertain or even unknown, uncertainty needs to be considered in every single scientific research process. This means in its final consequence that optimal solutions to a problem can never be attained – only optimal under certain assumed conditions. Hence, we have to start looking for robust solutions that deliver acceptable outcomes for many possible boundary conditions with a high degree of uncertainty.

Finally, in order to do a good job as scientists in the support of watershed decision making, a transdisciplinary approach is absolutely essential. Only by involving decision makers and stakeholders in the research process can a sufficiently realistic picture of the institutional watershed context be attained as a starting point for research, and only such a participatory procedure opens up the possibility to generate a solid basis for science-policy cooperation.

These lessons learned can also be considered as guides for further research needs in the ecological economics of watershed management. As the contributions of this book show, some progress has been achieved in interdisciplinary research and modeling, context specificity, multi-criteria approaches, uncertainty analysis, and transdisciplinarity. However, we are not yet at the end of this research path and many challenges are ahead – among them is the foreseeable trend that water resources in the future will become a more and more contested good in many parts of the world, which will require new, innovative, and context-adjusted scientific approaches and policy solutions.

PART II: WATERSHED MANAGEMENT

RIVER BASIN MANAGEMENT IN GERMANY: PAST EXPERIENCES AND CHALLENGES AHEAD

Daniel Petry and Ines Dombrowsky

ABSTRACT

Given that the European Water Framework Directive (WFD) calls for the management of water resources at the river basin level, the German water sector, which has historically been dominated by the federal states and has been organized along administrative borders, is now challenged to be reorganized. The article introduces the German water sector, reviews past experiences with river basin management such as North Rhine–Westphalia's water associations, the river basin organizations of the former German Democratic Republic, and international river commissions, and addresses current challenges in connection with the implementation of WFD.

1. INTRODUCTION

In recent years, the concepts of river basin management (RBM) and integrated water resources management (IWRM) have gained increasing attention as strategies for sustainable resource use within a complex

Ecological Economics of Sustainable Watershed Management
Advances in the Economics of Environmental Resources, Volume 7, 11–42

ISSN: 1569-3740/doi:10.1016/S1569-3740(07)07002-2

multi-institutional regulatory context (e.g., Newson, 1992; GWP, 2000). RBM calls for the management of water resources at the catchment or river basin level, and thus mainly refers to a spatial or natural system integration of water management functions. In contrast, IWRM mainly focuses on a sectoral or social system integration among the various water using sectors, while it leaves the spatial organization of water management open. For instance, the global water partnership defines IWRM as "a process which promotes the coordinated development and management of water, land and related resources, in order to maximize the resultant economic and social welfare in an equitable manner without compromising the sustainability of vital ecosystems" (GWP, 2000, p. 22). The emphases of the two concepts differ; however, they clearly incorporate elements of each other (see also Huppert, 2005). A recent attempt to implement these concepts in the European Union is the European Union Water Framework Directive (WFD) of December 2000 (European Communities, 2000). The WFD requires its member states to implement a river basin approach in order to reach common environmental objectives for all surface waters and groundwater bodies in the EU.

Germany is a country in which the main responsibility for water resources management is with the federal states and municipalities. Hence, the German water sector is primarily organized along administrative and not along hydrological boundaries and within hierarchical orders. In view of Germany's federal structure, the WFD's RBM approach represents a significant challenge for Germany's water sector.

However, despite the federal organization of Germany's water sector, in a number of instances experiences with river basin approaches could be gained in the past. Due to the industrialization in the 19th century, water availability in sufficient quality and quantity became key factors of public health and economic development. This led to the foundation of water associations in the territory of today's North Rhine–Westphalia at the beginning of the 20th century. Until 1990, in the former German Democratic Republic river basin authorities were responsible for safeguarding water availability. Experiences with RBM were also gained in the management of international watercourses, such as the Rhine, the Elbe, the Odra, and the Danube rivers.

In the implementation of the WFD's RBM approach, the German federal states have opted for the so called 'cooperation model', which leaves the main responsibilities with the states, but sets up coordination mechanisms at the river basin level. The result is a complex multi-level institutional structure that seeks to satisfy jurisdictional and hydrological requirements.

This article introduces the organization of Germany's water sector (Section 2), assesses past experiences with RBM (Section 3), and discusses opportunities and constraints Germany faces in the implementation of the EU WFD (Section 4). In doing so, it also assesses the integrative character of the WFD and seeks to come to a first evaluation of the German implementation process. The article highlights the following aspects:

- While the WFD is ambitious and its implementation remains challenging for the EU member states, it still does not fully incorporate important water policy fields.
- The federal political structure of Germany hampers the implementation of the WFD's RBM approach.
- The evolving institutional arrangement based upon inter-state coordination mechanisms only partly bridges hydrological and institutional misfits, but might yet represent an appropriate adaptation to the existing water management system.

2. WATER MANAGEMENT IN GERMANY

2.1. Structures, Institutions, and Instruments

In line with Germany's federal constitution, the main responsibility for water resources management in Germany is with the federal states. The states have the legislative and institutional power in most fields of water policy, such as water supply, wastewater treatment, flood protection, environmental monitoring, and emission control, and each state has its own State Water Act. The national or federal level – led by the Federal Ministry of Environment – only provides the framework as laid out in the Federal Water Act (*Wasserhaushaltsgesetz*), the Waste Water Charges Act, the Drinking Water Ordinance, and the Waste Water Ordinance. Any standards, emission limits, or quality objectives defined under federal framework legislation have to be taken over by state legislation and are specified and differentiated according to the more specific water management needs at state level. To a growing extent, the federal level prepares the implementation of binding European directives in the field of water policy for the federal states. Table 1 provides an overview of the institutional structure of water management in Germany prior to the introduction of the EU WFD.

Water resources management is hierarchically organized, comprising the federal, state, county, and municipal level. Some federal states feature an

Table 1. Institutional Framework of Water Management in Germany prior to the EU Water Framework Directive.

Level	Legislation	Organizations (Examples)	Duties and Instruments
European Union	Nitrates Directive, Drinking Water Directive, Bathing Water Directive, Urban wastewater treatment Directive	European Commission European Parliament Council of the European Union	Legal initiatives by the European Commission are negotiated and adopted by the Council and the Parliament. The Commission controls member state implementation of EC legislation
Other international levels	Bi- or multilateral treaties, conventions and programs	International commissions for the protection of the large transboundary rivers Rhine (ICPR), Elbe (ICPE), Danube (ICPD)	Commission secretariats prepare, facilitate, and monitor action plans and programs in various fields of water resources management, e.g., water quality, flood prevention on an international level. Plans and programs are non-legally binding but strengthen political commitment of partners
National or federal level	Federal Water Act, Waste Water Charges Act, Waste Water Ordinance, Drinking Water Ordinance	Federal Government; Federal Ministries (e.g. for environment, transport infrastructure or public health)	In many fields of WRM framework legislation and regulations; more specific legislation and regulations under responsibility of federal

		Subordinate agencies with executive and/or advisory functions to the Ministries (e.g., Federal Environment Agency, Federal Waterways, and Shipping Administration)	states. Exemptions: e.g., Federal Waterways and Shipping Administration with far-reaching responsibilities for the management of navigable rivers
		Professional associations of practitioners and scientists in the water sector (DWA, DVGW, and BWK)	Provision of technical and scientific advice; close collaboration with the LAWA
Interstate level			
	Interstate treaties and provisions for common activities of federal states	Working Group of the Federal States on Water Problems (LAWA) as an official body with coordinative and advisory duties	Coordination of interstate cooperation in WRM; development and definition of non-binding standards, procedures, and guidelines
		Working groups for certain rivers, e.g., for the purification of the Elbe river (ARGE Elbe) as partly as statutory bodies with coordinative and facilitative duties	Coordination of federal state activities on specific issues in shared rivers (e.g., monitoring programs, action programs on pollution mitigation)
Federal states	State Water Acts, ordinances, and statutory orders in all fields of water management	State Government State Ministry with responsibility for environmental issues – the 'Supreme Water Authority' Subordinate environment or water agencies with executive functions (planning, monitoring, and preparation of authority decisions)	Supreme Water Authorities as key decision-makers (preparation of acts and ordinances, definition of quality and emission standards, and strategic management and planning); Subordinate agencies with advisory and executive functions (monitoring, licenses, and charges)

Table 1. (*Continued*)

Level	Legislation	Organizations (Examples)	Duties and Instruments
Local level	Regulations, e.g., on abstraction rights and charges, wastewater management	District, County and City councils – the 'Upper' and 'Lower Water authorities'	Upper and Lower Water Authorities decide on licenses, authorizations and charges, control water supply and wastewater treatment
		Technical departments of town or county administration with executive functions	
		Public, public–private or private bodies for water supply and wastewater treatment	
		Water and soil associations (landowners, municipalities, and/or other actors along a certain watercourse or within a certain territory)	Statutory bodies under supervision of the Lower Water Authorities; responsibility for land drainage, maintenance of smaller water courses, local flood protection measures, partly also for public water supply, and wastewater disposal

additional regional district level. At all levels, water authorities need to be distinguished from water agencies. While the first have decision-making responsibilities, the latter are charged with executive duties such as environmental and hydrologic monitoring. They are also provided with powers to sanction the authorities' decisions. In addition, there are special administrative bodies not mentioned in Table 1 for many different tasks such as reservoir management or navigation (see below).

A specific feature of the institutional arrangements within the water sector is the management of navigational issues. Whereas in most fields of water policy the federal level has framework responsibilities only and powers are in the hands of the federal state environmental administration, navigation, and navigability of rivers is under control of the Federal Waterways and Shipping Administration under the responsibility of the Federal Ministry of Transport. With 19,000 employees and an annual budget of half a billion Euro (excluding budgets for watercourse and canal development) the Waterways and Shipping Administration forms the largest branch within the water management administration (Kahlenborn & Kraemer, 1999, p. 128). Therefore, decisions on navigational issues are made under the responsibility of the Federal Ministry of Transport and thus outside the institutional arrangement of water resources management dominated by the state ministries of the environment. Very often this leads to conflicts, for instance when watercourse construction and improvement measures collide with objectives to improve or maintain the hydromorphological integrity of rivers. Conflicts are likely to expand in the future, due to the binding environmental objectives of the WFD.

Coordination among the different federal states in regulatory and management matters is institutionalized within the 'Working Group of the Federal States on Water Problems' (LAWA), which was established in 1956. With its annually shifting presidencies and offices as well as an inscrutable variety of permanent and ad hoc committees or boards, the LAWA is a typical product of Germany's federal structure. The LAWA is judged as being a rather successful water management institution considering the fact that 16 independent and equal actors, the federal states, have to agree on common strategic, legislative, and technical issues. (Kahlenborn & Kraemer, 1999, p. 130).

At the regional and local level there are two main public actors in the water sector. In most cases, the municipalities are responsible for drinking water supply and wastewater management. They are in turn controlled in terms of emission and quality standards by local and district water authorities. Municipal responsibility results in an extremely fragmented organizational structure. More than 7000, mostly municipally operated, enterprises are in charge of wastewater management in Germany (UBA, 2001a, p. 47). In recent years a deregulation and liberalization of the water sector is taking place in many European countries, leading to different forms of public–private partnerships. According to recent reforms, municipalities may now sell water services and the infrastructure in international tenders. Especially in metropolitan areas, more and more private companies run formerly public water services and hold shares of formerly municipally owned enterprises. Global players emerge in the field of water services such as the French company Veolia (former Vivendi), which runs the wastewater treatment and drinking water supply of Berlin, Praha, and other European cities. Veolia's worldwide revenues in the water sector totaled more than 12 billion Euros in 2000. However, unlike developments in the deregulated European electricity and energy sector, there is no national water grid emerging from the current developments. Drinking water distribution mainly operates within closed local and regional networks. Long-distance water transfer schemes are in place to ensure drinking water supply in the metropolitan regions (e.g., Hamburg, Frankfurt, Stuttgart, and Leipzig), but rarely operate over longer distances than a few hundred kilometers.

2.2. Past and Current Challenges of Water Resources Management in Germany

For decades, the main challenge of Germany's water sector has been the control of point sources of pollution. Recently, the reduction of non-point pollution, the ecological and hydromorphological restoration of rivers, and flood control play an increasing role.

The peak of organic and chemical pollution of rivers and streams was reached in the second half of the 20th century: salmon, formerly an important part of the diet of the people living along the river, became extinct around 1950 in the river Rhine. In the 1970s, the Rhine river had gained the doubtful reputation of being 'Europe's sewer'. In 1971, only five species of insects from the macrozoobenthic aquatic community were found in the river Rhine, while at the beginning of the 20th century river Rhine had been habitat for more than 100 insect species. Within the same period, the number of fish species indigenous to Rhine declined from 46 to 23 (UBA, 2001b, p. 10). West German water policy in the 1970s and 1980s concentrated on hazard prevention and the control of point sources of pollution from industrial and household effluents. Sewer networks were expanded considerably and enhanced technical standards for sewage treatment works came into practice.

A very powerful monitoring instrument in this context was and still is the biological quality classification, a qualitative assessment of all river and stream reaches on a seven-step-scale from unpolluted to excessively polluted. The classification system is based on the monitoring of aquatic macroinvertebrates, indicating hydrological, chemical, and physical stress to the freshwater ecosystems. Developed as a monitoring system in the 1920s, the biological quality maps became a key instrument of water managers to draw public and political attention to water pollution and the necessity of pollution control. With growing public attention, enforcement of emission standards, and increasing investments in purification technologies from the 1970s, water quality improved considerably until 1990. Following German unification in 1990, water quality improved in eastern Germany as well due to the decline of polluting heavy and chemical industries and technical improvements of sewage treatment (see Table 2).

Today, severe eutrophication and resulting oxygen depletion are no longer a problem in most surface waters, because technical measures like substitution of phosphorous in detergents or denitrification techniques in sewage treatment were successful in mitigating point sources of nutrients (NH_4-N and PO_4-P). A considerable reduction was also realized for salt loads, some heavy metals, pesticides, and hydrocarbons. But for other heavy metals, e.g., cadmium and zinc, as well as for nitrate the situation remains critical (UBA, 2001b, p. 14ff).

With decreasing pressures from point sources of pollution, other sources of ecological deterioration came into focus. Pollutants from non-point sources, especially nitrogen and phosphorous from agriculture (see Table 3), represent a major unsolved problem for water management. At the end of the 20th century about two thirds of the nutrient load in surface waters were coming from non-point sources of pollution, mainly agriculture[1] and partly from urban surface runoff and sewage overflows (UBA, 2001b; Behrendt et al., 1999; De Wit, Behrendt, Bendoricchio, Bleuten, & van Gaans, 2002).

In recent years, new substances came into the focus of water management, e.g., pesticides, organic chlorine compounds, polycyclic aromatic hydrocarbons (PAH). Rather new challenges for the protection of waters are the so-called endocrinically active substances: industrial chemicals and pharmaceuticals with hormonal effects on aquatic species and humans (UBA, 2001b, p. 57f).

Furthermore, the hydromorphological structure of most rivers and streams is modified, for instance by weirs, bank protection, and straightening. This

Table 2. Time Series of Nutrient and Heavy Metal Pollution of the Rivers Elbe and Rhine.

			1975	1980	1985	1990	1995	1999	2001
Rhine (German–Dutch border)	Flow rate in m^3/s		2,320	2,580	1,990	1,930	2,850	2,920	
	Nutrients in mg/l[a]	Ammonium-*N*	1.20	0.60	0.50	0.20	0.20	0.10	0.10
		Nitrate-*N*	3.00	3.60	4.20	3.90	3.10	2.60	2.60
		o-Phosphate-*P*	0.69	0.22	0.35	0.11	0.09	0.07	
	Heavy Metals in mg/kg[b]	Lead		110	105	71	76	63	63
		Cadmium		1.80	1.70	1.20	1.00	0.80	1.00
		Mercury		0.60	0.72	0.39	0.47	0.49	0.50
Elbe (100 km above tidal limit)	Flow rate in m^3/s				558	447	908	578	548
	Nutrients in mg/l[a]	Ammonium-*N*			3.60	1.50	0.20	0.10	0.10
		Nitrate-*N*			3.20	5.10	5.10	4.00	3.40
		o-Phosphate-*P*			0.25	0.19	0.07	0.06	0.09
	Heavy Metals in mg/kg[b]	Lead	153	215	164	178	145	160	160
		Cadmium	9.70	11.50	14.70	13.00	11.10	9.10	9.00
		Mercury	16.30	21.10	11.90	7.50	5.30	3.60	3.50

Sources: UBA, (2001b) and UBA, (2005).
[a]mean values
[b]50-percentiles; suspended particular matter (dry matter).

Table 3. Time Series of Total-*N* and Total-*P* emissions into German Surface Waters in the 1980s and 1990s.

	Nitrogen				Phosphorous			
	1983–1987		1993–1997		1983–1987		1993–1997	
	kt/a	%	kt/a	%	kt/a	%	kt/a	%
Total emissions	1,082	100.0	818	100.0	94	100.0	37	100.0
Point sources (sewage treatment works and industrial effluents)	429	39.7	232	28.4	60	63.8	12	32.4
Non-point sources (agriculture, atmospheric deposition, and urban runoff)	653	60.3	586	71.6	30	36.2	25	67.6

Source: Behrendt et al. (1999).

leads to increased runoff velocities and flood risks, as well as to the destruction of water-related habitat and ecosystems. In a densely populated country such as Germany, most flood plains are used for agricultural, infrastructural, and other purposes. Therefore, the hydrological interaction between the river and the floodplain is restricted or prevented by dykes for flood protection. This not only results in increasing flood risks downstream, but also in an ecological degradation of the river systems.

In order to establish a monitoring system of the hydromorphological structure of surface waters and to gain public and political attention for the ecological consequences of morphological degradation of rivers, in the 1990s, a structural or hydromorphological quality classification was established by all federal states. Recently, water management activities have shifted from pollution reduction to the implementation of river rehabilitation schemes and the restoration of longitudinal continuity (see international programs like 'Salmon 2000' by the ICPR). In many cases, the limiting factors for these activities are the economic interests of water users and riparian land owners. Other conflicts emerge in connection with competing goals of other water uses like navigation, either existing, such as in the Rhine basin as one of the most heavily used shipping lanes in the world, or planned, such as the planned extension of the navigability of the Elbe and Saale rivers (Petry & Klauer, 2005). Another limiting factor is the fact that water managers have little influence on policies in the agriculture sector.

This means that after successfully mitigating pollution from many point sources, the reduction of non-point source pollution and the ecological and hydromorphological restoration of rivers play an increasing role today. In this context, navigation and agriculture turn out to be constraining factors. Flood control and prevention have always been important duties of German water authorities. The severe floods during the mid 1990s at the river Rhine and in 2002 at the rivers Elbe and Danube set flood control high on the political agenda.

In general, conflict resolution in water management is becoming more complex. In the past, severe pollution could be mitigated by building sewage treatment plants, and flood risks were reduced by dyke constructions. In view of increased flood risks, unsolved pressures from agriculture, and a growing awareness for up- and downstream interactions, single actor, single level, and single technology solutions reveal their limitations and the need for more integrated approaches in water management becomes evident. But for the adaptation of RBM or IWRM approaches, the current institutional structure of water resources management in Germany seems inadequate:

- The organization of water resources management along administrative borders hinders the adoption of integrated *RBM* approaches. RBM requires – until today – cooperation of independent political and administrative bodies, relying on expensive coordination procedures.
- There is a need for greater *intrasectoral cooperation* within the water sector, as management duties such as navigation, flood protection, water supply and wastewater disposal, and water protection are in the responsibility of different ministries and authorities at different hierarchical levels.
- In addition, there is a need for greater *intersectoral cooperation* of the water sector with other policy fields such as agriculture, land use planning, and infrastructure and housing development. While industrial discharges into a river are comparatively easy to control within the existing institutional arrangements, the control of nutrients from non-point sources such as agriculture goes beyond the established institutional scope of water management (an exception are the so-called water protection areas). The same holds true for flood protection when it is extended beyond technical measures such as dyke and polder construction toward integrated schemes for the land use of flood plains.

The EU WFD has initiated fundamental changes in water resources management in Germany as described in more detail in Section 4. This reform process can build upon some experiences with IWRM and RBM, which will be presented in Section 3.

3. PAST EXPERIENCES WITH RIVER BASIN MANAGEMENT

Examples of experiences with IWRM include the water associations in North Rhine–Westphalia, the river basin organization of the former German Democratic Republic, and various international river commissions.

3.1. Water Associations in North Rhine–Westphalia

The history of the water associations in North Rhine–Westphalia dates back to the second half of the 19th century and is closely linked with the industrialization of the Ruhr district. Within a few decades the Ruhr district became the industrial heart of Germany, based on coal mining, related heavy industries and a rapidly growing population. Water supply in sufficient quality and quantity and wastewater disposal became key factors for further economic development. At the turn of the century, two landmark decisions were made, affecting the river network until today. The Emscher river in the northern part of the Ruhr district was chosen to serve as the 'main sewer' of the region's wastewater. On the other hand, the Ruhr river became the major source for drinking and industrial water supply. In order to organize the different interests, a special law was passed which allowed for the creation of corporate water associations. The *Emschergenossenschaft* was established in 1904 to coordinate waste water disposal into the Emscher river. In 1913, the *Ruhrverband* and the *Ruhrtalsperrenverein* were founded to provide a sufficient runoff of the Ruhr river. Upstream reservoirs were built to guarantee a constant water supply for downstream water works and hydropower generation.

Today, in North Rhine–Westphalia all major tributaries to the Rhine river have water associations in place under the supervisory control of the Federal State Ministry of Environment (Holm, 1988, p. 59). The water associations are self-governing public bodies, assembling municipalities and districts, public and private water users, and other stakeholders like agriculture within the particular river basin that forms the association's territory. Financial expenditures for water management are carried by the members of the associations, which in turn regulate fees for water abstractions and wastewater discharges.

Historically, the responsibilities of these water associations were rather restricted. Legal competencies and management objectives were limited to water supply or wastewater management, and to the river instead of the entire river basin. Nevertheless, long before the WFD came into being, the water associations developed more integrative management practices, giving more regard to environmental concerns.[2] However, today the water associations play an active role in the implementation of the WFD not only at a technical but also at a strategic and conceptual level. Their long tradition of involving all relevant stakeholders serves as an example of good practice in participation processes.

3.2. River Basin Authorities in the former German Democratic Republic

In comparison to the federal structure of West Germany, the centralized political and administrative structure of the German Democratic Republic (GDR) facilitated the institutionalization of the river basin approach in the water sector. In 1958, seven Water Management Authorities (*Wasserwirtschaftsdirektionen*) were established, each of them responsible for one or several river basins within the GDR. They served as head institutions within a hierarchical structure of subordinate authorities responsible for sub-basins and consistent river reaches. Water management was thus organized in a hydrologically defined hierarchical spatial context, separately from the regular administrative system at national, regional, and district level. Cooperation with other policy fields such as land use planning or agriculture was partly realized in regional or district level commissions and working groups (Apolinarski, 2003, p. 72).

This early and consequential RBM structure was constrained in its effectiveness by centralistic political obligations. Water management in the GDR was part of the socialistic economy and had therefore primarily a service function for the economic development. The main purpose of water management was to ensure a stable water supply for all parts of the national economy, namely agriculture and industry, whereas environmental concerns played a subordinate role. However, as Moss (2001) observes, it is not without irony that just a decade after dissolving the river basin organization of water management, the WFD calls for the reorganization of management activities within watersheds.

3.3. International Water Commissions

Most of Germany's larger rivers cross international borders. The Danube river is shared by thirteen riparian countries, the Rhine by nine, the Elbe by four, and the Odra by three. Over the course of time, these transboundary rivers have repeatedly given rise to transboundary negotiations on issues relating to borders, navigation, flood control, hydropower development, fisheries, and pollution control.[3] Often negotiations took place on a bilateral basis, and bilateral border commissions exist with France, the Netherlands, Poland, and the Czech Republic. An early concern leading to multilateral negotiations was navigation. In 1815, the Vienna Congress established the principle of free navigation for all signatory states. In 1816, Switzerland, France, the German states, Belgium, and the Netherlands established the first multilateral water commission, the Central Commission for Navigation on the Rhine, which exists to date. In 1948, the Soviet Union, Bulgaria, Czechoslovakia, Hungary, Romania, Ukraine, and Yugoslavia established the Danube Commission with the aim of securing free navigation on the Danube river. In 1960, Austria acceded to the Danube Commission.

A second set of issues giving rise to multilateral negotiations in the Rhine basin was fisheries and water pollution.[4] Due to the loss of habitat and increasing pollution by mining, industries, and municipalities, fish stocks in Rhine declined drastically since the second half of the 19th century. In 1885, the riparian states agreed on the regulation of salmon fisheries, and in 1887 on a general fisheries agreement. However, the pollution of Rhine further increased and the fishing profession was not able to enforce their interests against those of the growing industries. By the mid 20th century, salmon were effectively extinct.

As a response to rising pollution levels, in 1950, five of the Rhine riparian countries, Switzerland, France, Luxembourg, Germany, and the Netherlands established the International Commission for the Protection of Rhine against Pollution (ICPR). ICPR was to deal with Rhine's 'main stem' between lake Constance and the delta in the Netherlands. In the beginning, the collaboration was still on an informal basis, but in 1963 the five countries signed the Convention for the Protection of the Rhine against Pollution (Bern Convention). In 1975, the European Economic Community acceded to ICPR. Already in 1959, Switzerland, Austria, Liechtenstein, and the German federal states of Baden-Württemberg and Bavaria had founded the International Commission for the Protection of lake Constance, and in 1961, Germany, France, and Luxembourg established the International Commissions for the Protection of the Moselle and Saar, two major tributaries of the Rhine. Already in 1892, Switzerland and Austria had created the International Commission for the Regulation of the Alpine Rhine. While ICPR covers the largest part of the river, regimes for the transboundary protection of the Rhine and its tributaries developed in a de-central fashion, and historically there was no overall coordination authority in place.

The initial objectives of ICPR included the exchange of information and the coordination of national contributions toward the protection of the river. Despite these efforts, there were no immediate improvements, but the Rhine's water quality continued to decline, until public pressure led to national pollution abatement programs that started to become effective by the mid 1970s. In parallel to these national efforts, the countries sought to harmonize emission standards at the international level. Given that the decisions of ICPR only had the character of recommendations to the national governments, the countries sought to negotiate separate legally binding international agreements. In 1976, the ICPR members concluded two conventions, the Convention for the Protection of the Rhine against Chloride Pollution (Chloride Convention) and the Convention for the Protection of the Rhine against Chemical Pollution (Chemical Convention). Both conventions had the character of framework conventions that were supposed to be implemented through further protocols. This approach turned out to be extremely cumbersome. Concrete steps toward the implementation of the Chloride Convention did not start until 1987 (Bernauer, 1995, 1997). In the case of the Chemical Convention, by 1986, emission standards had only been ratified for two out of a great number of relevant pollutants, for mercury in the chlorine production in 1982, and for cadmium in 1986 (Bernauer & Moser, 1996). It can thus be concluded that until the mid 1980s, the contribution of transboundary water management efforts toward the improvement of Rhine's water quality remained negligible. By then, the significant improvement of Rhine's water quality was mainly the result of national initiatives (Oterdoom, 2002).

The benefits of international coordination, including greater flexibility, were only realized after the Sandoz spill. In November 1986, a fire broke out in a storehouse for pesticides at the Sandoz Company in Schweizerhalle close to Basle. The fire-fighting water discharged into the Rhine led to the extinction of fish and other water organisms over a distance of 500 km. The ecological catastrophe led to a public outcry and a drastic shift in national and international water policies. In response, the ICPR member countries developed the Rhine Action Program (RAP) that was based on four main objectives: (1) the reintroduction of higher species to Rhine (symbolized by the vision 'Salmon 2000'); (2) ensuring the continued use of Rhine water for drinking water production; (3) the reduction of the contamination of sediments in order to enable the use or disposal of dredged material without causing environmental harm; and (4) the protection of the North Sea. In the choice of objectives, ICPR also reacted to concerns that had long been upheld by non-governmental and local organizations, such as the International Association of Water Works on the Rhine, and the Port of Rotterdam (Durth, 1996).

At the international level, the countries agreed on joint goals, identified possible measures for implementation, and monitored progress on the basis of national reporting. In doing so, the member states set up a number of thematic working groups, which involved technical experts of the respective national bureaucracies. This implied that those who were responsible for the implementation were directly involved in the negotiation process. Also, for the first time, non-governmental organizations were admitted as observers. While the targets were not legally binding, they were negotiated and implemented in a highly participatory approach, and reinforced by high-level political commitment. The selection and implementation of measures to reach the agreed targets, however, was left to the respective national governments. By the year 2000, salmon has returned to Rhine, most of the water quality goals of the RAP had been reached. Further efforts will be needed to reduce the discharge of heavy metals, nitrogen, and pesticides from non-point sources, and to allow for a natural reproduction of salmon (IKSR, 2002).

With the RAP, a new, flexible instrument had been found to promote integrated RBM in a transboundary context. A main reason for the success of RAP is being seen in the political character of the program (e.g., Durth, 1996; Holtrup, 1999). The combination of high level political commitment to a publicly visible and verifiable objective ('Salmon 2000') and intense collaboration at the working level allowed for flexible management based on a rapid and dense exchange of information. At the same time, given that implementation was left to the member states, they maintained enough leeway to implement measures in their own way, thus speeding up implementation.

In the 1990s, ICPR broadened its scope by including flood control in its agenda, as reflected in the Action Plan on Flood Defence of 1998, and the Programme Rhine 2020 of 2001, which combines and continues the work of RAP and the Action Plan on Flood Defence. In 1999, the ICPR countries signed a new Rhine Convention that formally extended ICPR's scope from a pure water quality perspective to include broader environmental concerns and flood control, explicitly covering groundwater, aquatic, and terrestrial water-related ecosystems, and the protection of the North Sea. While ICPR's geographical mandate was not formally extended, the wording of the convention still provides room for interpretation as it applies to the 'Rhine catchment area' insofar as the Rhine is affected by pollutants or flooding (see also Epiney & Felder, 2002).

After the fall of the Iron curtain, ICPR served as model for the establishment of the International Commission for the Protection of the Elbe (1990), the International Commission for the Protection of the Danube (1994), and the International Commission for the Protection of the Odra (1996) (Holtrup, 1999). Also, many experiences of ICPR were considered in the design of the EU WFD.

With the WFD, there are now two international regimes in place for transboundary water quality management in the Rhine Basin. Both the ICPR and the WFD regime set water quality targets at the international level and require close collaboration in the implementation among the riparian countries. While the two regimes do overlap to a great extent, there is no complete congruence. While the WFD for the first time covers the entire river basin, it remains narrower in scope by not including flood control. Significant efforts are underway to coordinate activities under the two regimes, and the ICPR secretariat officially supports the implementation of the WFD at the international level.

4. THE NEW EU FRAMEWORK DIRECTIVE

4.1. Objectives and Instruments

In 1995, the legislative and executive institutions of the European Union – the parliament, the council, and the commission – agreed that a new directive was needed for establishing a sustainable water policy (CEC, 1996). After tedious negotiations between the different European institutions and the member states, at the end of the year 2000 the European WFD entered into force (e.g., Kallis & Nijkamp, 2000). The WFD forms a milestone in the environmental and water policy of the European Union with far reaching consequences for all member states. In Article 4, the WFD defines binding environmental objectives:

- a good ecological and chemical status for all surface waters;
- a good ecological potential and good chemical status for all heavily modified or artificial water bodies;
- a good quantitative and chemical status for all groundwater bodies.

Annexes provide precise definitions of the so-called quality components indicating the ecological, chemical, and quantitative[5] status. By the year 2015, the good status has to be achieved for all waters, apart from those under specific derogations (certain regulations such as the designation of heavily modified and artificial water bodies, under which exemptions from the environmental objectives are justified).

In order to fulfill these demanding objectives, the WFD sets out several integrative principles for water-related planning and policy:

(a) RBM approach: Implementation of the WFD and achievement of the environmental objectives shall be addressed through the set up of compulsory programs of measures (PoMs) and river basin management plans (RBMPs) for all river basin districts. A river basin district may consist of one large or several smaller river basins. In case of transboundary basins, international river basin districts have to be assigned and appropriate cooperation between concerned countries needs to be established. The first PoMs and RBMPs shall become operational by 2009, and shall be reviewed and updated every 6 years thereafter.

(b) Combined approach of emission limits and environmental quality objectives: The improvement of water quality shall be achieved in two complementary ways by emission control and quality objectives for the status of waters: Member states shall identify significant pressures and responsible driving forces and implement appropriate measures (e.g., best available technology, BAT) to mitigate negative impacts on water systems (emission approach). On the other hand, binding objectives for the ecological, chemical, and in the case of groundwater, quantitative quality of waters are being defined and monitored (immission approach). All measures taken to mitigate negative impacts on water bodies and to meet the environmental objectives have to be registered in the PoMs by 2009 and implemented within 3 years thereafter under the responsibility of the so-called 'competent authorities', which are the State Ministries for Environment in most cases.

(c) Point and non-point sources of pollution control: The pressure and impact analysis and the measures shall cover all causes of pollution and deterioration, including point sources of pollution such as industrial effluents as well as non-point sources such as nitrogen and phosphorous surpluses from agricultural land use.

(d) Combined management of groundwater and surface waters: The WFD is the first European regulation that calls for an integrated protection and management of groundwater and surface waters. Therefore, a classification system for groundwater is being established and hydrological as well as chemical interactions between groundwater and surface waters have to be documented. An impact analysis has to estimate the effects of groundwater bodies on associated surface waters and terrestrial ecosystems.

(e) Polluter pays and cost recovery principle: While the polluter pays principle is long established in environmental policy, it is rarely strictly enforced in legislation. The WFD specifies that water services such as drinking water supply or wastewater management shall work on a full cost recovery basis. The costs of water services have to include environmental and resource costs emerging from the services. Currently, cost recovery is achieved via prices for delivered goods such as drinking water and subsidies or other transfer mechanisms. Environmental and resource costs are rarely taken into account (e.g., through abstraction charges). The WFD now requires that all water uses – in terms of the WFD water services, their customers as well as all human activities causing significant pressures on water systems – have to contribute to the cost recovery of water services in an adequate fashion.

(f) Cost-effectiveness of measures: The PoMs shall consist of the most cost-effective combination of measures available for achieving a good quality of all water bodies. This requires the development and implementation of robust assessment procedures for the ecological as well as economic effects of measures.

(g) Regionally differentiated regulations: Environmental objectives as well as the measures and instruments to achieve them have to be adjusted to the specific geographic, economic, and social conditions within the European Union. What a good status for a specific water body means is being defined with regard to the geologic, climatic, and hydrological conditions of the water body. Management strategies can be adapted to the regional institutional arrangements.

(h) Combined top-down and bottom-up approach: It is often argued that policy and legislation of the European Union follows a strict top-down approach. The WFD may serve as an example of a new generation of EU-directives. Instead of being drafted by the European Commission, the directive was developed in a participative process in which the member states, the European Parliament and NGOs played a key role (Kaika & Page, 2003). For the first time, a so-called 'Common Implementation Strategy' is being developed (CIS WFD, 2001), which hands the implementation process over to the member states, which is supported by working groups on specific implementation issues.

Members of the expert groups are recruited from national and regional water authorities, NGOs, private consultancies, and the scientific community. Guidance papers delivered by the CIS working groups are not legally binding but strengthen the member states commitment to WFD implementation. The European Commission controls the in-time achievement of compliance with the environmental objectives, but not the way compliance is achieved.

(i) Increased role of participation: The WFD calls for a strong participative approach in RBM. The public has not only to be informed, but the member states shall encourage the active involvement of stakeholders in the preparation of RBMPs. To ensure participation early in the planning process, participation has to start at least 3 years before the plans become operational.

The above list demonstrates that the WFD is ambitious not only in its environmental objectives but also in its integrative and broad approach covering many aspects of the water sector. A particular characteristic of the WFD is that for the first time, an RBM approach is imposed from above. The WFD is certainly one of the most ambitious attempts to implement the concepts of RBM and IWRM. As such, the directive poses significant challenges for the different member states, which vary considerably in their organization of their respective water sectors as the following examples demonstrate. While German water quality management was traditionally based on emission standards, other countries followed an immission approach based on quality objectives. In Spain 66% of all water abstractions are used for agricultural purposes (25 billion m^3 per year out of 38 billion m^3), while in Germany only 0.7% of its total water abstractions of 43 billion m^3 per year are used for agricultural purposes, but 56% are used for industrial cooling processes (power plants). Hence, the use of economic incentives in order to stimulate an efficient use of water may require different forms of adaptation in different member states. Furthermore, the current practice of levying water abstraction charges varies widely, not only within the European context but also among German federal states (Unnerstall & Köck, 2004). Similar arguments apply to the institutionalization of RBM. While RBM is a current practice in some member states such as France and the UK, others have to reorganize their water sector. In doing so, it is up to the member states whether to establish river basin authorities with far-reaching responsibilities and powers or to organize RBM through cooperative arrangements among the relevant authorities.

In conclusion, the WFD is a hybrid type of an EU Directive that represents a new policy style in the EU (see also Knill & Lenschow, 2000; Moss, 2004). For the first time, the implementation process was designed in cooperation with the member states, as laid down in the Common Implementation Strategy (EC, 2001). Within the directive 'command and control' elements such as binding objectives and procedures are combined with soft elements such as the increasing importance of participation and the acknowledgement of regional physical, social, and economic diversities.

4.2. The Implementation of the WFD in Germany

The WFD has initiated fundamental changes in Germany's water sector, affecting the institutional arrangements as well as the kind of measures, instruments, and planning processes required. In 2002, the Federal Water Act was amended to accommodate the requirements of the WFD and to provide the framework for the ongoing amendments of the 16 Federal State Water Acts. In view of the challenges identified in Section 2.2, the implementation process will be discussed for the transposition of the river basin approach, and the question of intra- and inter-sectoral cooperation.

4.2.1. Transposition of the River Basin Approach

During the negotiations of the WFD, the member states considered the obligatory introduction of river basin authorities with clearly defined responsibilities for the implementation of PoMs and RBMPs. Such a regulation was favored by France and the UK, both of which have a comparatively long institutional tradition in RBM.[6] Germany was among the countries, which – in the end successfully – prevented such binding regulations, arguing that the European Commission may set objectives, but may not define the means toward their achievement (e.g., Reinhardt, 2001; Unnerstall & Köck, 2004).

For the implementation of the WFD in Germany, ten river basin districts have been defined, six of them being international catchments for which PoMs and RBMPs are developed until 2009. Due to its federal structure, Germany decided to stay away from the establishment of river basin authorities, but to set up cooperative arrangements among those federal states that share a river basin, allowing for coordinated water management at the basin level (Strathenwerth, 2002). For the implementation of the river basin approach, the federal states established 'coordination groups' at a sub-basin level, and 'river basin associations' with coordinative functions for the national river basin districts (*Flussgebietsgemeinschaften*) (see Fig. 1). LAWA serves as an overall coordination body, e.g., by developing a national implementation guidance document (LAWA, 2003).

This means that the federal states maintain their legislative as well as executive autonomy and continue to be the key players in water resources management. At the same time, their decisions have to be coordinated at the river basin level. In consequence, all management activities have to be coordinated at the state as well as at a river basin level. Based on the first national and international reports of the Elbe river basin district to the European Commission (e.g., FGG Elbe, 2005 (Flussgebietsgemeinschaft); IKSE, 2005),[7] it can be concluded that the resulting implementation and management process is rather convoluted. The emerging practice shows that while the EU procedures are formally harmonized within basins and sub-basins, e.g., in common reports, the methods, models, and assessment procedures, for instance used for pressure analysis and impact assessment, may still differ from federal state to federal state sharing a basin or sub-basin. For example, surface water or groundwater bodies in similar conditions might be at risk of failing the good ecological or chemical status in one federal state while meeting them in another for methodical reasons only. Risk assessments carried out according to Article 5 WFD for the first report to the European Commission in 2005 are not fully consistent and comparable throughout a river basin district. Another observation on insufficient inter-state cooperation can be drawn from the participatory processes initiated according to Article 14 WFD, which requires that information and consultation of the public shall be ensured and active involvement encouraged. While the WFD is not prescribing the spatial scale at which participation shall take place, given the complexity of issues and stakeholders concerned, it is clear that different scales have to be addressed of both administrative and hydrological nature (see also CIS Working Group 2.9, 2002). In the Elbe river basin district for instance, participatory boards and procedures are mainly established by federal state authorities within political rather than basin or sub-basin boundaries.

From these preliminary observations, it may be concluded that the German federal states have still some way to go until they achieve meaningful RBM. On the other hand, one of the benefits from the emerging institutional regime is that RBMPs and PoMs can rely on the legislative and executive power of the established federal state authorities. At the same time, the 'hydrological management track' (right column in Fig. 1) clearly increases transaction costs due to the requirement of double coordination within administrative and hydrological boundaries. The German Advisory

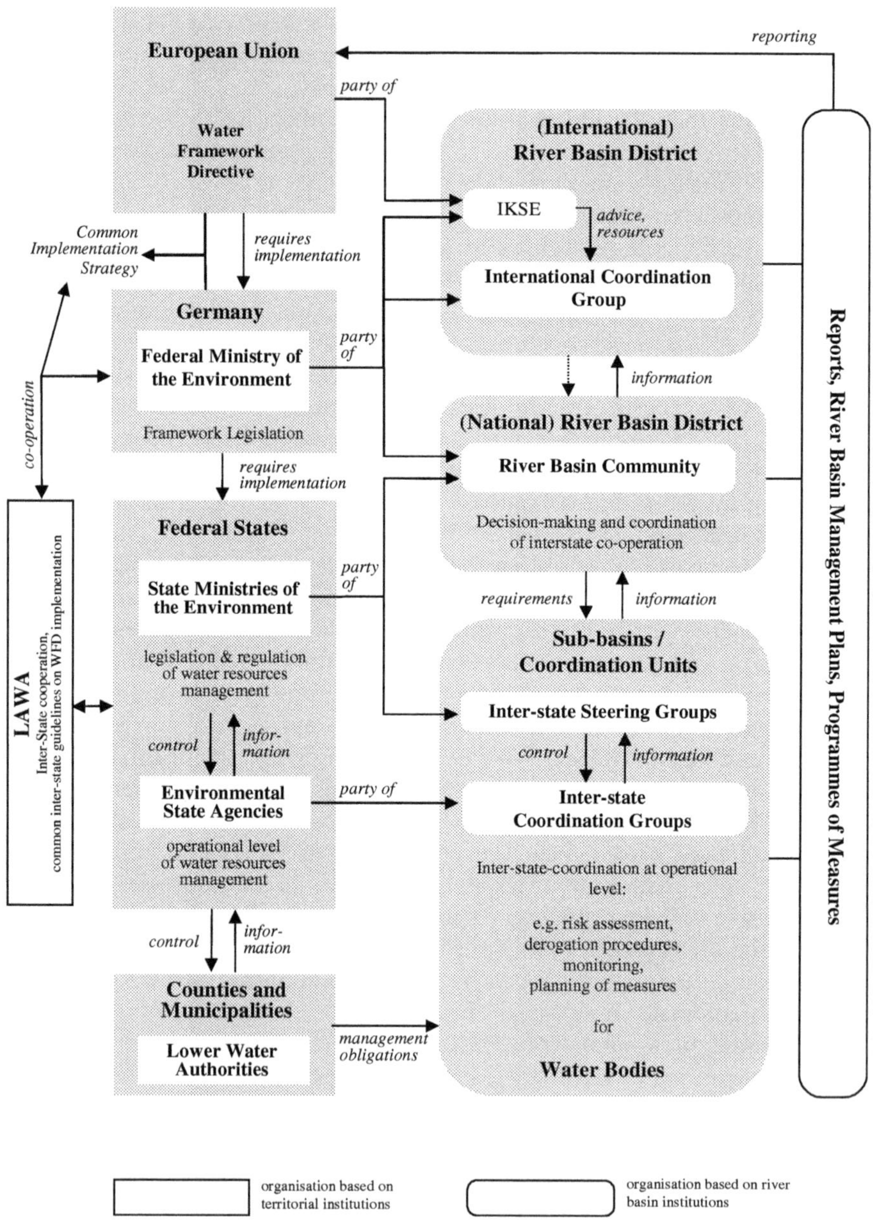

Fig. 1. Preliminary Institutional Arrangements for the Implementation of the WFD in the Elbe River Basin. *Source:* Modified after Petry (2005).

Council on the Environment, an advisory body for the federal government, has criticized the decision to pursue the cooperation model as inappropriate for the realization of adequate RBM. The Council recommends that

> "the federal government be given concurrent legislative powers in order to ensure the coherent transposition and subsequent implementation of EU legislation on the protection of surface waters. In parallel with changing legislative competences, alternatives to the planned, or already agreed upon, interstate cooperations should be considered, as the current administrative structures are, in the opinion of the Environmental Council, not compatible, in the final analysis, with the effective and efficient management [of] surface waters in river basin sections." (SRU, 2004, p. 36).

4.2.2. Intra- and Inter-Sectoral Cooperation in the Implementation of the WFD

The set up of an RBM track raises the question of what the implications are for the intra-sectoral and inter-sectoral coordination and institutional interplay with the relevant actors in the water sector. Moss (2003) argues that the introduction of RBM in Germany might solve problems of 'spatial fit' at the expense of problems of 'institutional interplay'. In other words, given that the water authorities put more emphasis on the coordinated action at the river basin level in order to gain a better spatial fit, they may have (even) less capacity to ensure the intra-sectoral or inter-sectoral institutional interplay that is necessary to solve pressing conflicts and to meet the environmental objectives of the WFD.

Despite the WFD's ambitious and far-reaching objectives and approach, the scope of the directive remains restricted compared to the definition of IWRM referred to in Section 1. The focus is on achieving environmental objectives, while other aspects of the water sector are not covered by the binding regulations of the WFD. Given the WFD's almost exclusive focus on environmental objectives, the authorities and agencies in charge of the implementation of the WFD are mainly located within the environment ministries and administration. Other water-related policy fields, such as the shipping administration, are not formally included in the implementation process. Thus, the environmental objectives of the WFD are prone to give rise to intra-sectoral conflicts between water supply, wastewater treatment and nature conservation on the one hand, and power generation, navigation, and flood protection on the other.[8] On the one hand, with the WFD the environment agencies have a formal instrument to push their agenda. On the other hand, the question remains how powerful this instrument really is. As the emerging practice demonstrates, there appears to be a tendency to exclude highly conflicting or political issues such as navigation (see last paragraph of Section 2.1 of this chapter) or flood protection from the implementation process. It is likely that in this context, the derogation procedures of the WFD that allow for less stringent environmental objectives (e.g., by designating a river as heavily modified due to its morphological modifications enabling navigational use but degrading the river's ecological status) will be applied when conflicts cannot be resolved.

At the same time, inter-sectoral cooperation of water management with other policy sectors such as agriculture, nature conservation, land use planning, and infrastructure and housing development will become more important for meeting the WFD's objectives. The achievement of the environmental quality objectives affords measures in the field of land use policy where the scope of necessary action goes beyond the scope of emission standards or where the morphological rehabilitation of watercourses is conflicting with other land use objectives (e.g., infrastructure). Apart from spatially and financially limited powers to establish riparian buffer strips or to designate water protection areas, water authorities and river basin managers have little legal or executive responsibilities outside the watercourse. In addition, the reduction of nutrient pressures of surface waters and groundwater will have to include measures on the agriculturally used land within the basin. Legal, and more importantly, financial instruments influencing agricultural practices are mainly determined by the Common Agricultural Policy of the EU and the agro-environmental schemes of the federal states – not by the PoM or RBMP for a certain river basin district.

Thus, it is certainly true that intra- and inter-sectoral cooperation and 'institutional interplay' will at least be as important as 'solving problems of fit' for the achievement of the WFD's objectives. In this regard, the WFD has little 'teeth'.

5. CONCLUSIONS AND OUTLOOK

Given that the European WFD calls for the management of water resources at the river basin level, the German water sector that has historically been dominated by the federal states and organized along administrative borders is now challenged to reorganize. In doing so, it can build upon selected experiences within international river commissions and the North Rhine–Westphalian water associations.

The implementation of the WFD in Germany can be characterized as an institutional learning process on how to establish RBM within a hierarchical administrative system with strong federal states. The evolving institutional arrangements can be characterized as a 'double-track' multi-level governance structure that provides linkages between different administrative levels (federal government, federal states, counties, and municipalities) and hydrological scales (basin, sub-basin, and water body). The two-track structure consists of:

- the existing, hierarchically organized jurisdictions with strong legislative and executive powers within political and administrative boundaries; and
- new coordination mechanisms at various hydrological scales, pulling expertise, budgets, and political commitment into hydrological units.

As such, the evolving arrangements can be considered as a pragmatic approach toward accommodating the realities and obligations of a federal system with an RBM approach.

The question is whether this rather complex and susceptible regime represents an appropriate way to establish the required 'spatial fit' and the 'institutional interplay' in all directions: horizontal coordination at the river basin level, vertical coordination among different administrative levels and hydrological scales, intra- and inter-sectoral coordination with other policy sectors. With respect to horizontal coordination, an advantage of the current approach is that the implementation of RBMPs and PoMs can rely on the legislative and executive power of the established authorities. A disadvantage is that the transaction costs for coordination at the river basin level are likely to remain high. Furthermore, the results of the first phase of the implementation of the WFD, i.e., the characterization of river basins, the risk assessment of water bodies, as well as the initial participation processes, demonstrate the difficulties to harmonize assessment and management procedures and data at the river basin level. However, it is still too early to conclude whether the respective differences are an expression of a positive competition of ideas or rather of a poorly managed river basin approach.

How powerful the established coordination mechanisms will be will become evident when it comes to the set up of PoMs and RBMPs with real conflicts of interests between the jurisdictions involved. However, these medium-to-long term operational deficiencies and conflicts are apparently being seen as the 'lesser evil' compared with a complete reorganization of the system.

While the German states have at least taken the first step toward an RBM approach, tools for an improved intra- and inter-sectoral coordination with other water policy fields and sectors remain weak. Improved coordination with navigation, flood protection, agriculture and land use policy will be important for achieving the WFD's environmental objectives. In that sense, Moss (2003) may, at least to a certain extent, be right that the WFD may solve 'problems of fit' at the expense of 'institutional interplay'. However, in that respect, the German approach, where the federal states continue to play a strong role in water issues, may even have some advantages, as it might be easier for the existing jurisdictions to further their dialogue with other policy sectors than it is for newly established river basin authorities.

This notwithstanding, there is certainly room for further adaptation over time, if the need for stronger basin orientation becomes obvious to the decision makers. This might become the case during the establishment of PoMs and RBMPs, which will start in 2005 and is bound to be a continuous process over the next two decades. Should the German states fail to achieve the environmental objectives, interstate arrangements may become inevitable in order to avoid fines by the European Commission. Especially in the case of mitigating quantitative, chemical, and nutrient pressures, locations of impact (i.e., the water body failing to meet the objectives) and origin of pressure may be situated in different federal states. Therefore, one federal state may depend on the action of another upstream state to achieve the objectives in its territory. Such a situation may require much closer basin-oriented cooperation than is realized today. Conflict may furthermore evolve around financial arrangements among the federal states of a river basin district. Representatives from federal authorities may possibly serve as mediators in conflict resolution.

Irrespective of these future developments, in particular the experiences of the ICPR show that the future success of RBM in Germany depends on additional factors, which are only partially influenced by the way RBM is institutionalized:

- *Political commitment* by the federal states within a river basin unit for the common objectives and management duties. The ICPR experience shows that a problem-oriented approach toward the WFD's PoMs and RBMPs might be more effective than legal regulations.
- The ability of river basin managers to *communicate* with actors from different fields of the water sectors and those from other policy fields such as agriculture, nature conservation, infrastructure, and land use planning in order to put the WFD's objectives on their agenda.

NOTES

1. With emission to groundwater and drainage in the case of nitrogen and erosion in the case of phosphorous as major pathways.

2. In a comparative study of the French river basin agencies and the North Rhine–Westphalian water associations, Holm (1988, p. 257) concluded that the water associations are of limited effectiveness in reaching ecological goals. She argued that this is due to the powerful role the emitting industries play as members of the water associations. It would have to be assessed whether this still holds true today.

3. The Food and Agriculture Organization (FAO, 1978; FAO, 1984) lists about 500 international treaties pertaining to Rhine between 805 A.D. and 1984.

4. A more detailed analysis of experiences with transboundary water management in the Rhine river basin can be found in Dombrowsky and Holländer (2004).

5. According to Article 2 of the WFD "Quantitative status is an expression of the degree to which a body of groundwater is affected by direct and indirect abstractions."

6. Examples are the National Rivers Authority in England and Wales (since 1996 part of the Environment Agency) and the river basin agencies (*Agences de bassin*) in France.

7. The reports fulfill the requirements of Article 3 WFD to characterize river basin districts in terms of pressures, impacts, and economics of water uses by 2004. This comprises the determination of the likelihood of every single surface water body of failing to meet its environmental quality objectives (risk assessment; see also CIS Working Group 2.1, 2002).

8. Where the protection of natural flood plains is concerned, nature protection and flood control are complementary activities.

REFERENCES

Apolinarski, I. (2003). Wasserwirtschaft in der DDR. In: T. Moss (Ed.), *Das Flussgebiet als Handlungsraum. Institutionenwandel durch die EU-Wasserrahmenrichtlinie aus raumwissenschaftlichen Perspektiven* (pp. 47–87). Münster: LIT-Verlag.

Behrendt, H., Huber, P., Ley, M., Opitz, D., Schmoll, O., Scholz, G., & Uebe, R. (1999). *Nährstoffbilanzierung der Flußgebiete Deutschlands*. Berlin: UBA-Texte 75/99.

Bernauer, T. (1995). The international financing of environmental protection: Lessons from efforts to protect the river Rhine against chloride pollution. *Environmental Politics*, *4*(3), 369–390.

Bernauer, T. (1997). Protecting the Rhine river against chloride pollution. In: R. Keohane & M. Levy (Eds), *Institutions for Environmental Aid: Pitfalls and Promises* (pp. 201–232). Cambridge: Massachusetts Institute of Technology.

Bernauer, T., & Moser, P. (1996). Reducing pollution of the river Rhine: The influence of international cooperation. *Journal of Environment and Development*, *5*(4), 389–415.

CEC (Commission of the European Communities) (1996). European Community Policy: Communication from the Commission to the Council and the European Parliament of 21 February 1996. Brussels.

CIS WFD (2001). Common Implementation Strategy for the Water Framework Directive. Strategic Document as Agreed by the Water Directors under Swedish Presidency, 2 May 2001.

CIS Working Group 2.1 (2002). Analysis of Pressures and Impacts. Guidance for the Analysis of Pressures and Impacts in Accordance with the Water Framework Directive.

CIS Working Group 2.9 (2002). Best Practices in River Basin Planning. Guidance on Public Participation in Relation to the Water Framework Directive.

De Wit, M., Behrendt, H., Bendoricchio, G., Bleuten, W., & van Gaans, P. (2002). The contribution of agriculture to nutrient pollution in three European rivers, with reference to the European Nitrates Directive. European Water Management Online.

Dombrowsky, I. & Holländer, R. (2004). Erfahrungen im integrierten Management grenzüberschreitender Flüsse in Europa: Das Beispiel des Rheins. *Zeitschrift für Angewandte Umweltpolitik 15/16* (*Special edition of issues 3–5* (2003/04)): 443–459.

Durth, R. (1996). *Grenzüberschreitende Umweltprobleme und regionale Integration: Zur Politischen Ökonomie von Oberlauf-Unterlauf-Problemen an internationalen Flüssen.* Baden Baden: Nomos.

EC (European Commission) (2001). Common Implementation Strategy for the Water Framework Directive (2000/60/EC). Strategic Document as Agreed upon by the Water Directors under Swedish Presidency, 2 May 2001.

Epiney, A., & Felder, A. (2002). Überprüfung Internationaler Wasserwirtschaftlicher Übereinkommen im Hinblick auf die Implementierung der Wasserrahmenrichtlinie, Umweltbundesamt, Texte 17-02, Berlin.

European Communities (2000). Directive 2000/60/EC of the European Parliament and the Council of 23 October 2000 establishing a framework for Community action in the field of water policy. *Official Journal of the European Communities*, 22 December 2002, L327, 1–72.

FAO (Food and Agriculture Organization of the United Nations) (1978). Systematic Index of International Water Resources Treaties, Declarations, Acts and Cases by Basin. Agriculture and Food Organization of the United Nations. Rome.

FAO (Food and Agriculture Organization of the United Nations) (1984). Systematic Index of International Water Resources Treaties, Declarations, Acts and Cases by Basin (Vol. II). Agriculture and Food Organization of the United Nations. Rome.

FGG Elbe (Flussgebietsgemeinschaft Elbe) (2005). Bericht an die EU-Kommission nach Art. 3 Wasserrahmenrichtlinie für die Flussgebietseinheit Elbe.

GWP (Global Water Partnership) (2000). Integrated Water Resources Management. Technical Advisory Committee Background Papers No. 4. Stockholm: GWP.

Holm, K. (1988). *Wasserverbände im Internationalen Vergleich: Eine ökonomische Analyse der Französischen Agences Financières de Bassin und der Deutschen Wasserverbände im Ruhrgebiet.* München: ifo-Institut für Wirtschaftsforschung.

Holtrup, P. (1999). Der Schutz grenzüberschreitender Flüsse in Europa: Zur Effektivität internationaler Umweltregime. Dissertation Universität Bonn, "Berichte des Forschungszentrums Jülich – 3642" – Programmgruppe Technologiefolgenforschung – Kenn-Nr.: Jül-3642.

Huppert, W. (2005). Was ist IWRM? Plädoyer für ein differenziertes Verständnis des Konzepts Integriertes Wasserressourcen-Management. In: S. Neumann, W. Scheumann, A. v. Edig & W. Huppert (Eds), *Integriertes Wassermanagement: ein Konzept in die Praxis überführen* (pp. 15–30). Baden Baden: Nomos.

IKSE (Internationale Kommission zum Schutz der Elbe) (Eds). (2005). Internationale Flussgebietseinheit Elbe. Merkmale der Flussgebietseinheit, Überprüfung der Umweltauswirkungen menschlicher Tätigkeiten und wirtschaftliche Analyse der Wassernutzungen. Bericht an die Europäische Kommission.

IKSR (Internationale Kommission zum Schutz des Rheins) (2002). Vergleich des Istzustandes des Rheins 1990 bis 2000 mit den Zielvorgaben. Bericht Nr. 123-d.doc. 68. Plenarsitzung, Luxemburg, 2./3.7.2002.

Kahlenborn, W., & Kraemer, R. A. (1999). *Nachhaltige Wasserwirtschaft in Deutschland.* Berlin, Heidelberg: Springer-Verlag.

Kaika, M., & Page, B. (2003). The EU water framework directive: Part 1. European policy making and the changing topography of lobbying. *European Environment*, *13*, 314–327.

Kallis, G., & Nijkamp, P. (2000). Evolution of EU water policy: A critical assessment and a hopeful perspective. *Zeitschrift für Umweltpolitik und Umweltrecht (ZfU)*, *3*, 301–335.

Knill, C., & Lenschow, A. (2000). On deficient implementation and deficient theories: The need for an institutionalist perspective in implementation research. In: C. Knill & A. Lenschow (Eds), *Implementing EU Environmental Policy. New Directions and Old Problems* (pp. 9–35). Manchester, NY: Manchester University Press.

LAWA (Länderarbeitsgemeinschaft Wasser) (2003). German Guidance Document for the Implementation of the EC Water Framework Directive, April 30, 2003, at http://www.lawa.de/pub/kostenlos/wrrl/Arbeitshilfe_englisch.pdf, visited April 4, 2005.

Moss, T. (2001). Solving Problems of 'Fit' at the Expense of Problems of 'Interplay'? The Spatial Reorganisation of Water Management following the EU Water Framework Directive (draft). Discussion paper 5/2001. Institute for Regional Development and Structural Planning, Erkner.

Moss, T. (2003). Solving problems of 'fit' at the expense of problems of 'interplay'? The spatial reorganisation of water management following the EU water framework directive. In: H. Breit, A. Engels, T. Moss & M. Troja (Eds), *How Institutions Change: Perspectives on Social Learning in Global and Local Environmental Contexts* (pp. 85–121). Opladen: Leske und Budrich.

Moss, T. (2004). The governance of land use in river basins: Prospects for overcoming problems of institutional interplay with the EU water framework directive. *Land Use Policy*, *21*, 85–94.

Newson, M. D. (1992). *Land, water and development: River basin systems and their sustainable management*. London: Routledge.

Oterdoom, H. (2002). The International Commission for the Protection of the Rhine. In: *Deutsche Stiftung für Internationale Entwicklung (DSE)*, The Nile: Sharing Experiences, Sharing Visions. Berlin: DSE.

Petry, D. (2005). Institutional change in river basins induced by WFD implementation – A success story? A case study from the Elbe river basin. In: EWRA European Water Resources Association (Ed.), *Sharing a common vision for our water resources, 6th International Conference*, September 2005 in Menton, Conference Proceedings.

Petry, D., & Klauer, B. (2005). *Umweltbewertung und politische Praxis in der Bundesverkehrswegeplanung*. Marburg: Metropolis-Verlag.

Reinhardt, M. (2001). Deutsches Verfassungsrecht und Europäische Flussgebietsverwaltung. *Zeitschrift für Umweltrecht, Sonderheft*, *12*, 124–128.

SRU (The German Advisory Council on the Environment) (2004). Environmental Report 2004 – Ensuring Environmental Protection Capacity. Summary, (p. 115). Berlin.

Strathenwerth, T. (2002). Die Bewirtschaftung nationaler und internationaler Flussgebiete. In: S. von Keitz & M. Schmalholz (Eds), *Handbuch der EU-Wasserrahmenrichtlinie: Inhalte, Neuerungen und Anregungen für die nationale Umsetzung* (pp. 324–342). Berlin: Erich Schmidt Verlag.

UBA (Federal Environmental Agency). (Ed.) (2001a). *Water Resources Management in Germany, Part 1: Fundamentals*. Berlin: The Federal Ministry for the Environment, Nature Conservation and Nuclear Safety.

UBA (Federal Environmental Agency). (Ed.) (2001b). *Water Resources Management in Germany, Part 2: Quality of Inland Surface Waters*. Berlin: The Federal Ministry for the Environment, Nature Conservation and Nuclear Safety.

UBA (Federal Environment Agency) (Eds). (2005). Online access to water quality monitoring data at: http://www.umweltbundesamt.de/index-e.htm, visited February 15, 2006.

Unnerstall, H., & Köck, W. (2004). The implementation of the EU water framework directive into federal and regional law in Germany. *Journal of European Environmental Policy and Legislation*, *1*(3), 207–217.

THE EVOLUTION OF WATERSHED MANAGEMENT IN THE UNITED STATES

Austin Troy

ABSTRACT

The United States today boasts of a complex and extensive set of public and private institutions and arrangements for managing its water resources. Today's system of watershed management is neither entirely top-down nor bottom-up. It is not entirely planned, nor is it entirely laissez-faire. Rather it is a hybrid. This chapter analyzes through a historical lens how American watershed management evolved to this state. It looks at two driving factors: technological change and trends in American political culture. Technology provided the reason for water resource and watershed management to evolve because of the conflicts provoked by its unintended and negative side effects, such as pollution. American political culture mediated the way that individuals and government reacted to these conflicts and spurred the evolution of new institutions.

Ecological Economics of Sustainable Watershed Management
Advances in the Economics of Environmental Resources, Volume 7, 43–66

ISSN: 1569-3740/doi:10.1016/S1569-3740(07)07003-4

1. THE AMERICAN CONTEXT: PRIVATE PROPERTY AND THE PUBLIC INTEREST

While it may seem parenthetical to a chapter on watershed management, an understanding of American political culture is both highly relevant and explanatory. The framers of the United States' Constitution were greatly divided over the appropriate level of governmental control in civil and economic affairs. At the heart of the disagreement lay differing levels of distrust in government. Thomas Jefferson perhaps best embodied the side most wary and distrustful of government. At the time, his distrust was well warranted, given the preponderance of despots, tyrannies, and otherwise corrupt governments around the world. He and his disciples, who, following the Revolution, were known as *Republicans*, until their party disintegrated in the early 19th century and was reconstituted as the *Democrats*,[1] felt that big government would untowardly benefit capitalists and large mercantile interests, because of its inherent corruptibility. Because they saw the lack of strong government as a check on the power of monopolies, trusts, and other large monied interests, they were against government involvement in major infrastructural improvement projects, seeing them as unfair support for the wealthy. In others words, if a project was so big that individual citizens could not do it on their own, it was likely only to benefit the powerful. As an example, one of the earliest attempts at a natural resource management bill, a rivers and harbors bill, was struck down for similar reasons in the 1840s with a veto by Democrat James K. Polk (Benson, 1961). Put simply, this political movement attracted those who believed in individuals and distrusted institutions.

On the other side of this socio-political divide was a group of largely Northeastern people who saw government as an essential means towards the progress and economic development of the nation. This group, initially known as *Federalists* and later *Whigs* and then *Republicans* (no relation to the original Jeffersonian Republicans), first led by the likes of Alexander Hamilton, John Adams, and Henry Clay, saw a much greater role for corporate bodies, community institutions, and higher level coordination (for the sake of consistency, I will refer to them from hereon as Hamiltonians and to the other group as Jeffersonians). This group saw a far larger role for government in encouraging and regulating trade, mobilizing industrial development, and providing services and improvements that required economies of scale and that might otherwise not be provided, such as roads, harbors, and railroads. They also saw a role for government in resolving conflict that would otherwise be irreconcilable among individuals.

Moreover, Hamiltonians tended to be highly technocratic in their thinking. They believed strongly in enlightenment ideals of science, technical progress, and education, and further believed that it was necessary to have well-educated people in government and its agencies or commissions to make technically informed decisions that would benefit the greatest number.

The differences between these two camps were largely rooted in different understandings of human nature. Jeffersonians felt that, released from the tyranny and class divisions of "Old Europe," people would naturally come together in a loose and classless federation without power and money being consolidated in the hands of a few.[2] The Hamiltonians, on the other hand, believed that humans acting as individuals were inherently corruptible and that the same forces that led to accumulation and abuse of money and power in the Old World could just as easily take root in the New, unless some intermediary – a large federal government, elected by the people – was given the power and resources to intervene. As Federalist John Adams wrote in his early 19th century correspondence with Jefferson "as long as Property exists, it will accumulate in Individuals and Families ... the SNOW ball will grow as it rolls."[3]

2. EARLY GOVERNMENT INVOLVEMENT IN WATERSHED CONFLICTS: THE CASE OF FARMERS AND MINERS IN CALIFORNIA'S CENTRAL VALLEY

Throughout the early 19th century, Jeffersonians dominated the federal government. Even when 19th century Republicans began to be elected in the mid-1800s, they inherited a weak central government with weak institutions by today's standards. Moreover there was a civil war brewing and, until its finale and the ramping down of the subsequent reconstruction of the South in the 1870s, fairly little attention was paid towards the development of governmental institutions for managing and regulating land and natural resources.

The few resource-related policies from this era are telling of an extreme laissez-faire attitude. The Green Act, passed in California in 1868 and written by Will Green a fierce Jeffersonian Democrat in the California State Assembly, allowed individuals to buy up swampland and join a "reclamation" district which could be comprised of just one owner and one property (Kelley, 1989). These districts coordinated reclamation, draining, and creation of levees. Experience has since shown that large-scale reclamation of land by disparate, uncoordinated owners within the same floodplain can lead to disastrous outcomes. But as Kelley (1989) deftly demonstrated, it took decades for the independent minded farmers of the state to realize how inappropriate this strategy was and the law remained as an established public policy for almost 50 years. As Kelley writes "The result was that for most of the next half-century, the Sacramento Valley would be scissored into a crazy-quilt of small reclamation districts whose levees followed property lines, not the Valley's natural drainage pattern. Flood control anarchy, and therefore massive flood control failure would be the result" (Kelley, 1989, p. 63).

The Green Act was passed with the best Jeffersonian ideals in mind. Will Green (who was not a major owner of swampland) wanted to see a utopia of small farmers, cultivating reclaimed bottomland throughout the Valley. Instead, the law's provisions for essentially giving land away if "reclamation efforts" could be demonstrated, resulted in the amassing of large tracts of land in the hands of a few wealthy speculators through corrupt means, in some cases in transactions of hundreds of thousands of acres at a time.

The change to more governmental involvement in land use planning came slowly and incrementally through the late 19th and early 20th centuries. Even in the face of overwhelming evidence that intervention was beneficial, it took years before many stakeholders were able to overcome their innate abhorrence of centralized government and reluctantly ask for such intervention. For example, in the late 19th century miners in the Sierra Nevada Mountains of California began using hydraulic mining[4] to remove ore. Along the Feather, Yuba, American, and Bear Rivers, enough debris was produced to bury the city of Washington DC under 19 feet of rubble (Brechin, 1999). This highly erosive practice led to massive debris flows that destroyed farmland and structures, filled in river channels and resulted in terrible floods for the downstream farming communities of the Central Valley. These debris flows were so catastrophic that on the Yuba River, the bed of the river rose 16 feet in height until it was above the level of the surrounding land. Whether for ideological reasons or economic (their farming economy was strongly tied to the state's mining economy) farmers put up with this practice for many years before they demanded intervention (Steinberg, 2002). While today it would seem surprising, the response of most of the bottomlands farmers in the early years of hydraulic mining was not to put up a fight, but to propose simply abandoning their lands (Kelley, 1989).

At the time, the only mode of intervention in nuisance disputes was litigation. This option was not practical since debris that reached the valley bottoms was thoroughly commingled, and hence it was impossible to attribute blame to any given individual or company. Moreover, the notion of regulating land uses because of nuisance was not yet well established, nor was the problem of nuisance given much sympathy (Kelley, 1989). To a large extent, the technology that could produce large-scale nuisances with the ability to impact such large groups of people was a new phenomenon, and the law could not keep up with these changes in technology. There was also a carryover in attitudes from an earlier day of hands-off government, even after it became evident that new industrial technologies had impacts that the law was ill-equipped to deal with. The pre-Civil War "instrumentalist" attitude of the judiciary held that the court should refrain from interfering with commerce so as to promote its development.

An example of the prevailing pro-commerce attitudes is found in one of the earliest cases of pollution litigation. In the 1886 case *Pennsylvania Coal Company v. Sanderson* (113 Pa. 126, 6 A. 453), a downstream plaintiff sued, claiming that discharge from a coal mine rendered their water supply unusable for domestic use. The Pennsylvania Supreme Court found that if someone bought property downstream of a known polluter it was their problem and not the polluter's, not just because the polluter was a known use, but also because the polluter had no intent to harm. Interestingly, another case in the same year showed the extent to which the country was split on this doctrine. Lux v. Haggin (69 Cal. 255, 335, 337, 10 Pac. 674), argued before the California Supreme Court, attempted to settle rights to water between riparian users and irrigators with "prior appropriation rights" (Worster, 1985). It found that the riparian user has primary rights to the water "so long as he does no material damage to his neighbors below him." In other words, the judiciary, and likely much of America, was split at this time as to what were reasonable expectations from upstream users.

Given the failure of conflict resolution in the courts, it was becoming apparent that only the intervention of the legislative branch of government in the use of private property could alleviate such water conflicts. However, such intervention was so unprecedented and outside of the American political tradition that it was many years, and thousands of acres of lost farmland, before anything was done (Kelley, 1959, 1989). In 1875, a massive flood inundated the Sacramento Valley (the Northern part of the Central Valley) and, in the process, deposited vast amounts of mining debris on towns such as Marysville and their surrounding farmland. This extreme event finally galvanized the residents of Sutter and Yuba Counties, who started to question their initial assumptions about the nature of private property and came together for what might be considered the first watershed association meetings. These people, recognizing that they shared a common hydrologic boundary and a common destiny with those many miles away, organized public meetings and began a crusade against the upstream hydraulic mining industry.

For years they tried varied strategies to fight the mining industry, but to little avail. Lawsuits in the late 1870s were unsuccessful, but were followed by the beginning of state involvement. In 1880, the new State Engineer commissioned a study, concluding that the two uses were compatible so long as large debris dams were built. This was quickly discredited in 1881 when many of the recently constructed debris dams failed in a major flood. The farmers returned to court thereafter, but now a shift was becoming evident in the judiciary from hands-off instrumentalism towards formalism – a doctrine of greater judicial involvement in reigning in economic interests. This ideological shift was acted upon when the Ninth Circuit Court issued an injunction against mines discharging any further debris into streams in the 1883 case *Woodruff v. North Bloomfield Gravel Mining Company* (16 F. 25). In his decision, Judge Sawyer split with his predecessors by arguing that miner's property rights are not so extensive that they give them the right to inflict such extensive damage on others' property (Kelley, 1959).

This effectively shut down the mines for several years. However, by the 1890s there were strong economic pressures to reopen the mines so as to increase precious metal production and inflate the currency (which, ironically, would help farmers). The economic importance of the mines, in combination with the building of a movement at the federal level towards stronger and more centralized regulatory institutions (Skowronek, 1982) in the Hamiltonian tradition, combined to create ideal conditions for legislation that would allow mines to reopen under the auspices of a new regulatory agency. A related change was the dawning recognition among many that commissions independent from a legislature and formed of technically educated experts were becoming necessary given the increasing complexity of many regulatory problems. These largely Hamiltonian Republican advocates argued, in the case of railroad regulation, for instance, that no legislation could be detailed and far-sighted enough to anticipate all the stratagems that railroad entrepreneurs would employ to gain advantage, and hence independent commissions were needed that gave informed individuals enough leeway and latitude to enforce the overall intent of the law (Kelley, 1989).[5] These principles would soon be seen to apply to natural resources as well.

In this context, the U.S. Congress passed the Caminetti Act of 1892, which created the California Debris Commission (CDC), one of the earliest federally chartered environmental regulatory agencies dealing with watershed issues. In the best Hamiltonian tradition, this act called for the creation of an agency led by engineers and scientific experts, who, in its authors' minds, would dispassionately guide the process of reconciling the heretofore seemingly incompatible uses of hydraulic mining and agriculture. The act proposed that the CDC would be given authority to license individual mining operators to restart their operations, provided that they built proper debris dams – according to CDC regulations – to impound their tailings.

The passage of this bill did not sit well with the farmers who, after many years of fighting for higher-level governmental intervention, now were suspicious that such intervention was merely an excuse to allow miners to begin again and the regulatory or punitive aspects were merely empty fronts to gain approval (Kelley, 1989). Again, the lowlanders formed citizen's groups (such as the Anti-Debris Association) and again they were successful, this time in assuring that an amendment was put in the bill stipulating a $5000 fine or year of imprisonment for anyone violating the act. Republican President Benjamin Harrison gladly signed the act. Hence, 80 years before similar debates would rage again during the drafting of key environmental legislation, local constituencies were fighting with the government over the enforceability and punitive measures of environmental legislation designed to protect the health of watersheds.

The Caminetti Act had far-reaching implications for governmental involvement in natural resource management. For the first time, an expert-driven federal agency existed for managing problems that occurred on a watershed-scale – a scale too large to be dealt with by local authorities or private individuals. Beyond this, though, the Act had a further – and almost unnoticed at the time – provision that was groundbreaking in respect to the government's role in natural resources. Previously, under the doctrine articulated in *Gibbons v. Ogden*, the federal government was believed to be responsible only for navigability when it came to managing inland water ways. The Caminetti Act, on the other hand, gave the CDC the jurisdiction not just to regulate mining and facilitate navigation, but also to mitigate flood hazards, at the time a revolutionary role for the federal government, and its first large-scale foray into watershed management. Hence, what was sold as a bill to help out miners, ended up being a paradigm shift in American government as to how it managed the environment, representing one of the first[6] expert-driven agencies independent from the legislature.

If the CDC had actually developed into a major and successful agency, this bill would have been remembered as one of the critical moments of American environmental legislative history. Instead, the CDC got off to a slow start and slowly fell into obscurity amongst other rising agencies, until it finally dissolved in the 1980s. Part of its slow start was due to a change in the political landscape in Washington. The period of Republican ascendancy that had allowed for the Caminetti Bill to pass was soon followed by the ascendancy of strongly Jeffersonian, anti-big government Democrats, many of whom saw the CDC as simply a tool meant to serve the interests of the hydraulic mining companies and who starved the infant CDC of the financial resources it needed to perform its tasks. Much of the rationale for the CDC was further undercut when massive floods in 1896 breached debris dams that had been built under the auspices of the CDC. If previous evidence had not been enough, it was now becoming clear that hydraulic mining and downstream farming were by definition mutually incompatible. In addition to reinforcing many stakeholders' distrust of governmental commissions, this event also underscored a pattern common among many early "expert commissions" of egregious errors in judgment. Although the framework for scientific inquiry was well established, vast gaps still existed in the empirical knowledge needed to make such informed decisions. In hindsight, many decisions often seemed wrong, over-confident, and arrogant. Hence, it is not surprising that it was extremely difficult for early environmental commissions to gain either the financial or the popular legitimacy they needed to enact their visions.

By the late 1890s it was clear that as long as the *North Bloomfield* decision remained the law of the land the hydraulic industry was doomed and, in fact, all large-scale mining had ceased by 1900 (Kelley, 1959). Just as the CDC looked moribund, it was rescued by the arrival of $250,000 in federal and matching state money to undertake a massive flood control and navigation project. From here on, the CDC would cease to regulate mining and be a flood management and river navigation agency, among the first agencies of this scale and expertise in the nation. Shortly thereafter, the federal government funded the CDC to start building among the first large-scale, governmentally-backed structural flood control works in the United States.

The importance of the CDC in the context of future watershed management did not end there. After the arrival of Thomas H. Jackson, an activist and ingenious engineer, as the de facto leader of the CDC, it began to espouse new ideas of multiple use of hydrologic systems that would not attain full acceptance for years to come. These ideas found their way into his 1910 "Jackson Report" which eventually won Congressional approval and stands as the primary foundation for the Sacramento Flood Control Project, which continues to this day (Kelley, 1989). His report proposed the revolutionary idea of tackling multiple problems within a hydrologic system in an integrated fashion, as opposed to the separable task-by-task approach previously espoused by the CDC and the U.S. Army Corps of Engineers (an agency which, in subsequent years, would institutionally entrench this multiple use concept as it became the dominant national actor in the engineered management of watersheds). Unlike many before him, Jackson thought at the watershed level and saw the close associations between debris, navigation, and flooding. The CDC proposed dredging rivers to increase flow, building levees to protect communities, and building bypasses and overflow basins to direct high floodwaters.

From this modest beginning, multiple use approaches to basin and watershed management went from being derided by players such as the US Army Corps of Engineers, to being the dominant paradigm at the highest levels of government by the New Deal. This paradigm culminated in agencies adopting the combined system of levees, bypasses, dredging, dams, and reservoirs that yield flood control, power, navigation, irrigation, and recreation, and that, until very recently, constituted the structural core of most large-scale river-basin management. Although this mixed-purpose, mixed-tool doctrine would likely have taken hold eventually, it is likely that the CDC, inheritor of the historic farm–mine conflict, played a critical role in moving this forward.

3. INDUSTRIAL POLLUTION AND THE EMERGENCE OF POINT SOURCE MANAGEMENT: AN EAST COAST STORY

Meanwhile, New England was shifting away from a natural resource-based economy towards industrialization. With the increasing power of technology to mass produce, new and previously unimagined types and levels of air and water pollutants were being generated, including newly synthesized toxic chemicals that had never before been released into the environment in quantity before. Also, indoor plumbing and sewers that had been developed to carry away the untreated solid wastes of growing cities served to maximize the impact of those wastes by delivering them in massive quantities into rivers.

New England was a very different political landscape from the West Coast. California's need for government intervention in its otherwise insurmountable environmental conflicts was at odds with the extremely Jeffersonian, anti-interventionist sentiments that predominated at the time.

It was only after decades of tragically failed attempts at other approaches that citizens finally acknowledged the necessity of governmental intervention from entities such as the CDC. In New England, on the other hand, while the geography did not lead to conflicts on such a grand scale, the predominating Hamiltonian political attitude created a situation far more amenable to government involvement in management of the environment.

During the mid to late 19th century, New England saw the development of a new generation of early "environmental pioneers," including scientists, engineers, and naturalists. Through their official capacities, many were among the first in the nation to bring the power of the state government to bear upon industry to mitigate its environmental and, most importantly, its public health impacts (Cumbler, 2001). Among this group of pioneers were, on one hand, moderates who sought to find scientifically based compromises that would not adversely inconvenience industry. On the other hand were public health advocates who attempted to confront – usually unsuccessfully – economic interests. Ironically, the public health-based argument for cleaning up rivers failed because of the discovery of the engineering solution of filtering drinking water, which did away with the immediate necessity of treating effluents.

Cumbler (2001) and Steinberg (1991) both detail these early years in Massachusetts, showing how, starting in the 1860s, a few scientists working in the framework of newly created state-level institutions began making inquiries about the causes of fish decline. This science would eventually lead to a far better understanding of how pollution affects people and the environment, as well as change the way government intervention in such matters was conceptualized. In 1865, one of the most thorough examinations of inland fisheries up to the time was undertaken, led by Theodore Lyman, soon to be commissioner of inland fisheries, and his colleague Alfred Read (Lyman was considered to be one of the environmental "moderates" by Cumbler). Initially concerned with dams, and later with sawdust in the Merrimack River, they soon found they were quite concerned with the large amounts of industrial dyes. At the time, rivers were seen as legitimate repositories for wastes, even by the experts (Steinberg, 1991).

The Merrimack Report qualified, perhaps for the first time in an official government correspondence, that not all types of river pollution were equally excusable. However, rather than decrying such dumping on environmental grounds, the report authors were doing so on utilitarian, economic grounds, suggesting that such pollution reduced the overall productivity of the river and reduced its investment value. According to Lyman and Read, given that the return on fish harvesting was high relative to the return on producing additional textiles, they felt that industry leaders would surely perceive that allowing for greater fish production was clearly in society's and their best interest, a view not shared by the mill owners (Cumbler, 2001). While the report did not deter the mill-owners sentiments, it did lead to the passage of one of the nation's earliest river management bills (albeit on a state level). While the bill did call for the formation of an official state commission (as opposed to the ad hoc committee that initially put the findings together), the commission's powers were largely confined to construction of "fishways" and had little to do with pollution.

Shortly thereafter, in 1869, the Massachusetts State Board of Health (MSBH) was created, resulting in one of the first public bodies to study the links between environmental and human health. However, it would still be many years before pollution became an important public policy issue. The 1870s saw a shift towards larger and more polluting types of industries in eastern Massachusetts. Among the types of industrial facilities that underwent the largest growth were paper mills, which then used lime chloride and sulfuric acid to bleach pulp, and woolen mills, which spewed out dyes and detergents.

By the late 1870s, water quality was so bad in many rivers that mills had to filter the water before using it. The pollution problem, while fairly minimal by 20th century standards, was still significant enough to get the attention of the MSBH. While a report did find the pollution problem to be significant, it was cautious in its recommendations, suggesting that pollution was, at least for the moment, an unavoidable side-effect of progress (Folsom, 1877). In other words, it was still too early in the evolution of the interventionist framework for a recommendation to be made as radical as demanding that industry alter its behavior. This mentality was furthered by the fact that there were still enough rural parts to the watershed, and enough flow, to dilute and attenuate the pollution levels in many rivers to the extent where, with some filtration, its effects were tolerable.

These factors kept active management of pollution off the political agenda for some years to come. Nevertheless, a core group of "activists" such as MSBH Commissioner Henry Bowditch were, at the time, articulating a vision for increased governmental intervention in managing the pollution problem – a vision that, in combination with the worsening of the pollution problem, would lay the groundwork for the eventual creation of governmental pollution-management institutions. The MSBH, which was becoming a repository for such radical thought, was increasingly proposing radical and unprecedented Hamiltonian ideas of governmental intervention. In their 1874 annual report they wrote, "Though hitherto the Massachusetts

Board of Health has judiciously abstained from the that general exercise of its authority which would prematurely cut short its usefulness, it is to be hoped that at no distant day the use of apparently arbitrary measures may become so common, and manifestly so beneficial, that the people at large will regard the existence of the Board as necessary and as immutable as the judiciary system'' (Massachusetts State Board of Health, 1874, quoted from *The North American Review*, 1974, *119*(245), 447).

Pollution levels grew steadily more intolerable throughout the late 19th century. As pollution got worse, and urbanization increased, the toll from sanitation-related diseases grew. For instance, in Hartford, CT 32 people died of typhoid in 1890 followed by 41 the next year. Moreover, the linkages between water pollution and disease became increasingly acknowledged among both scientists and citizens (Cumbler, 2001). At the time, however, no institutional framework existed for addressing such large-scale nuisance problems. In the past, most tort cases that involved water centered on rights to water flow. A few pollution tort cases had made it to the courts, but the current situation did not lend itself to traditional methods of nuisance litigation because of the diffuse and multiple nature of the offending agents. That is, as a downstream riparian land owner, who would you sue given that the pollution was being caused by such a vast array of users?

Given the inability of the courts to resolve this situation, it was becoming clear to officials that a more pronounced form of governmental involvement was imminent. As the Connecticut Board of Health noted in 1880, ''It is only a question of time how long it will be before each state … must provide some official means to also protect the public at large'' (Connecticut State Board of Health, 1880, p. 25 as quoted in Cumbler, 2001, p. 111). Not only was it becoming evident in New England that coordinated intervention was necessary, but also that the institutional level at which such action would have to take place was the state.

These conditions presented a window of opportunity in which the radical ideas of the ''environmental reformers'' such as Bowditch could finally see realization. Bowditch believed that, where public health was concerned, industry should be subject to the decisions of governmental bodies such as his, known informally as ''state medicine.'' Given that these ideas were entirely new to the United States at the time and had only recently been tentatively adopted in Europe, Bowditch's ideas were groundbreaking. Addressing this problem meant mitigating with two very different source types: industrial pollution and municipal sewage. It was the growing acceptance of industry's free-ride on the public trust that led the Massachusetts legislature to pass the Act Relative to the Pollution of Rivers, Streams and Ponds in 1878.

This act, borne of the necessity of Massachusetts' heavy state of industrialization, represented one of the earliest and most radical examples of governmental intervention in watershed management. It prohibited discharge of sewage into streams and rivers and it required that all industrial pollution be "cleaned and purified" before dumping into rivers. It also created a standing rivers pollution commission with sweeping authority to monitor pollution, review pollution plans, permit new actions, and issue nuisance injunctions. However, recognizing the burden this would place on industry, the act did exempt two of the most densely industrialized rivers (the Connecticut and Merrimack) and grandfathered polluting rights for some corporations. Despite these concessions, industrialists were irate. In arguments foreshadowing 20th century environmental discourse, industrialists complained that such regulations would strangle them, forcing them to leave the state and causing unemployment (Rosenkrantz, 1972).

What was critical about this act in the context of American watershed management was not just that it was an early example of governmental intervention in industrial affairs, but also that it represented the dawning realization in government that when it came to the environment, the public and private good were not mutually promoting but rather, "may even stand for the time in direct opposition to each other" (Massachusetts State Board of Health, 1886, p. 278). Despite its strong symbolic importance, the efficacy of the act, however, was considerably less. In one of the earliest examples of a conservative backlash against environmental regulation, the MSBH was merged with several other, largely unrelated commissions,[7] such that it soon lost most of its power and the voice of its reformers was drowned out. Former MSBH Chair Bowditch resigned in protest.

Thus, one of the earliest official Commissions with environmental regulatory power was disabled almost as soon as it was created in one of the earliest examples of agency cooptation. Bowditch came to believe, foreshadowing similar arguments to be made against environmental agencies in the 20th century, that the emasculating of his Commission had not come about through necessity of cost-cutting, as was claimed, but through the deliberate machinations of the industrialists in alliance with government officials who had purposefully weakened and coopted the Commission (Cumbler, 2001). In fact, the new Chairman, who was a laissez-faire conservative Democrat, had family connections to textile manufacturing and had been an industry lawyer (Conway & Cameron, 1909).

With the reformers on one side, and Donnelly, the governor and industrialists on the other, a considerable controversy grew over the staffing of the Commission and over the larger issue of the role of the state in public health. This soon drew such public attention that the *Boston Herald* did a multi-part series on the story. On the one hand were the reformers who, in their Hamiltonian tradition, saw that only a strong state could provide for the public health where pollution was concerned. On the other was industry and its allies, such as Donnelly, who stated to a *Boston Herald* reporter that "It is contrary to American ideas for the state to take care of the health of the people. By doing so self reliance is taken away. The average citizen needs no state Board" (Cumbler, 2001, p. 124). Rather than representing the opinion of the majority, these statements inflamed public opinion such that the governor had no choice but to create a new and independent board of health with reformer Henry Walcott as its chief.

Moreover, the anti-industrialist public sentiment led to the passage of a second major pollution act in 1886, which gave the board authority over the care of all inland waters (Steinberg, 1991). With its purpose to protect the purity of drinking water sources, including streams and ponds, this act represented one of the earliest and, for the time, most progressive watershed protection laws in the country. However, even more importantly, the scientific bodies that it empowered now had the power and resources to truly begin to understand the linkages between human and environmental health and to highlight "more powerfully than ever before, the extraordinary interdependence of the watershed" (Steinberg, 1991, p. 239). The gradual improvements in scientists' understanding of these linkages, as well as the dispelling of myths, such as the idea that rivers purified themselves, not only led to new technical solutions, such the development of sand filtration, but also had a wider effect in changing consciousness throughout the nation about waste and pollution. In fact, in the 1890s the MSBH was being contacted by public health officials from throughout the country (Rosenkrantz, 1972).

While many industries would avoid cleaning up for some years to come, and the improvements to water quality stemming from the development of new municipal sewage treatment methods would take attention away from industrial pollution for some time, the writing was on the wall; state intervention in the regulation of industrial pollution was there to stay and industry would have to learn to adapt, especially as new pollution mitigation technologies became available. In fact, after the example set by Massachusetts, this model spread and soon almost all new England states had their own Boards of Health, all with the statutory ability to monitor and regulate pollution in the interests of public health. By the turn of the 20th century, just as the Progressive moment was sweeping Washington, the Hamiltonian approach to environmental management appeared to be the new paradigm.

4. GROWTH OF GOVERNMENTAL INTERVENTION IN WATER RESOURCE MANAGEMENT

The 20th century would see the role of state and federal governments in regulating water quality and quantity grow considerably. In the early Progressive years of the century, there was a profusion of new government agencies, commission, and bodies, many with powerful legislative backing and appropriations. In the case of water, this era saw vast growth in agencies such as the Army Corps of Engineers which, as described by Kelley (1989), came to play a vital role in re-engineering the hydrology for vast areas of the country. Initially the Corps had limited most of its activities to navigation, a matter of interstate commerce and hence beyond state or local jurisdiction. However its shift, in the early 20th century, towards active involvement in structural flood control, and later damming and reservoir building, represented a radical departure towards a new doctrine of federal involvement in non-interstate commerce-related matters.

As Kelley (1989) and Worster (1985) point out, flood control and irrigation were initially thought of as individual tasks. When the limits of the individual approach became evident, they began to be thought of as community-level tasks. When those limitations became evident, they began to be thought of as county or state-level tasks. When the magnitude of the funding needed for massive basin-wide projects became apparent, they became thought of as state–federal partnerships. However, through the early 20th century, the power of the states in those partnerships relative to the federal government waned until agencies such as the Corps and the U.S. Bureau of Reclamation became entirely dominant in the areas of irrigation, flood control, hydro-power, wetlands draining, and damming.[8] Among the key policy mechanisms through which these powers and the necessary funding were granted were the Rivers and Harbors Acts of the late 19th and early 20th centuries.

This trend towards increasing concentration of power in a few agencies produced many unintended consequences. As the immense political power inherent in management for multiple uses became evident, these agencies attempted to increase their scope as far as they could. Often one agency eventually collided with competing agencies, resulting in massive inefficiencies, the most classic example of which is the clash between the Corps and the Bureau of Reclamation (Reisner, 1993). Many natural resource agencies took the Hamiltonian model to an extreme by developing large, top-down, expert-driven hierarchies that tended to isolate and alienate stakeholders (Thomas, 1999), a prime example being the US Forest Service (Kaufman,

1960). However, this model did not apply to all agencies. In particular, the Bureau of Land Management is much more decentralized, in part because it simply does not have the staff to operate as a large centralized bureaucracy and, hence has been viewed by many as more responsive to local concerns (Foss, 1960). The flip side of this, though, is that the BLM has also been considered by many to be a "captured agency," or one whose priorities lie more with its chief economic actors than with the public (Clarke & McCool, 1996).

Environmental values were seldom considered in the early years of these agencies, except inasmuch as they directly concerned human welfare. At the time, the nature of the indirect linkages between human welfare and the environment were poorly understood. As a result, many water and flood projects ended up having extremely deleterious environmental consequences and, in turn, long-term human consequences. For instance, the Army Corps of Engineers were the single greatest filler of wetlands yet, because wetlands attenuate floods, they were making their other job of reducing flood risk that much more difficult.

It was only in the latter half of the 20th century that science had advanced to the level where it could begin to adequately address the linkages between human health and welfare and the environment, as well as the intrinsic value of the environment. Policies shortly thereafter followed suit. While it is not the purpose of this chapter to provide a detailed inventory of all these pieces of legislation, something that has been done well in other studies (Pontius, 1997; National Research Council Committee to Review the New York City Watershed Management Strategy, 2000, Chapter 3), several deserve mention. Although neither particularly groundbreaking nor powerful, the Federal Water Pollution Control Act (FWPCA) of 1948 represented one of the earliest federal examples of this newly developing sentiment. Unlike previous acts whose wording was largely in terms of direct human welfare, this act called for consideration of public water supplies, propagation of fish and aquatic life, recreation, and agricultural and industrial uses.

Later amendments to this Act in 1961, 1966, 1970, 1972, 1977, and 1987 would result in more stringent standards and greater enforcement capability. The 1972 amendments (known as the Clean Water Act or CWA) were of particular importance because it was there that Congress established the National Pollutant Discharge Elimination System, which, for the first time at a federal level, required point source polluters to obtain a permit (something that the MSBH was trying to impose on polluters one hundred years prior). Another revolutionary concept established by the CWA was the Total Maximum Daily Load (TMDL) program. This requires state waterways out of compliance with Section 303 to develop TMDLs, or quantitative assessments of the maximum amount of a pollutant that a waterbody can receive and still meet water quality standards, along with a specification of the amount of pollution reduction that must be undertaken to meet those standards. Under TMDLs, pollution-control responsibilities are then allocated among polluters.

As this Act was being amended incrementally, additional water-related acts were being passed, including the Safe Drinking Water Act of 1974, which contained provisions for assessing, managing, and preventing the biological and chemical contamination of drinking water supplies, including surface water and groundwater. A 1989 amendment (the Surface Water Treatment Rule) required that all surface-drinking-water systems filter their water unless proven to not exceed certain coliform and turbidity levels.

The early 1970s also saw the creation of a variety of new public agencies designed to oversee the management of water and air resources and to implement, enforce, and administer this new suite of legislation. Among the most notable was the Environmental Protection Agency (EPA), created by President Nixon in 1970. The same reorganization plan that created the EPA also created the National Oceanic and Atmospheric Administration. That plan[9] stated that new organizations were needed because of the complexity of environmental problems. It recognized that to identify pollutants, trace them, determine exposures, examine interactions, and identify remedial actions, the newly formed EPA would require a highly complex organizational structure involving research, monitoring, creation of standards, and enforcement which, at the time, were function that were scattered about through a variety of federal and state agencies.

5. NON-POINT SOURCE POLLUTION AND THE WATERSHED PARTNERSHIP APPROACH

This suite of legislation began to successfully contain the problem of point-source pollution. However, the overall levels of many pollutants still remained high. It became increasingly clear that point sources only accounted for a portion of the total pollution load and that this regulatory approach was not successfully dealing with the more pervasive problem of non-point source pollution (John, 1994; Davies & Mazurek, 1998). This was partially because the command and control mechanisms set in place were inadequate for dealing with diffuse pollution sources crossing administrative and jurisdictional lines (Marsh & Lallas, 1995). Hence, while the spirit of this suite of legislation represented a major paradigm shift towards managing for environmental values, its end-of-pipe approach was only appropriate for dealing with certain types of problems. It was becoming increasingly clear that watershed-level problems required watershed-level solutions.

This concept was not new. The watershed approach to managing natural resources had been proposed to the federal government over 125 years ago, but was largely ignored at the time. John Wesley Powell, civil war veteran and explorer of the Colorado Plateau, was commissioned to write a report for Congress about approaches for settling the west and managing its natural resources (Geographical and Geological Survey of the Rocky Mountain Region (U.S.) & Powell, 1879). He saw, presciently, water as the critical limiting factor to western settlement. He recommended that all subsequent settlement should be planned based on watershed boundaries and that each hydrologic planning unit should be self-governing. Moreover, the planning of settlement should be based on meticulous surveys of water resources so that all units were within their carrying capacity.

The regulatory framework of the 1960s and 1970s gave little statutory authority for the federal government to regulate non-point source pollution on a watershed basis. There are no federal laws specifically mandating watershed management for source water (National Research Council Committee to Review the New York City Watershed Management Strategy, 2000). In fact, it was only in a court decision in 2000 that TMDLs were finally considered to apply to both point and non-point source pollution under Section 303d of the CWA,[10] a viewpoint that differed from previous interpretations (Larson, 1999) and that was politically unpopular among many interests groups.

This court case was important because it dealt with a largely rural watershed, with few point source polluters but with serious sediment-loading problems, caused largely by a forestry operation and the roads servicing it. The EPA TMDL, which called for a 60% reduction in sediment output, was upheld in court. In language reminiscent of the 19th century debate on mining debris, the court qualified the definition of "pollutant" under the CWA, which already included dredged spoil, rock and sand, as also including the term "sediment."

This critical decision implied that the EPA had indirect statutory influence over land use, generally considered the realm of states and municipalities. The influence was only indirect because EPA could only set the required outcomes through TMDLs, while states were expected to determine how to attain them. This proved to be a highly divisive issue, and came to a head when Congress included a "rider" (an addition to a legislative bill) prohibiting the EPA from spending on any new TMDL regulations anticipating that President Clinton would attempt to include a wider range of uses and sources, including non-point sources, under those rules. Opponents of the rules, representing the gamut of economic interest groups, loudly voiced their opposition using rhetoric strikingly familiar to that used by mill owners in Massachusetts 100 years earlier. Clinton's rules were never allowed to go into effect. In March of 2003, the Bush administration, as part of their promise to make environmental regulation more voluntary, withdrew the July 2000 rules drafted by Clinton.

While the role of the federal government in water-quality management was and is highly contested, there has recently been increasing consensus over the value of taking a watershed approach to regulation. This stemmed partially from the increasing importance of the watershed as an organizing principle in ecology, in particular "watershed ecosystem analysis" (Hornbeck & Swank, 1992) and partially from the recognition of the disconnect between socio-political boundaries and the spatial hierarchy of ecosystems (Keiter, 1994).

A few pieces of legislation tacitly recognized the importance of watershed management without directly requiring it. Among them were the 1987 amendment to the CWA,[11] establishing Section 319, under which states became eligible to receive federal grants to support non-point source management activities, and the 1996 Safe Water Drinking Act amendments, in which states were required to assess watershed conditions and create watershed control programs for unfiltered surface water supplies. As the 20th century drew to a close, the watershed approach gained attention and advocates within the EPA and elsewhere. This was reflected in EPA's 2003 Watershed Initiative program which funded roughly $15 million in grants to 20 watershed associations around the country to undertake community-based approaches to managing non-point source pollutants. This program was recently reauthorized for $20 million in 2004. In announcing the program, EPA Assistant Administrator G. Tracey Mehan emphasized his agency's commitment to the watershed approach: "The watershed approach should not be seen as merely a special initiative, targeted at just a selected set of places or involving a relatively small group of EPA or state staff. Rather, it should be the fulcrum of our restoration and protection efforts, and those of our many stakeholders, private and public. Failure to fully incorporate the watershed approach into program implementation will result in failure to achieve our environmental objectives in many of our nation's waters."[12]

These are not particularly powerful legislative tools, but they do represent the legitimation and validation of watershed-based thinking at the federal level. Today at least 17 federal resource agencies have officially adopted the watershed approach to some extent (McGinnis, 1999). Moreover, in 1998 a commission created by Congress to review federal water-resource policy released a report in which they, like Powell, proposed adopting new governmental structures based at least partially on hydrologic geography (Western Water Policy Review Advisory Commission, 1998).

While the role of government in watershed management today is important, its influence has been overshadowed by the development of a coordinated movement of grass-roots watershed associations. Over the last 20 years, hundreds[13] of watershed associations and councils have formed across the country, attempting to do much of the work that Powell had intended for government to do at the watershed level (although Powell was more concerned with water distribution and these modern groups tend to be more concerned with water quality). These organizations are generally locally directed and feature diverse private and non-governmental actors, but generally work in an officially sanctioned partnership with governmental institutions.

This multi-actor approach allows them to simultaneously deal with a large array of interlinked issues, such as timber harvests, salmon runs, agricultural best-management practices and ecosystem restoration, in a comprehensive way. To reach these ends, they utilize collaborative, voluntary, and consensus-based decision-making mechanisms, rather than traditional regulatory approaches (Kenny, 2000).These decisions are supposed to be informed by local knowledge and stakeholder preferences as well as by expert-driven science, although the role of science can be compromised if the appropriate checks and balances, such as independent reviews, are not included (Johnson & Campbell, 1999).

However, not all watersheds have their own associations and, where they exist, not all are the same in terms of effectiveness or approach. Recent research suggests that watersheds with severe pollution problems from agricultural and urban sources, as well as those with the lowest levels of command and control intervention, are the most likely to see watershed associations/partnerships develop (Lubell, Schneider, Scholz, & Mete, 2002). The study adds, however, that income is positively associated with the incidence of associations, while percent Black or Hispanic variables are negatively associated. This may be because higher income, more enfranchised communities have the resources necessary to secure intergovernmental grants and to overcome the transaction costs that stand in the way of enabling a successful watershed partnership (Yaffee, Philips, & Fretz, 1996). This result implies that watershed associations are not nearly as likely to develop in high minority watersheds. Hence, if the government were to place too much reliance on watershed associations relative to regulatory mechanisms to resolve pollution problems, they run the risk of disproportionately favoring the environmental quality of higher income, whiter communities.

6. CONCLUSION

This chapter has shown how political culture has, to a large extent, shaped the way that watersheds and water resources are managed in the United States. The two case studies outline what are among the earliest examples of governmental involvement in watershed management in the United States. Each was borne of an extreme resource conflict that was nearly irreconcilable without some higher level intervention. Each was also borne out of a unique socio-political landscape. In the case of California, the conflict started many years earlier but took much longer to militate for government involvement because of the highly independent, Jeffersonian nature of the rural populace at the time. In the case of Massachusetts, the conflict became intolerable much later, but intervention was quicker to come because of the greater prevalence of the interventionist and technocratic Hamiltonian political philosophy. However, both these cases set an important precedent in demonstrating that governmental involvement in natural resources matters, especially where related to water, was not only useful, but was here to stay.

Both of these conflicts were a result of technology. New, more destructive technologies increasingly underscored the inadequacy of the pure Jeffersonian approach to mitigating environmental conflict. These two examples illustrate the early years of that realization, while the Progressive era saw the further development of new governmental institutions designed to harness these new technologies for the "greater good" and, in some cases, contain their effects. But it was not until the 1960s and 1970s that the pure Hamiltonian approach fully enveloped the environmental realm when sweeping legislation finally gave the federal government the authority and tools to address water pollution.

Among many factors that likely led to the great success in the growth of the watershed partnership approach, two stand out. First, while the regulatory approach was adequate for dealing with point-source pollution, it proved inadequate for dealing with diffuse non-point sources. Second, the United States was and continues to be a divided nation in terms of political philosophy. Many voters disagreed strongly with the heavily top-down environmental regulations of the late 20th century, as expressed in periodic "backlash" periods where environmental regulations were weakened or repealed.

As the backlash grew, it became clear that local, grassroots initiatives would play an increasing role in watershed management. While not perfect, watershed partnerships represented not only a good way to deal with nonpoint source pollution, but also a new hybrid approach towards management of water resources, falling somewhere in between the Hamiltonian and Jeffersonian extremes and providing institutional arrangements that both could stomach.

NOTES

1. To avoid confusion, it should be noted that a changing economic context in the mid 20th century led to a flip-flop in the roles of the parties such that the Democrats became the party of greater government intervention as a check against the power of capital, while Republicans increasingly espoused the rhetoric of reducing the size of government.
2. A somewhat hypocritical perspective given that Jefferson and his fellow white Southerners, most of whom were Democrats, prospered off the labor of slaves (Jefferson owned about 200).
3. From John Adams' letter to Thomas Jefferson, 15 November 1813. Quoted from Ellis (2000, p. 235).
4. This is a method in which water is blasted through high-pressure hoses against the earth surface, freeing ancient deposits of former river-bottom gold.
5. One of the first major commissions created in this spirit was the Interstate Commerce Commission which was created in 1887 but remained essentially powerless until Theodore Roosevelt gave it additional powers in the early 20th century.
6. Similar River Commissions had been established for the Missouri and Lower Mississippi shortly before.
7. The Boards of Charity and of Lunacy.
8. The Corps maintained for decades that reservoirs did not work in flood control until about the 1920s, at which point it did such a rapid about face that it "built them at a pace that would have left the most ambitious pharaoh dazzled – something like six hundred in sixty years". Reisner (1993, p. 172).
9. Reorganization Plan of No. 3 of 1970. *35 F.R. 15623, 84 Stat. 2086, as amended Pub. L 98-80*, § *2(a)(2), (b)(2), (c)(2)(C)*.
10. Pronsolino v. Marcus, No. C 99-01828-WHA (N.D. Cal. March 30, 2000).
11. It read "Each management program proposed for implementation under this subsection shall include … development on watershed basis. A State shall, to the maximum extent practicable, develop and implement a management program under this subsection on a watershed-by-watershed basis within such State."

12. Memo from G. Tracy Mehan, III to EPA Directors. "Committing EPA's Water Program to Advancing the Watershed Approach" Dec. 3, 2002.

13. As of 1997, 958 watershed organizations had been identified by the Conservation Information Center. As of publication, the number is likely much higher.

REFERENCES

Benson, L. (1961). *The concept of Jacksonian democracy. New York as a test case*. Princeton, NJ: Princeton University Press.

Brechin, G. A. (1999). *Imperial San Francisco: Urban power, earthly ruin*. Berkeley, CA: University of California Press.

Clarke, J. N., & McCool, D. (1996). *Staking out the terrain: Power and performance among natural resource agencies*. Albany: State University of New York Press.

Connecticut State Board of Health. (1880). Second Annual Report, Connecticut State Board of Health.

Conway, K. E., & Cameron, M. W. (1909). *Charles Francis Donnelly: A memoir, with an account of the hearings on a bill for the inspection of private schools in Massachusetts, in 1888–1889*. New York: J.T. White & Co.

Cumbler, J. T. (2001). *Reasonable use: The people, the environment, and the state, New England, 1790–1930*. Oxford, New York: Oxford University Press.

Davies, J. C., & Mazurek, J. (1998). *Pollution control in the United States: Evaluating the system*. Washington, DC: Resources for the Future Press.

Ellis, J. J. (2000). *Founding brothers: The revolutionary generation*. New York: Alfred A. Knopf.

Folsom, C. (1877). The Pollution of Streams, Disposal of Sewage, etc. Massachusetts Boards of Health. Eighth Annual Report.

Foss, P. O. (1960). *Politics and grass: The administration of grazing on the public domain*. Seattle: University of Washington Press.

Geographical and Geological Survey of the Rocky Mountain Region (U.S.), & Powell, J. W. (1879). *Report on the lands of the arid region of the United States, with a more detailed account of the lands of Utah*. With maps. Washington: Govt. Print. Off.

Hornbeck, J. W., & Swank, W. T. (1992). Watershed ecosystem analysis as a basis for multiple-use management of eastern forests. *Ecological applications: A publication of the Ecological Society of America, 2*(3), 238–247.

John, D. (1994). *Civic environmentalism: Alternatives to regulation in states and communities*. Washington, DC: CQ Press.

Johnson, B. R., & Campbell, R. (1999). Symposium: Watershed policy – ecology and participation in landscape-based planning within the Pacific Northwest. *Policy Studies Journal, 27*(3), 502–529.

Kaufman, H. (1960). *The forest ranger: A study in administrative behavior*. Baltimore: Johns Hopkins Press (published for Resources for the Future).

Keiter, R. B. (1994). Beyond the boundary line: Constructing a law of ecosystem management. *University of Colorado law review, 65*(2), 294–333.

Kelley, R. L. (1959). *Gold vs. grain, the hydraulic mining controversy in California's Sacramento Valley; a chapter in the decline of the concept of laissez faire*. Glendale, CA: A. H. Clark Co.

Kelley, R. L. (1989). *Battling the inland sea: American political culture, public policy, and the Sacramento Valley, 1850–1986*. Berkeley: University of California Press.
Kenny, D. S. (2000). *The new watershed source book: A directory and review of watershed initiatives in the Western United States*. Boulder, CO: University of Colorado School of Law, Natural Resources Law Center.
Larson, R. L. (1999). Court rules the CWA does not regulate non-point source. *Water, Environment and Technology*, *11*(1), 72.
Lubell, M., Schneider, M., Scholz, J., & Mete, M. (2002). Watershed partnerships and the emergence of collective action institutions. *American Journal of Political Science*, *46*(1), 148–163.
Marsh, L., & Lallas, P. (1995). Focused, special-area conservation planning: An approach to reconciling development and environmental protection. In: D. R. Porter & D. Salvesen (Eds), *Collaborative planning for wetlands and wildlife: Issues and examples* (pp. 7–33). Washington, DC: Island Press.
Massachusetts State Board of Health. (1874). *Fifth Annual Report*. Boston.
Massachusetts State Board of Health. (1886). *Eighteenth Annual Report*. Boston.
McGinnis, M. V. (1999). Symposium: Watershed policy – making the watershed connection. *Policy Studies Journal*, *27*(3), 497–598.
National Research Council Committee to Review the New York City Watershed Management Strategy (2000). *Watershed management for potable water supply: Assessing the New York City strategy*. Washington, DC: National Academy Press.
Pontius, F. W. (1997). Overview of SDWA source water protection programs. *Journal American Water Works Association*, *89*(11), 22–24.
Reisner, M. (1993). *Cadillac desert: The American West and its disappearing water*. New York: Penguin Books.
Rosenkrantz, B. G. (1972). *Public health and the state: Changing views in Massachusetts, 1842–1936*. Cambridge: Harvard University Press.
Skowronek, S. (1982). *Building a new American state: The expansion of national administrative capacities, 1877–1920*. Cambridge [Cambridgeshire], New York: Cambridge University Press.
Steinberg, T. (1991). *Nature incorporated: Industrialization and the waters of New England*. Cambridge [England], New York: Cambridge University Press.
Steinberg, T. (2002). *Down to earth: Nature's role in American history*. New York: Oxford University Press.
Thomas, C. W. (1999). Symposium: Watershed policy – linking public agencies with community-based watershed organizations: Lessons From California. *Policy Studies Journal*, *27*(3), 544–564.
Western Water Policy Review Advisory Commission (1998). *Water in the West: The challenge for the next century*. Report of the Western Water Policy Review Advisory Commission. Arlington, VA.
Worster, D. (1985). *Rivers of empire: Water, aridity, and the growth of the American West*. New York: Pantheon Books.
Yaffee, S. L., Philips, A., & Fretz, I. C. (1996). *Ecosystem management in the United States: An assessment of current experience*. Washington, DC: Island Press.

PART III: SCENARIO ANALYSIS AND INTEGRATED WATERSHED MODELING

SCENARIO ANALYSIS IN THE ELBE RIVER BASIN AS PART OF INTEGRATED ASSESSMENT

Frank Messner

ABSTRACT

This chapter describes the scenario technique used for the integrative methodological approach (IMA) of the German global change project GLOWA Elbe. It is outlined how regional scenarios are systematically derived to analyze water use conflicts and their resolution in the context of global change for the German Elbe river basin. Through the combination of frameworks of development and policy strategies a consistent set of developmental scenarios can be generated that makes it possible to examine the regional impact of policy strategies under conditions of different future global change development paths. The scenario analysis of the framework of development starts on the global level with qualitative IPCC storylines, translates them to the regional level, and quantifies their regional effects by means of modeling and statistical estimation methods. The policy strategies are derived in close cooperation with regional stakeholders.

Ecological Economics of Sustainable Watershed Management
Advances in the Economics of Environmental Resources, Volume 7, 69–96

ISSN: 1569-3740/doi:10.1016/S1569-3740(07)07004-6

1. INTENTION AND APPROACHES OF SCENARIO ANALYSIS

Scenario analysis is a scientific tool to explore the future and to integrate future developments into scientific analysis. In order to reflect the high degree of uncertainty about the future and the momentous changes in nature and society that may happen, a bunch of scenarios with different sets of assumptions is constructed to get an idea of the range of possible future developments. Scenarios are often based on scientific modeling (e.g., Alcamo, Leemans, & Kreileman, 1998), but sometimes qualitative reasoning is also used as a technique to explore possible future developments in a thinking experiment (Aligica, 2005). In the context of scenario analysis, the term scenario can be defined as follows. A scenario comprehends three elements: first a delineation of the current state of the world or the region under consideration, second a description of a possible or desirable future state, and third a bundle of events and policy actions over time that determines the development path (Veeneklaas & van den Berg, 1995). In scenario analysis, two different kinds of scenarios are distinguished, projective and prospective scenarios. Projective scenarios are based on the developments of the past and the present and extrapolate into the future. Prospective scenarios are more normative in character. They start up with a desirable future state and, in order to "backcast" the developments that may lead to this future state they contain explanatory variables of change (Schoonenboom, 1995).

In contrast to forecasting, scenario analysis does not intend to predict "what will happen" in the future, but it explores possible futures or the scope of future developments in connection with a specific question in order to clarify "what might happen". While forecasting efforts build on historical regularities and well-known dynamics to make a best guess on future developments, scenario analysis intends to open up the future perspective and to include unexpected changes in the future (Schoonenboom, 1995).

When used to aid political decision making, scenarios have two main functions. Firstly, given that politicians are generally committed mainly to the present, scenarios can make them aware of possible future situations. Especially, the fact that the future scale of various effects could differ enormously from the present can be elucidated. Secondly, scenario analysis can be used to reveal how widely the effects of political action may vary under different natural, economic, and legal future conditions (cf. Veeneklaas & van den Berg, 1995). The second function will be highlighted in this chapter.

In the following, the scenario analysis technique used in the context of the integrative assessment approach IMA will be presented, which was the central methodology applied in the project "Global Change and the Hydrological Cycle of the Elbe River" (GLOWA Elbe) financed by the German Ministry of Education and Science (BMBF). After a theoretical delineation in Section 2, the application of the technique is portrayed in Section 3 for the analysis of the water allocation problem in the Elbe river basin under conditions of global climate and societal change. An outlook in Section 4 will complete this chapter.

2. SCENARIO DERIVATION WITH THE INTEGRATIVE ASSESSMENT APPROACH IMA

IMA stands for integrative methodological approach of the German global change project GLOWA Elbe. It is a scenario-based and participation-oriented integrative assessment approach to evaluate policy strategies and future scenarios in accordance with the principles of sustainable development. It provides an overall framework to structure participatory evaluation processes on public decision issues under explicit consideration of global change processes. This approach has been conceptualized to support public decisions on complex environmental problems and conflicts in the context of global change affecting many people, large regions, and long periods of time, involving considerable social, ecological, and economic effects, and comprising significant uncertainty issues. The major goal of IMA is to improve the quality of environmental decision making. IMA can be described by four major research elements. Element 1 consists of problem and stakeholder analysis as well as scenario derivation in close cooperation with stakeholders. Element 2 focuses on identification of indicators and criteria to evaluate the scenarios. Element 3 deals with impact analysis in order to quantify scenario effects. And, finally, element 4 encompasses evaluation of scenarios using benefit–cost and multicriteria analysis in close cooperation with stakeholders (Messner, Zwirner, & Karkuschke, 2006; Becker et al., 2001; Klauer, Drechsler, & Messner, 2006; Klauer, Meyer, Horsch, Messner, & Grabaum, 2001; Wenzel, 2001; Horsch, Ring, & Herzog, 2001). The IMA approach is described in more detail in Chapter 12 (Messner, in this volume). In this section, only the scenario analysis technique used in the IMA context is presented.

In order to analyze complex public decision processes in the context of global change different specific notions of the term "scenario" are needed. In the IMA approach, four specific types of scenarios are distinguished:

global change scenarios, policy action scenarios, developmental scenarios, and baseline scenarios. They are defined in the following.

As Fig. 1 exhibits, the four different types of scenarios feature a close hierarchical relationship. Broadly speaking, scenarios of global change that contain exogenous boundary conditions for the study region are combined with political action scenarios inside the study region to identify developmental scenarios that are at the focus of scenario analysis.

Global change scenarios describe the development of important driving forces of natural and societal global change (like climate change, population development, water demand, etc.). A bundle of global change scenarios that contains scenarios for all important driving forces is called *Framework of Development* (FoD). From the perspective of a regional decision maker an FoD contains scenarios with all important external change processes that are relevant for the public decision and that cannot be influenced by the decision maker. Hence, even changes in policy at the national level are external boundary conditions for him and therefore national policy change scenarios must be considered in the FoDs for a regional analysis. From the perspective of a national decision maker, however, national policies can be influenced and therefore they do not belong to the FoD for an analysis with a national focus. Thus, it is important to state that the bundle of global change scenarios necessary to be considered in the FoDs depends on the issue to be analyzed as well as on the regional scale of the decision process. One problem in the building of FoDs reads that a large number of global change scenario combinations can be constructed using the decision tree technique (Messner et al., 2001). Hence, it is necessary to select a limited number of scenario combinations. In the IMA approach the storyline technique is used to select scenario combinations for the FoDs. *Storylines* tell a qualitative story about how a future might look like under different circumstances and different dominant driving forces. Such storylines can be designed by interdisciplinary scientific working groups in cooperation with important representatives of the public and the political sphere. For these storylines, FoDs are worked out that contain quantitative future change scenarios that match the qualitative assumptions of the underlying storylines.

A *policy action scenario* represents the actual execution of a policy strategy over time in order to resolve specific problems or conflicts in the region under study or to respond to potential challenges of global change. A policy strategy is defined as a combination of policy options from one or several policy fields (like water policy, agricultural policy, etc.) that also includes a specific attitude regarding policy adjustment in times of societal and/or

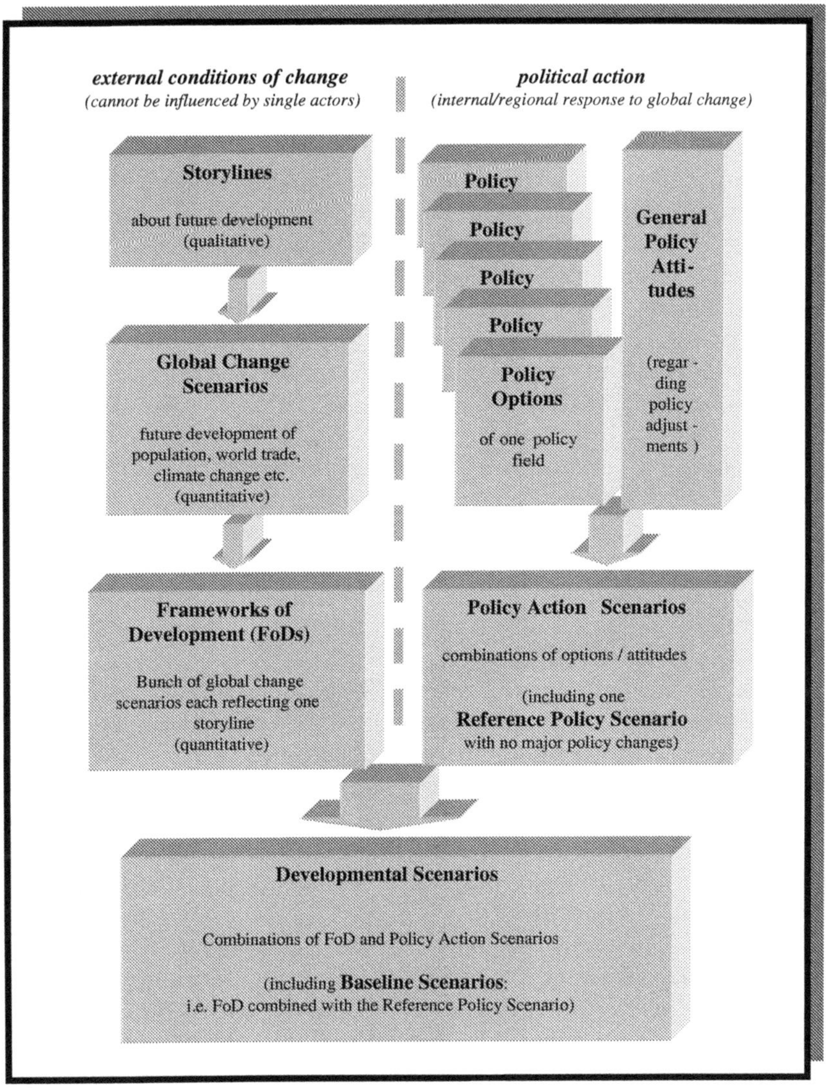

Fig. 1. Types of Scenarios in the IMA Approach.

natural change events (e.g., a risk averse attitude or an affirmative attitude toward policies that are very market oriented). A policy action scenario is characterized by a set of chosen policy options with a timetable for their practical implementation. In IMA, one of the policy action scenarios, which usually is characterized by no major policy changes, is defined as a *reference policy action scenario* to which other policy action scenarios can be compared to examine the impact of policy change.

Developmental scenarios are eventually those scenarios that contain both, an FoD with several global change scenarios as external conditions of change and a policy action scenario reflecting the internal factors of change. These scenarios are subject to impact analysis and evaluation in the IMA research elements 3 and 4 and they are in the center of scientific attention in the interdisciplinary analysis of processes of global change and political action.

Finally, *baseline scenarios* are developmental scenarios with any of the FoDs combined with the reference policy action scenario. These scenarios serve as reference scenarios in the overall analysis. On the one hand, they are used to compare the different future visions reflected in the different FoDs without any change in policy action. On the other hand, in the evaluation of policy strategies for one FoD, a baseline scenario serves as a reference to evaluate the alternative policy strategies.

Since developmental scenarios are subject to the evaluation in IMA they are at the focus of analysis. These scenarios are both, projective and prospective in character. This hermaphrodite disposition is due to the fact that in IMA global change scenarios are projective in the sense that current global change dynamics are projected into the future. On the other hand, policy action scenarios, which are worked out together with stakeholders, are prospective because they are tied to certain policy goals that reflect specific desirable future states.

Choosing this way of analyzing policy strategies with policy action scenarios and FoDs means to explicitly include future uncertainty into scientific analysis. However, the price of proceeding this way is high, because in most cases there is no single policy action performing best. Even if a well-performing policy solution can be identified for every FoD, a favorable solution for *all* FoDs is only existent, if a policy strategy performs best for all FoDs.[1] However, this case will not occur very often. In most instances, certain policy strategies are proper solutions for the various FoDs. Based on this information, decision makers have to select the policy they prefer. If they are risk averse they would tend to select a solution with an average yet acceptable performance in all FoDs. If they are prepared to take a risk they would tend to select a solution that performs best in an FoD they deem to be most realistic and accept the risk that this policy could perform poorly if another FoD becomes reality in the future.

3. SCENARIO DERIVATION RELATING TO A WATER ALLOCATION CONFLICT IN THE ELBE RIVER BASIN

3.1. The Water Allocation Conflict

In the river basins of the river *Spree* and the river *Schwarze Elster*, which are sub-basins of the Elbe river basin in eastern Germany, a water quantity and quality problem arose in recent years due to 100 years of excessive lignite mining. In the early 20th century, lignite mining was initiated in the area, and in the last decades large-scale open-pit mining was practiced with production of up to 220 million tons per year in Lower Lusatia, consuming 2,000 ha of land every year (see Map 1). Since open-pit mining requires a lowered ground water table, the mining areas were dewatered. In view of the

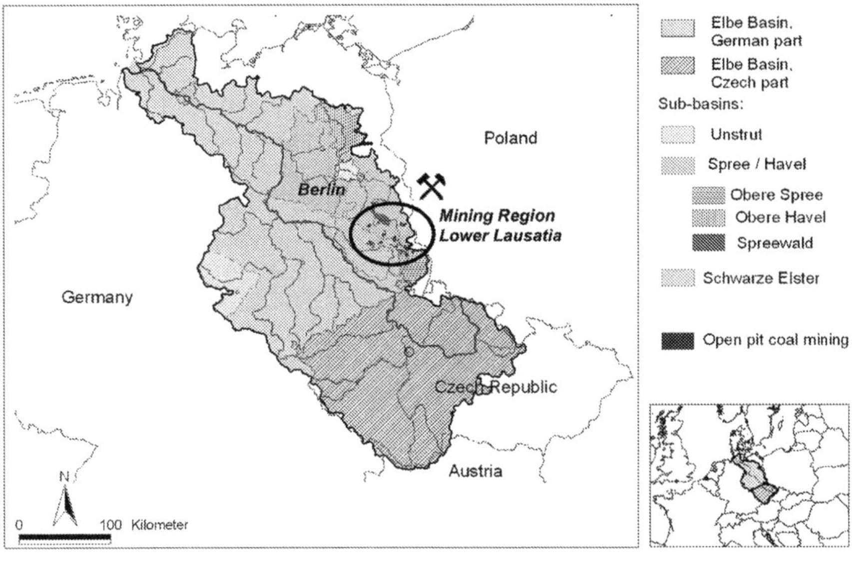

Map 1. The Lower Lusatia Mining Region in the Elbe River Basin.

fact that extracting one ton of lignite involves pumping of more than six tons of groundwater, about 1,220 million cubic meters of groundwater was pumped out of the subsurface per year in the late 1980s. As one result of this process, a groundwater depression cone of about 2,100 km^2 with a cumulative groundwater deficit of 13 billion cubic meters was generated, which will call for restoration after the closure of the pits (Grünewald, 2005; Koch et al., 2005). Another result was that a relatively dry region with precipitation of about 550–600 mm per annum was changed by man into a surface water abundant region. Consequently, economic use of surface water increased during the 20th century: among others, the Spreewald wetland area and the related tourism could evolve better, fish farming activities benefited from high surface water supply and the capital city of Berlin and its about three million inhabitants did never experience problems with water provision.

However, things changed drastically after German reunification in 1990 when, as a result of economic re-structuring in eastern Germany, several unprofitable mines were closed and the production of lignite decreased down to 50–60 million tons per year. As a consequence, the pumping of mining water into the Spree and Schwarze Elster rivers was reduced by about 50%. At the same time of declining surface water availability a new and water-intensive type of water use came into existence: the restoration of the closed mining pits and the plans for future tourist use of the mining lake landscape. Millions of cubic meter of surface water is needed to countervail the groundwater depression cone and to fill the lakes. Quick action is also needed, because a slow filling of the mining lakes can lead to acid mine drainage (Grünewald, 2005). And a new lake landscapes with acid lakes would not be attractive for tourism, which is one of the few economic activities with a potential to advance economic development in the Lusatia region. On the other hand, reduced surface water in the rivers and the huge water demand to fill the mining pits also involved the consequence that essentially less water was available for other water users. A variety of traditional water users (energy production, fish farming, inland navigation, tourism at reservoirs, and in the Spreewald wetland) located in three different German federal states (Saxony, Brandenburg, and Berlin) had to fear that they would loose off in the dispute over water allocation. Consequently, a typical water allocation conflict threatened to arise.

Intensive negotiations started with decision makers from the three German federal states involved. Due to the many interests and asymmetric power constellations in this water conflict it took about 10 years to agree on a common water management strategy for the Spree and Schwarze Elster rivers. This strategy is described in the report on the *Principles of the transboundary Water Management of the Rivers Spree and Schwarze Elster* (AG Flussgebietsbewirtschaftung, 2000). Major goals stated in this report are: not to deteriorate the position of traditional regional water users and achieving a quick filling of the lignite pit mines. In accordance with these goals, water use priorities were formulated, stating that traditional water users (including instream flow requirements) are to be served first with surface water of the river basin, refilling of reservoirs being the second priority, and filling the pit mines being the third priority. In order to prevent water scarcity at all, it was decided to complement the water availability of the affected region by water transition from the nearby Neiße river and by optimizing water distribution by means of computer-based operational water control. For the context of the scientific scenario analysis, this water management strategy is called *basic strategy*.

3.2. Frameworks of Development (FoDs)

An important prerequisite to generate FoDs for the future analysis of this conflict was to define the geographical boundaries of the regional study area and a time horizon for the analysis. With regard to the geographical boundaries this task seemed to be easy done from a water resources management point of view, because the river basin is the unit of water resources management analyses. Therefore, to analyze the conflict described in Section 3.1 the hydrological boundaries of the Spree and Schwarze Elster river basins were considered to be the best choice for the geographical specification of the study area. However, in the context of the GLOWA Elbe project additional conflicts and problems in other parts of the Elbe river were planned to be analyzed by means of scenario analysis, too. As a consequence, it was decided to choose the Elbe river basin as a super ordinate geographical reference area. For this reference area, overall datasets relating to global change trends were to be drawn up. Based on this super ordinate reference area for the whole project, problem-specific sub-basin reference areas could be defined to create a more specific data pool with disaggregate data for the analysis of regional conflicts like in the Spree and Schwarze Elster river basins.

Another problem related to the geographical boundaries was the lack of socio-economic data at the river basin level. Usually, socio-economic data and trends are only available for administrative units like counties or states, which, however, are often only partly located in one river basin or sub-basin. Therefore, relating to socio-economic data all administrative county units overlapping with the hydrological boundaries of the Elbe river basin and its sub-basins, respectively, were used together as a socio-economic reference area. That way it was possible to crudely approximate the hydrological boundaries of the reference areas. If even on this county scale data were not available, the five East German federal states plus the capital city of Berlin were considered to be the socio-economic super ordinate geographical reference area relating to future scenarios in the Elbe river basin. From a socio-economic point of view even this very rough geographical approximation of the Elbe river basin is still appropriate for the definition of the FoDs. The reason for this is that this region of the former socialist German Democratic Republic plus the capital city of Berlin – called Eastern Germany in the following – can be characterized by similar economic, cultural, and social trends during the past decades.

Regarding the general time horizon, the involved scientists chose the period 2003–2052 – i.e., about the first half of the 21st century – as the common reference period of interdisciplinary analysis. This was not an easy choice due to different time scales of global change processes and their scientific examination. For example, climate change processes are relatively slow and are usually analyzed and modeled over the time scale of centuries. However, considering half a century can already reveal first changes and impacts due to climate change in the reference areas. Contrary to this, changes in society and economy are usually much quicker in their evolution and a robust analysis of these processes cannot comprise many decades. Nevertheless, if the objects of analysis are long-lived capital stocks like power plants, mining pits, and tourist regions or if some processes are very stable in their evolution – like developments in population – crude propositions can be made for 50 years. An exception was made with regard to agriculture, because its regulatory framework strongly determines production figures. Therefore, only a time horizon until 2020 was chosen to reflect the relatively stable institutional conditions related to the EU agricultural policy, which is given by the AGENDA 2000 (EC (European Commission), 2006).

After having defined the geographical and time boundaries of the regional investigation, the next work to be done for the derivation of the FoDs was the identification of the most important external driving forces and boundary conditions of global change and regional development with essential direct and/or indirect impacts on the regional water cycle of the Elbe river basin and its sub-basins. In this context, eight exogenous factors with a significant influence on regional development and the water cycle of the

Spree and Schwarze Elster river basins were identified. They are listed and their choice is motivated shortly in the following:

- Climate change is an important driving force, because it determines the quantities of naturally available water.
- Economic development in general is a crucial driving force due to the relation between water demand, economic production, and regional income.
- Population development was chosen as a driving force, because it governs the availability of regional manpower and the water demand of households.
- EU water policy represents an important boundary condition in the form of the legislatory framework for water use and water quality in all kinds of water bodies.
- EU agricultural policy is an important boundary condition with regard to the legal requirements for the usage of nutrients in agricultural practice and its impact on non-point pollution of rivers.
- Energy policy and electricity production are crucial driving forces, which determine the availability of surface water in lignite mining areas. Moreover, more than 50% of German surface water usage is used as cooling water in electricity production (StaBu, 2001).
- Development of fish farming and respective national and EU policies were chosen as boundary conditions, because the study region includes many fish farming activities (generating a turnover of several million Euro per year, BLE, 2004), which are very dependent on surface water and very vulnerable in case of reduced water availability.
- Development of tourism in the region is an important boundary condition, because tourism does represent an increasingly important source of water-related economic income in eastern Germany and the economic development of the study region, Lower Lusatia, will depend on future tourism trends.

For these exogenous factors, scenarios of change were to be generated and combined in FoDs in a way as to characterize potential paths of future development in the reference areas by means of quantitative data. To accomplish this, the following research activities were executed: first, defining qualitative storylines for a global scale; second, transferring these qualitative storylines to the regional sphere of the study areas; and third, quantifying the regional storylines by means of modeling techniques, statistical estimation methods, trend analyses, and the like. These three research activities are described in the following.

3.2.1. The Global Change Storylines A1 and B2 of the IPCC

In order to ease and harmonize the task of selecting assumptions for the global change scenarios two storylines of the IPCC (2000) were chosen. These storylines, called "A1" and "B2", were used to qualitatively describe sets of assumptions for different FoDs for global change in the Elbe river basin and its sub-basins. The storylines characterize the future paths mainly through different forms of economic, societal, technological, and policy developments. The storyline A1 characterizes future development of the world for the coming decades by rapid economic growth including increasing globalization and liberalization of markets as major features, rapid global population growth up to 2050 but decline afterwards due to global convergence of fertility rates, substantial reductions in regional peculiarities, quick development and diffusion of new technologies, and rather defensive environmental policy. The storyline B2 has a quite different focus: it describes future development with an emphasis on local solutions to economic, social, and environmental sustainability (regionalization). Economic and population growth are both less rapid than in A1, technological development and diffusion are also less pronounced, whereas environmental and climate policies are given political priority with precautionary policy solutions (IPCC, 2000, p. 4f). These two storylines must be understood as "empty tubes" that define the direction of global future development qualitatively, but which, in order to be applied in the regional case study, needed to be translated to regional circumstances and, finally, quantified using appropriate scientific methods.

3.2.2. Regionalizing the Qualitative Storylines to the Study Area

Transferring the qualitative global storylines to the regional sphere of the super ordinate reference area of the Elbe river means to derive appropriate regional assumptions that translate the global storylines into a regional story. This could be executed rather straightforward for most of the exogenous factors. The regionalization of the qualitative global storylines of A1 and B2 resulted in the following regional assumptions for the external factors, which significantly determine the regional features of global change.

Economic development. In general, the mean economic growth path of the EU was assumed to be about 2% per annum over the next decades. More important for the regional development is the question to what extent equalization of economic development, income, and living standards in eastern and western Germany will take place. For the globalization storyline A1, it was assumed that growth rates and income per inhabitant in eastern Germany will conform to western trends over time. On the contrary, for B2 it was supposed that current trends will continue, leading to an economic decoupling of eastern Germany with growth rates and income positions remaining distinctly below the western standards.

Population development. The population development in Germany is characterized by a decreasing trend. This is mainly due to an almost constant fertility rate of about 1.4 (1,400 children for every 1,000 women) since 1980. The German federal statistical office frequently executes population development prognoses. In its last prognosis, the population development trend for 2000–2050 was estimated (Statistisches Bundesamt, 2000) for two major variants. Both of them assume a constant fertility rate of 1.4 over time, but they differ in terms of net immigration, with one variant showing a presumed number of net immigration of 100,000 persons per year, the other with 200,000 persons per year. These prognoses were taken as a basis for the regional population scenarios. Regarding the classification of these scenarios to the storylines A1 and B2, the population scenario with the higher net immigration figure was selected to represent A1. This attribution was chosen, because globalization and liberalization imply an increasing migration activity compared to a regionalization storyline context.

EU Water policy. The most important water policy framework for Germany is the EU Water Framework Directive (WFD). This is a large body of innovative legislation that has been passed in 2000 by the EU and it will have a profound impact on national water policies in EU countries in the coming decades. The WFD prescribes very ambitious water quality and quantity goals and requires that every water body in the EU should attain good ecological status in the year 2015 (see Petry & Dombrowsky, in this volume). However, the WFD also includes several exceptional rules such that the attainment of its goals can be postponed if economic or social necessities prevail. It is assumed that the exceptional rules are much more applied in the globalization storyline A1 than in the regionalization context B2. As a consequence, the quality of water bodies in the A1 world is lower than in the B2 world. For the latter it is taken for granted that the WFD goals will be achieved in 2015.

Agricultural policy in the EU. The global storyline B2 was interpreted in a way that the EU in terms of agricultural production remains quite separated from world trade. As a consequence, the agricultural policy as indicated by the Agenda 2000 was taken as a basis for this storyline and its effects on the regional agricultural sector. In terms of environmental impacts, it was assumed that subsidy payments will successively change in character as to promote environmentally benign production methods. Regarding A1, it was assumed that far-reaching liberalization and globalization will take place in agricultural production with massive subsidy cuts and substantial deregulation of the current quota regime.

Sectoral development and policy relating to fish farming. Fish farming is highly protected in Germany and the EU, similar to the agricultural sector. For the regionalization of storyline B2 it was therefore assumed that this protection will prevail in the future due to the cultural and environmental value of fish farming activities and the related lake landscapes. Furthermore, while subsidies were assumed to remain stable, direct commercialization of fish products was supposed to increase to 50% of the total revenue with the consequence of small average price increases for regional fish. For A1, it was presumed that in the surge of liberalization subsidies will be cut by half and direct commercialization will decrease to 25%, indicating slightly decreasing average fish prices in the region.

Development of tourism in the region. Tourism will be an important future economic activity in the former and current mining areas of eastern Germany. The number of tourists visiting the new landscapes and the revenues of the sector will depend highly on the attraction of the regions on visitors from outside – nationally and internationally. Corresponding assumptions were linked to tourism experiences in a traditional German lake landscape region – the franconian lake region in southern Germany (Eckart-Montanconsult and Planung/IBA Fürst-Pückler-Land, 2002). For the regional storyline of A1, it was assumed for the tourism sector that the Lusatia lake landscape will be able to realize similar turnover results as they are usual in the franconian lake region 10 years after the start of tourism in the new landscape area. For B2, a more regional-oriented and ecologically soft tourism was supposed with less tourism from abroad and therefore with low turnover figures up to one third compared to the franconian region.

Energy policy and regional climate change. The translation of the global storylines into regional ones turned out to be more complicated for the future trends of energy policy and production as well as for regional climate change.

The most important assumption regarding the regional energy sector and its impacts on the regional water cycle related to the amount of regional lignite extracted and burned for electricity production in future times. Two variants were generated together with the major energy producing enterprise: for the first it was assumed that electricity production and regional lignite output will not decrease for the chosen time horizon and for the second one a declining trend in electricity production was adopted with a phasing out scheme for lignite use correlated to the life-time of current power plants (about 35 years). However, it was not possible to exactly classify these energy scenarios as either A1 or B2. Considering the variant with continued long-term lignite use, it could be argued that this should belong to A1 because of the large significance of lignite energy resources to sustain a higher level of growth and electricity production and an energy policy that does not limit the burning of lignite due to environmental reasons. Contrary to these arguments, it could also be claimed that this variant characterizes B2 because the use of the *regional* lignite resources is emphasized and, what is more, the environmental performance of the new eastern German lignite power plants in terms of energy efficiency and CO_2 emissions is much better than many old western German power plants. Therefore, this variant could also be attributed to be the one that has a higher emphasis on regional development as well as on environmental and climate policy. With accordant arguments, the variant with declining lignite use could also be attributed to A1 or B2. To resolve this problem it was decided that the variant with declining lignite consumption, which was an accepted variant by the stakeholders and which also involved a higher pressure on surface water availability, will be used for both FoDs – while the constant use variant is just used in sensitivity analysis.

Regarding regional climate change, regional climate models based on the global IPCC storylines were used to translate the global assumptions to regional impacts. As a consequence, different regional climate change scenarios were created for eastern Germany. However, these scenarios only start to show significant differences in precipitation and temperature developments in the second half of the 21st century, i.e., outside the time horizon chosen in the GLOWA Elbe project. As a consequence, it was decided to use only one climate change scenario (for A1 *and* B2). It was complemented by a status-quo climate scenario characterized by the absence of climate change, reflecting a climate according to the time phase 1951–2000. In order to combine these scenarios with the socio-economic factors described above, it became necessary to double the FoDs, being now:

1. status quo climate *plus* A1 in socio-economic terms (short: A1 + no climate change);
2. status quo climate *plus* B2 in socio-economic terms (short: B2 + no climate change);
3. climate change *plus* A1 in socio-economic terms (short: A1 + climate change);
4. climate change *plus* B2 in socio-economic terms (short: B2 + climate change).

3.2.3. Examples for the Quantification of Regional Global Change Scenarios
The quantification of the qualitative assumptions of the regional storylines was performed by using very different methods, including sophisticated sector or climate models as well as statistical methods to regionalize and downscale data of already existing national prognoses or estimations. Due to space restrictions, it is not possible to show all global change scenarios that were generated for all important driving forces and boundary conditions. However, the methods and results for the most important ones regarding their impact on the water cycle of the super ordinate reference area of the Elbe river will be presented. These are: energy policy and production, agricultural policy, population development, and climate change.

Energy policy and production. The general basis to estimate the use of primary energy resources as well as the development of electricity production and water use by the east German energy sector in the context of the European and global energy markets was the energy report III for Germany (Prognos, 2000). For this report, the energy sector model EIREM was used to approximate the power plant structure of Germany within the evolving EU for the coming decades until 2020. Using this information as input and complementing the assumptions on the regional storylines A1 and B2 regarding the future use of regional lignite, the IKARUS data base with specific micro data about all power stations in Germany and the KaSIM sector model (Wagner & Stein, 1999; Martinsen, Kraft, & Markewitz, 2001) were applied to calculate the electricity production in eastern Germany according to primary resources used for both energy scenarios and to compute the respective water use of energy production. The results are shown in Fig. 2a and b and Fig. 3a and b below.

Fig. 2a and b show the trend of energy production in eastern Germany for the two energy scenarios described above. First of all, it must be stated that the structure of electricity production in the eastern part of Germany is highly dominated by lignite power plants. In the case of the continued use of lignite in electricity production (Fig. 2a), the structure of the energy sector remains more or less the same. The calculations for the case of phasing out of lignite in electricity production (Fig. 2b) brings about a structural change with gas and renewable energy resources as substitutes for the declining use of lignite.

Fig. 3a and b show fresh water demand, used water discharge and mining water pumped from the subsurface related to the two regional energy scenarios. As the figures reveal fresh water input and used water discharge are declining over time in both scenarios, mainly due to the diffusion of

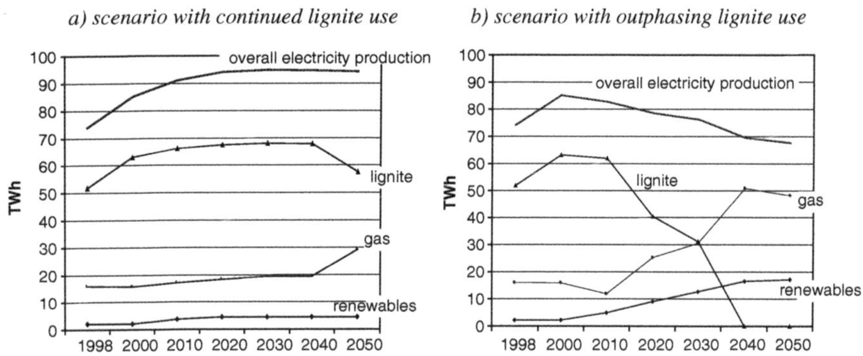

Fig. 2. Electricity Production by Primary Energy Resources Used for Eastern Germany Considering Two Regional Storylines (1998–2050, in TWh). *Source:* Adapted from Vögele and Markewitz (2001).

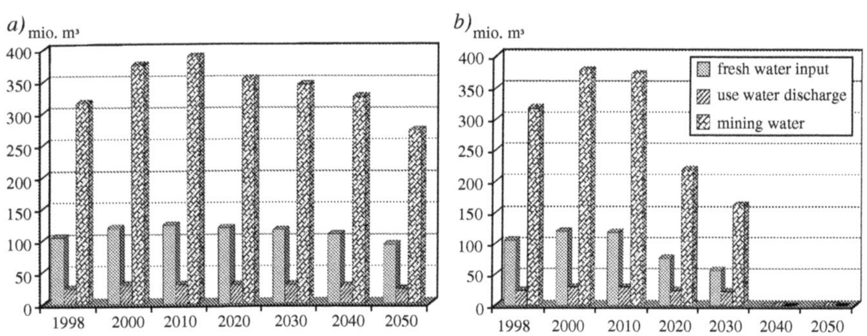

Fig. 3. Water Use for Lignite-Based Energy Production in eastern Germany Considering two Regional Storylines (1998–2050, in TWh and Mio. m^3). *Source:* Adapted from Vögele and Markewitz (2001).

technological advances in the management of water use in lignite-based electricity production.

Although fresh water demand is decreasing more explicitly in the phasing-out-of-lignite scenario, the general pressure on surface water availability is larger in this scenario. The reason for this relates to the huge amounts of mining water that significantly augment the surface water availability in the river basin. Since this water is generally used to satisfy the fresh water demand of lignite-based electricity production, there is never a water provision problem for energy production, because the amount of mining water in the study region is about three times higher than fresh water demand of energy production. The large impact on surface water availability in the phasing-out-of-lignite scenario is due to the fact that lignite mining ends in the 2030s and hence the additional mining water is no longer obtainable to sustain past levels of surface water discharge in the river basin.

These results are shown in an aggregated form for eastern Germany as a whole, but they are also available in disaggregated form with specific regard to geographical location of individual power plants (Vögele & Markewitz, 2001). Therefore, these disaggregated data on water use in the energy sector created for the super ordinate geographical reference area of the Elbe river basin (approximated by socio-economic data on eastern Germany) could directly be used in the case study to model the regional surface water availability in the Spree and Schwarze Elster river basins by means of hydrological and water management simulation models.[2]

Agricultural policy. The modeling of the impact of changing agricultural policy in the context of global change on agricultural production in eastern Germany was executed with the Regional Agriculture and Environment Information System, in short RAUMIS. This model system describes the agricultural sector by means of "region farms" with the German Agricultural Accounts as a framework of consistency. Basically one administrative district in Germany is considered to be one farm with a region-specific agricultural output. The input data used to describe the production of the agricultural sector and the different regional farms are taken from various sources, e.g., the Farm Accounting Data Network. Applying a mathematical programming approach and the profit maximization hypothesis, RAUMIS calculates for every "region farm" the amount of agricultural production by product types, the production technology used, capital inputs, remuneration for all factor inputs, as well as nutrient balances. Regarding the global world market context of agricultural production in Germany, simulation results from the world agriculture trade simulation model (WATSIM) were taken as external boundary conditions (Henrichsmeyer et al., 1996; Cypris, 2000; see also Kreins et al., in this volume).

The specification of RAUMIS according to the regional storylines A1 and B2 as portrayed above brought about a series of agricultural production data for the period 2000–2020. Fig. 4a and b below show the modeling results for the two storylines with regard to the nitrogen surplus for the year 2020, which are important pieces of information related to the non-point nitrogen emissions of the agricultural sector. As can be seen in the figures,

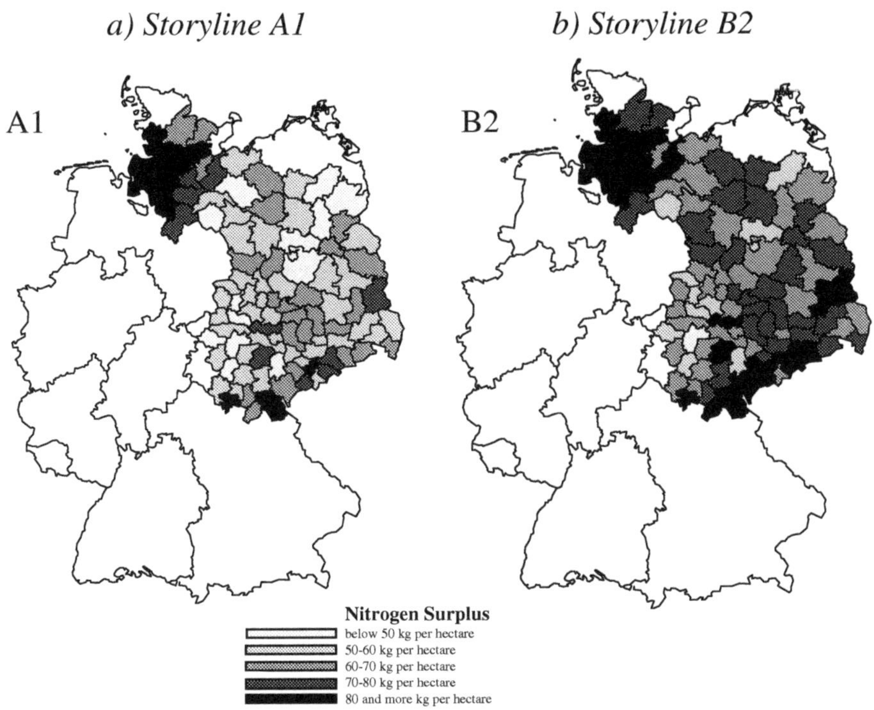

Fig. 4. Nitrogen Surplus in Agricultural Production for German Districts of the Elbe River Basin for the Storylines A1 and B2 (2020), in kg per Hectare and Year. *Source:* Gömann, Kreins, and Julius (2005).

the nitrogen surplus of agricultural production in 2020 is rather high (darker in Fig. 4b) for storyline B2 (average values of about 60–80 kg per hectare and year) with more or less stable policy conditions as they are determined by the regulations of the Agenda 2000. On the contrary, liberalization of agricultural policy would lead to overall reduced production with the consequence of reduced nitrogen surplus in storyline A1 (Fig. 4a: much more "region farms" with values below 60 kg per hectare). This GIS-based location-specific data were used as an important input for the hydrological model and the nutrient transport model, which together calculate the amount of nutrients and pollutants as well as the local pollutant concentrations in the Elbe river basin.

Population development. Having read the assumptions on population development above, it might be expected that no further work was necessary to characterize the population trends, because the two population scenarios of the Statistische Bundesamt (StaBu, 2000) were used. However, the data of the prognoses only referred to the German federal states as a whole without taking the population distribution within the states and their potential changes into account. Population data on such a high aggregation level could not be used as an input to calculate local and regional changes in water demand and wastewater generation. Therefore, the prognosis of the StaBu (2000) was complemented by regional population prognoses with a lower aggregation level. The following prognoses were available (Vassolo & Döll, 2002, p. 2ff):

- One population scenario for Germany for 97 regional areas called "Raumordnungsregionen" of the German Federal Office for Construction and Planning (Bundesamt für Bauwesen und Raumordnung – BBR) for the years 2005 and 2015 (BBR, 1999).
- One population scenario for Berlin and Brandenburg for all districts for 2005 and 2015 based on the prognosis of the authorities of the capital city of Berlin (Ströbl, 2001).
- Population scenarios for the district level for selected East German States until 2015.

A major problem in the downscaling of the StaBu (2000) population prognoses was that the different regional and local prognoses did not correspond properly to the overall German prognosis in many respects. First, all local or regional prognoses included only one future scenario. Second, the time horizon of the regional prognoses was very different. Third, the sum over all districts of one state in the regional prognoses did not equal any of the state prognoses in StaBu (2000). In order to resolve this problem of downscaling, it was decided not to include absolute population data of the regional prognoses to complement the StaBu (2000) prognoses, but only to include the information about the proportion of the population of a regional district compared to the population of the state. Proceeding this way, the population distribution within every state could be calculated in terms of percentage data for all districts and for all time periods available in the regional prognoses. Since most regional studies only computed data up to a year far from 2050, the development of the population distribution within the states was used until the final year of the regional studies and taken as constant afterwards. By this means, the specific information contained in the regional population studies could be used to downscale the StaBu (2000) prognoses.

Fig. 5a and b below show the results of the aggregated population scenarios for the eastern German states. For both A1 and B2 the trend is declining with B2 featuring a more distinct decline in population figures of about 25% in the period 2000–2050. The only exception of this trend relates to the suburban areas of Berlin, where population growth will occur. However, this means in general that population development is not a major pressure on water in the overall basin with the exception of the suburban areas of Berlin (Vassolo & Döll, 2002).

Climate change. Regarding the climate situation in the Elbe river basin for the period 1951–2000, it can be noted that most of the Elbe river basin has precipitation figures above 500 mm per year and only two areas in the western part of the basin show values below 500 mm. Using the data of 369 climate measuring stations in the Elbe river basin with daily records and applying the regional climate model system STAR of the Potsdam Institute for Climate Impact Research (PIK), a climate change scenario was generated with 100 realizations (Gerstengarbe & Werner, 2005). These different realizations reflect the uncertainty about the actual future course of regional temperature and precipitation figures related to the same climate change scenario. The generation of this scenario based upon the assumption that the mean yearly temperature will rise basin wide by 1.4 K until 2055. The temperature increase was postulated following results of the ECHAM4/OPYC3 global circulation model for the A1 emission scenario. According to the STAR climate change scenario the areas in the Elbe river basin with precipitation figures below 500 mm per year are considerably larger compared to the values measured in the period 1951–2000. Some areas even show precipitation figures below 400 mm per year. This scenario implies that climate change related to a global temperature rise of 1.4 K can have a significant impact on the regional water cycle of the Elbe river basin with considerable reductions in precipitation as well as an increase in regional evaporation and a reduction of discharge and groundwater regeneration (Gerstengarbe & Werner, 2005). The data for the climate scenarios are provided on a daily basis and they are an important input for the hydrological models.

3.3. Policy Strategies

After having presented the current water allocation conflict, the current policy strategy *basic* and after having defined four FoDs for possible future paths in the Elbe river basin until 2050, several alternative policy strategies

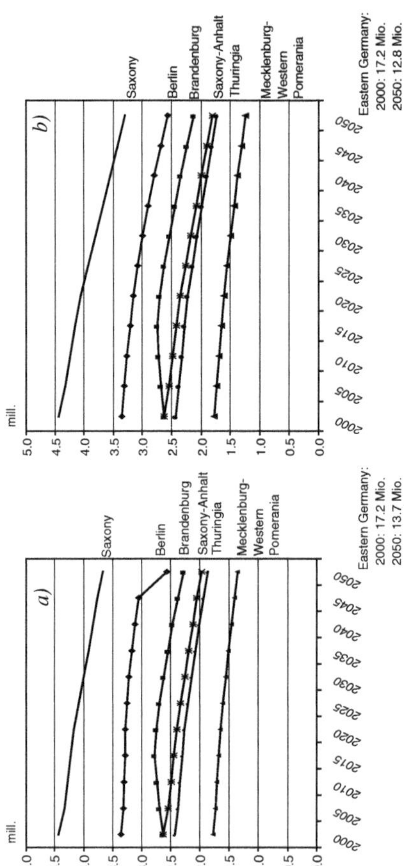

Fig. 5. Population Scenarios for the States in eastern Germany for A1 and B2 in Million Inhabitants, 2000–2050.

were to be characterized. These strategies were the objects of assessment in the following IMA analysis in order to identify the most appropriate one to deal with the current water allocation conflict and to take into account the policy challenge to adapt to global change.

In order to foster the identification of alternative strategies in cooperation with the decision makers and important water users, the scientists applied their hydrological and water resource management models to reveal the consequences of each of the four FoDs of global change under the circumstances of the *basic strategy*. The results showed that without climate change water availability problems will be moderate in the coming decade, whereas after 2030 the reliability of water availability for Berlin and the Spreewald wetland will decline due to reduced mining activity. Regarding the scenarios with climate change water scarcity will be much more pronounced with significantly reduced water flows and levels. In any case, the filling process of mining pit lakes is far from the environmental requirements to avoid a great amount of chemical lake water treatments. This is due to the low priority of water supply to the pits in times of water scarcity. After the presentation of these results to the decision makers and water users a discussion process started and, eventually, four additional policy strategies were identified to cope with the potential threat of global change (Koch, Kaltofen, Schramm, & Grünewald, 2006; Messner et al., 2006). They are characterized in the following:

- Filling strategy: Increasing the priority for the filling of lignite mining pits in order to secure their water quality without additional lake water treatment while ecological minimum flows remain unchanged.
- Reduced support: Based on the basic strategy but reducing the (high-cost) support of smaller streams near the mining pits.
- Odra–Spree Transition: Based on basic strategy with additional water through water transition of the Odra river over the Odra–Spree canal.
- Odra–Malxe Transition: Based on basic strategy with additional water through water transition of the Odra river over a pipeline.

The first two strategies aim at changing current water allocation priorities. In the *filling strategy* the filling of mining pits has no longer the lowest priority in water allocation, but will oust traditional water users like fish farmers and industry from their current priority positions. In *reduced support* significantly less water will be diverted to small streams nearby mining pits. A major argument for this strategy reads that the support of these streams is very costly in terms of pumping and opportunity costs, while it is not clear whether these streams will continue to exist under the circumstances of the new hydrological regime after the rehabilitation of the mining pits.

The *Odra–Spree* and *Odra–Malxe transition* strategies intend to redirect additional surface water from the Odra river basin into the Spree and Schwarze Elster river basins. The *Odra–Spree transition* seeks to pump water from the Odra river over the already existing Odra–Spree canal into the river Spree to support the water supply of the capital city of Berlin. In the Odra–Malxe transition strategy, it is planned to build a new transition system to divert additional water into the Malxe and, thereby, to support the water supply in the state of Brandenburg and to sustain water levels in the Spreewald wetland.

Thus, through the inclusion of decision makers and water users five alternative policy strategies were identified.

3.4. The Resulting Developmental Scenarios

As described in Section 2 above, the most important type of scenarios in the integrative assessment using the IMA approach are developmental scenarios. They are derived through combination of the FoDs with the policy strategies. Thus, since four FoDs and five policy strategies have been defined, twenty developmental scenarios were derived in the end. They are listed in Table 1 below. The name of each scenario was chosen as to characterize it properly with its two elements, the FoD and the policy strategy. Thus, e.g., the scenario *Filling_B2_climate change* indicates the *filling* policy strategy under global change conditions with socio-economic trends according to the IPCC storyline B2 with climate change.

4. OUTLOOK

This chapter describes how scenarios are systematically derived using the IMA approach to analyze water use conflicts and their resolution in the context of global change. Through the combination of FoDs and policy strategies a consistent set of developmental scenarios can be generated that makes it possible to examine the impact of policy strategies under conditions of different future global change development paths. Proceeding this way, the risk of policy making under uncertain future circumstances can be included into the analysis. In Chapters 11 (Messner et al.) and 12 (Messner, both in this volume) it is delineated how the assessment of the above-

Table 1. The Developmental Scenarios of the Conflict in the Spree and Schwarze Elster River Basins.

Developmental Scenarios (Abbreviation)	Policy Strategy	Framework of Development (FoD)
Basic_B2_no climate change	Basic	B2 + no climate change: (socio-economic development according to global change scenarios with B2 and no climate change)
Filling_B2_no climate change	Filling	
Reduced_B2_no climate change	Reduced support	
Transition OS_B2_no climate change	Transition Odra-Spree	
Transition OM_B2_no climate change	Transition Odra-Malxe	
Basic_B2_climate change	Basic	B2 + climate change: (socio-economic development according to global change scenarios with B2 and with climate change)
Filling_B2_climate change	Filling	
Reduced_B2_climate change	Reduced support	
Transition OS_B2_climate change	Transition Odra-Spree	
Transition OM_B2_climate change	Transition Odra-Malxe	
Basic_A1_no climate change	Basic	A1 + no climate change: (socio-economic development according to global change scenarios with A1 and no climate change)
Filling_A1_no climate change	Filling	
Reduced_A1_no climate change	Reduced support	
Transition OS_A1_no climate change	Transition Odra-Spree	
Transition OM_A1_no climate change	Transition Odra-Malxe	
Basic_A1_climate change	Basic	A1 + climate change: (socio-economic development according to global change scenarios with A1 and with climate change)
Filling_A1_climate change	Filling	
Reduced_A1_climate change	Reduced support	
Transition OS_A1_climate change	Transition Odra-Spree	
Transition OM_B2_no climate change	Transition Odra-Malxe	

described scenarios is executed using IMA and its specific approach to combine scientific modeling, benefit–cost analysis, multicriteria evaluation techniques, and participation of stakeholders.

NOTES

1. Another reason relates to multicriteria assessment that is used in the IMA approach. Since the weighting of criteria in the assessment of policies is of crucial importance and an "objective" weighting does not exist in societies with multiple value systems, it might even be impossible to find an unambiguous optimal solution for one framework of development. To deal with this problem a participatory

approach is chosen that shifts this question from the scientific to the political sphere (Messner et al., 2006).

2. For a description of the water management simulation model called WBalMo, which was used in the case study, see Section 2 in Chapter 11 (Messner et al., in this volume).

ACKNOWLEDGMENTS

The research was funded by the German Federal Ministry of Education and Research in the context of the GLOWA Elbe Project. Further support was received by all stakeholders and decision makers involved in the project, especially the members of the Cross-State Working Group River Basin Management. Furthermore, I would like to thank the interdisciplinary research team in GLOWA Elbe that worked together on the derivation of a consistent set of scenarios. Finally, I want to say a special thank you for helpful comments on early draft versions of this article to Hagen Koch, Michael Kaltofen, Oliver Zwirner, Frank Wechsung, Horst Gömann, and Petra Döll.

REFERENCES

AG Flussgebietsbewirtschaftung (2000). Grundsätze für die länderübergreifende Bewirtschaftung der Flussgebiete Spree und Schwarze Elster. In: LIWAG (Ed.), *Bericht zur Umsetzung des Positionspapiers des Landes Brandenburg und des Freistaates Sachsen zur Sanierung des Wasserhaushalts in der Lausitz*. Potsdam, Dresden.

Alcamo, J., Leemans, R., & Kreileman, E. (Eds). (1998). *Global Change Scenarios of the 21st Century – Results from the IMAGE 2.1 Model*. Oxford: Pergamon.

Aligica, P. D. (2005). Scenarios and the growth of knowledge: Notes on the epistemic element in scenario building. *Technological Forecasting and Social Change*, *72*(7), 815–824.

BBR (Bundesamt für Bauwesen und Raumordnung) (1999). *Perspektiven der künftigen Raum- und Siedlungsentwicklung*, Heft 11/12.

Becker, A., Behrend, H., Cypris, C., Dietrich, O., Feige, J., Grünewald, U., Hansjürgens, B., Hartje, V., Ipsen, D., Kaltofen, M., Klauer, B., Leinhos, S., Messner, F., Oppermann, R., Simon, K. H., Soete, B., Wechsung, F., Wenzel, V., & Wessolek, G. (2001). *Integrierte Analyse der Auswirkungen des Globalen Wandels auf die Umwelt und die Gesellschaft im Elbegebiet (GLOWA-Elbe) – Sozioökonomische Konzeption*. Unpublished document of the GLOWA Elbe project, Potsdam.

BLE (Federal Office for Agriculture and Nutrition) (2004). Karpfenbericht 2004 – Bericht über die Markversorgung und die Außenhandelssituation von Karpfen und anderen Süßwasserfischen in der Bundesrepublik Deutschland, Hamburg.

Cypris, C. (2000). *Positive mathematische programmierung (PMP) im Agrarsektormodell RAUMIS*. Unpublished doctoral dissertation, Universität Bonn.

EC (European Commission) (2006). Reform of the common agricultural policy. http://europa.eu/scadplus/leg/en/lvb/l60002.htm (as of September 2006).

Eckart-Montanconsult and Planung/IBA Fürst-Pückler-Land (2002). Erarbeitung der Stufe 1 des touristischen Wirtschafts- und Nutzungskonzeptes zur abgestimmten Entwicklung der entstehenden Wasserflächen der Lausitzer Seenkette. Regional study on tourism usage potentials of the Lausitz lake landscape area financed by the economic ministries of the German states of Brandenburg and Saxony, Görlitz/Großräschen.

Gerstengarbe, F.-W., & Werner, P. C. (2005). Simulationsergebnisse des regionalen Klimamodells STAR. In: F. Wechsung, A. Becker & P. Gräfe (Eds), *Auswirkungen des globalen Wandels auf Wasser, Umwelt und Gesellschaft im Elbegebiet* (pp. 110–118). Berlin: Weißensee Verlag.

Gömann, H., Kreins, P., & Julius, C. (2005). Perspektiven der Landbewirtschaftung im deutschen Elbegebiet unter dem Einfluss des globalen Wandels – Ergebnisse eines interdisziplinären Modellverbundes. In: F. Wechsung, A. Becker & P. Gräfe (Eds), *Auswirkungen des globalen Wandels auf Wasser, Umwelt und Gesellschaft im Elbegebiet* (pp. 119–128). Berlin: Weißensee Verlag.

Grünewald, U. (2005). Probleme der integrierten Wasserbewirtschaftung im Spree-Havel-Gebiet im Kontext des globalen Wandels. In: F. Wechsung, A. Becker & P. Gräfe (Eds), *Auswirkungen des globalen Wandels auf Wasser, Umwelt und Gesellschaft im Elbegebiet* (pp. 209–218). Berlin: Weißensee Verlag.

Henrichsmeyer, W., Cypris, Ch., Löhe, W., Meudt, M., Sander, R., von Sothen, F., Isermeyer, F., Schefski, A., Schleef, K.-H., Neander, E., Fasterding, F., Helmcke, B., Neumann, M., Nieberg, H., Manegold, D., & Meier, T. (1996). Entwicklung des gesamtdeutschen Agrarsektormodells RAUMIS96. Endbericht zum Kooperationsprojekt. Forschungsbericht für das BML (94 HS 021), vervielfältigtes Manuskript, Bonn/Braunschweig.

Horsch, H., Ring, I., & Herzog, F. (Eds). (2001). *Nachhaltige Wasserbewirtschaftung und Landnutzung – Methoden und Instrumente der Entscheidungsfindung und -umsetzung*. Marburg: Metropolisverlag.

IPCC (Intergovernmental Panel on Climate Change) (2000). *Emissions scenarios*. IPCC Special Report (http://www.grida.no/climate/ipcc/emission/index.htm; as of August 2006).

Klauer, B., Drechsler, M., & Messner, F. (2006). Multicriteria analysis under uncertainty with IANUS – Method and empirical results. *Environment and Planning C: Government and Policy*, *24*, 236–256.

Klauer, B., Meyer, B., Horsch, H., Messner, F., & Grabaum, R. (2001). Decision support for land use changes – A combination of methods for policy advising and planning. In: R. Krönert, U. Steinhart & M. Volk (Hrsg.), *Landscape Balance and Landscape Assessment* (pp. 281–297). Berlin: Springer-Verlag.

Koch, H., Kaltofen, M., Grünewald, U., Messner, F., Karkuschke, M., Zwirner, O., & Schramm, M. (2005). Scenarios of water resources management in the Lower Lusatian mining district, Germany. *Ecological Engineering*, *24*, 49–57.

Koch, H., Kaltofen, M., Schramm, M., & Grünewald, U. (2006). Adaptation strategies to global change for water resources management in the Spree River catchment, Germany. *International Journal of River Basin Management*, *4*(4).

Martinsen, D., Kraft, A., & Markewitz, P. (2001). *Szenarien zur CO_2-Minderung auf Basis des IKARUS-Optimierungsmodells*. Jülich: Bericht des Forschungszentrums Jülich.

Messner, F., Klauer, B., Horsch, H., Geyler, S., Volk, M., & Herzog, F. (2001). Ableitung von Szenarien für Entscheidungsanalysen: Methodik und beispielhafte Anwendung. In: H. Horsch, I. Ring & F. Herzog (Eds), *Nachhaltige wasserbewirtschaftung und landnutzung-methoden und instrumente der entscheidungsfindung und-umsetzung* (pp. 99–130). Marburg: Metropolis Verlag.

Messner, F., Zwirner, O., & Karkuschke, M. (2006). Participation in multicriteria decision support for the resolution of a water allocation problem in the Spree river basin. *Land Use Policy*, *23*, 63–75.

Prognos, A. G. (Hrsg.). (2000). *Energiereport III – die längerfristige Entwicklung der Energiemärkte im Zeichen von Wettbewerb und Umwelt*. Stuttgart: Schäffer-Poeschel Verlag.

Schoonenboom, I. J. (1995). Overview and state of the art of scenario studies for the rural environment. In: J. F. T. Schoute, P. A. Finke, F. R. Veeneklaas & H. P. Wolfert (Eds), *Scenario Studies for the Rural Environment* (pp. 15–24). Dordrecht: Kluwer Academic Publishers.

StaBu (Statistisches Bundesamt) (2000). *Bevölkerungsentwicklung Deutschlands bis zum Jahr 2050. Ergebnisse der 9. koordinierten Bevölkerungsvorausberechnung*. Wiesbaden.

StaBu (Statistisches Bundesamt) (2001). Wasserversorgung und Abwasserbeseitigung in der Industrie 2001. Data Series R.2.2. Wiesbaden.

Ströbl, B. (2001). Entwicklungsrahmen Bevölkerung für GLOWA Elbe: Skalierung insbesondere für Berlin-Brandenburg. Unpublished working paper, Potsdam Institute for Climate Impact Research.

Vassolo, S., & Döll, P. (2002). *Population scenarios for Germany and the Czech Republic with focus on the Elbe river basin*. Internal Report GLOWA-Elbe, Center for Environmental Systems Research, Kassel.

Veeneklaas, F. R., & van den Berg, L. M. (1995). Scenario building: Art, craft or just a fashionable whim? In: J. F. T. Schoute, P. A. Finke, F. R. Veeneklaas & H. P. Wolfert (Eds), *Scenario Studies for the Rural Environment* (pp. 11–13). Dordrecht: Kluwer Academic Publishers.

Vögele, S., & Markewitz, P. (2001). Die Analyse des deutschen Strommarktes mit Focus auf die neuen Bundesländer sowie die Ableitung von möglichen Strommarktentwicklungsszenarien bis zum Jahr 2050. Arbeitsbericht, Forschungszentrum Jülich, Programmgruppe Systemforschung und Technologische Entwicklung, Jülich.

Wagner, U., & Stein, G. (Hrsg.). (1999). *Das IKARUS Projekt: Klimaschutz in Deutschland. Strategien für 2000–2020*. Berlin: Springer.

Wenzel, V. (2001). Integrated Assessment and Multicriteria Analysis. *EGS Journal Physics and Chemistry of the Earth, Part B*, *26*(7–8), 541–545.

SCENARIO ANALYSIS OF ECONOMY–ECOLOGY INTERACTIONS IN THE HUDSON RIVER BASIN

Bongghi Hong, Karin E. Limburg, Myrna H. Hall and Jon D. Erickson

ABSTRACT

Our primary goal is to develop an integrated, quantitative assessment tool evaluating how human economic activities influence spatial patterns of urbanization, and how land-use change resulting from urbanization affects stream water quality and aquatic ecosystem health. Here we present a prototype of a holistic assessment tool composed of three "building blocks" simulating the social and economic structures, spatial pattern of urbanization, and watershed health as determined by various metrics. The assessment tool is applied to Dutchess County, New York and two of its largest watersheds, Wappinger and Fishkill Creek watersheds, demonstrating how an explicit link can be established between human economic activities and ecosystem health through changes in land use.

Ecological Economics of Sustainable Watershed Management
Advances in the Economics of Environmental Resources, Volume 7, 97–111

ISSN: 1569-3740/doi:10.1016/S1569-3740(07)07005-8

1. INTRODUCTION

Increasing human use of fresh water continues to reduce water quality for most of the world's population and threaten the health of aquatic ecosystems (Postel, 2000). In the short term, human water demand due to urbanization and economic development is a graver threat to water resources than global warming and climatic change (Vörösmarty, Green, Salisbury, & Lammers, 2000). In many parts of the world, urbanization is one of the most rapidly occurring changes to land cover, removing vegetation and increasing impervious surfaces. The loss of permeability has been linked to alterations in hydrology (Klein, 1979), sediment, nutrient, and toxicant loading, and general stream degradation (Forman & Alexander, 1998; Paul & Meyer, 2001; Center for Watershed Protection, 2003), with attendant loss of ecosystem function and biodiversity.

Ecosystem health may be defined as the maintenance of biotic integrity, resistance and/or resilience to change in the face of anthropogenic disturbance (Rapport, 1992; Shrader-Frechette, 1994; Rapport, Gaudet, & Calow, 1995), and includes physical and chemical environmental quality (e.g., stream temperature, conductivity, and element concentration), as well as biotic condition (e.g., diversity of fish and macroinvertebrate communities). Operationally, the urban–rural gradient is a useful framework for investigating how urbanization affects ecosystem health (McDonnell & Pickett, 1990; McDonnell, Pickett, Groffman, Bohlen, & Pouyat, 1997; Zhu & Carreiro, 1999). This gradient, along which population density and impervious surface area increase, typically promotes a suite of alterations, including such responses as increasing nitrate concentration (Zhu & Carreiro, 1999), conductivity (Limburg, Stainbrook, Erickson, & Gowdy, 2005), declines in fish fauna (Brown, Gray, Hughes, & Meador, 2005), and decreasing biodiversity and species richness (McKinney, 2002).

Land-use intensification, especially conversion into urban uses, is an important driver of stream health degradation. For example, the database from the National Water-Quality Assessment (NAWQA) Program, collected from many stations across the U.S., demonstrated decreasing water quality with increasing percent urban use (Meador & Goldstein, 2003). Past land-use change in the U.S. was dominated by large-scale conversion of forest and grasslands into agricultural use, but the expansion of urban and suburban areas is the most-important driver of land-use change at present (Naiman & Turner, 2000). The pressure for this change has been most acute around urban centers, via the process referred to as "urban sprawl," defined as "the spread of urban congestion into adjoining suburbs and rural sections in an irregular, unordered, and chaotic way" (Merriam-Webster Dictionary, 2002; see also Ewing, 1994). Thus, accurate prediction of future trends of urbanization is essential to the assessment of stream ecosystem health, and is a need voiced at local to national scales.

However, it has been a challenging task to evaluate and predict urbanization patterns resulting from urban sprawl due to the stochastic nature of the process (Polimeni, 2002). The conversion of land into urban use is the result of human decisions, often made one property, one home, and one business at a time (Erickson et al., 2005), primarily based on demographic and economic factors (Li, Sato, & Zhu, 2003). For example, it has been shown that the historic land-use changes in the Choptank River Watershed in Maryland and the Greater Yellowstone Ecosystem can largely be explained by socio-economic events that occurred in the region (Benitez & Fisher, 2004; Parmenter et al., 2003). Thus, the future course of urban sprawl, and its impact on ecosystem health, can only be appropriately evaluated with the help of a socio-economic model explicitly considering the social and economic structures of the region. Such a model should assess how these structures create the demand for new land development in response to anticipated socio-economic events. However, the explicit and quantitative link among socio-economic systems, land-use change, and ecosystem health has rarely been established, primarily because those systems, with their own complexities and non-linearities, have been studied by different academic disciplines and each has been considered at different temporal and spatial scales (Veldkamp & Verburg, 2004). Although the same physical landscape is shared by human and natural systems, the boundaries and scales delineating each system (e.g., counties and towns versus watersheds and subcatchments) are not the same (Fig. 1). We note, however, the need to recognize the reciprocal roles of human versus natural systems: that is, at some scales we can consider natural systems within the context of human ones (e.g., watersheds that fall within county boundaries), but ultimately, human systems are subsets of natural ecosystems.

Our primary goal is to develop an integrated, quantitative assessment tool evaluating how human economic activities influence spatial patterns of urbanization, and how land-use change resulting from urbanization affects stream water quality and aquatic ecosystem health. However, we acknowledge that the interactions of Fig. 1 are not unidirectional. Human choices affect and are affected by nature through various feedback links (Settle, Crocker, & Shogren, 2002), often mediated by changes in landscape features shared by humans and other organisms.

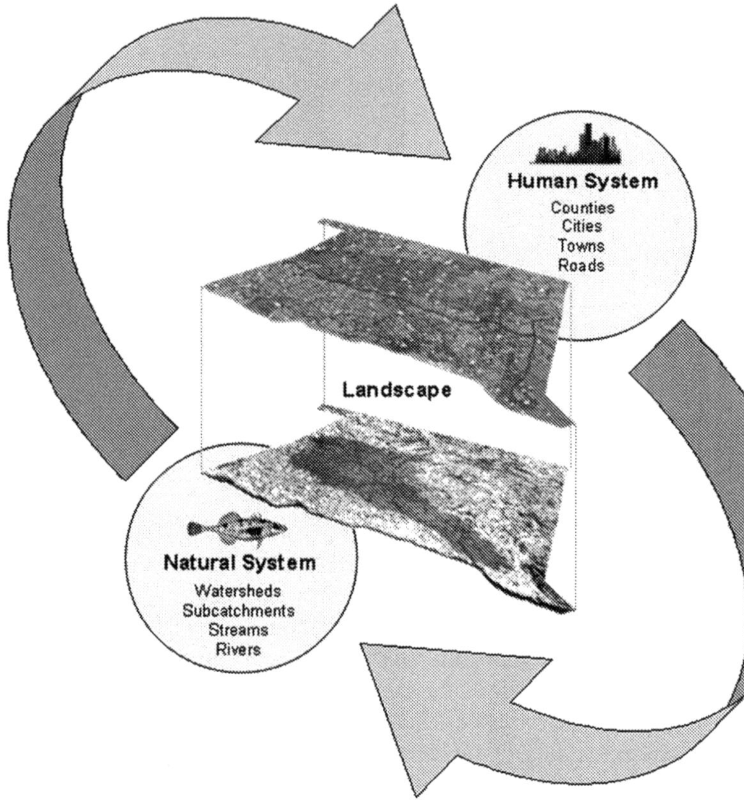

Fig. 1. Conceptual Diagram Showing Interactions among Landscape, Human, and Natural Systems. The Human System Alters the Landscape through Urbanization Processes, Affecting Ecosystems within the Landscape. As these Ecosystems Degrade, Resources Decline, and this May Send Direct or Indirect Signals Back to the Economy.

As an example, consider the following scenario: development to stimulate a local economy may include attracting an expanding industry. Aside from providing new employment for area residents, demand for new homes increases. Such demand will attract developers and increase the real-estate value of lands within some given distance. This may stimulate owners of large holdings (farmers and foresters) to parcelize, thus increasing building activity. However, over time, the increased development may in turn raise taxes, driving out farmers and foresters, resulting in even more land

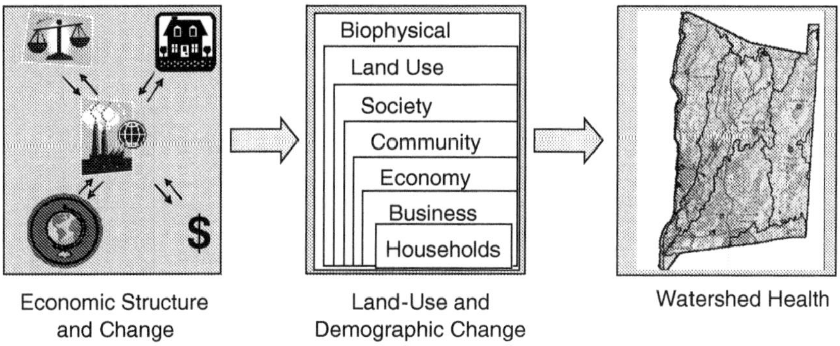

Fig. 2. Three "Building Blocks" Comprising the Current Assessment Tool.

available for development. Construction of homes, roads, and malls increase impermeable surfaces, causing flashier streams, washouts of plants and invertebrates, and unstable habitat for fish communities. Ultimately, development could achieve a scale at which local natural amenities decline, making it no longer attractive as a destination, leading to the decline of small lake-side resorts, bed and breakfasts, local restaurants, etc., while strip malls blossom. The character of the landscape is changed, supporting lower diversity of natural ecosystems, as well as lower economic diversity. Degraded ecosystems may produce downturns in the economy.

Over the past 4 years, we have developed a prototype (Fig. 2) of a holistic assessment tool composed of three "building blocks" simulating the social and economic structures (Nowosielski & Erickson, Chapter 8 in this volume), spatial pattern of urbanization (Polimeni & Erickson, Chapter 9 in this volume), and watershed health as determined by various metrics (Stainbrook, 2004; Limburg et al., 2005; Limburg & Stainbrook, in press). To implement these in decision making, we have begun to work with multi-criteria decision methods in order to bring together disparate perspectives and demands, and build consensus for environmental planning (Hermans & Erickson, Chapter 10 in this volume).

2. THE STUDY SYSTEM

Our study focused on Dutchess County, New York and two of its largest watersheds. The county (2,077 km^2) is located on the eastern side of the Hudson River estuary (Fig. 3). The Wappinger Creek (546.5 km^2) and

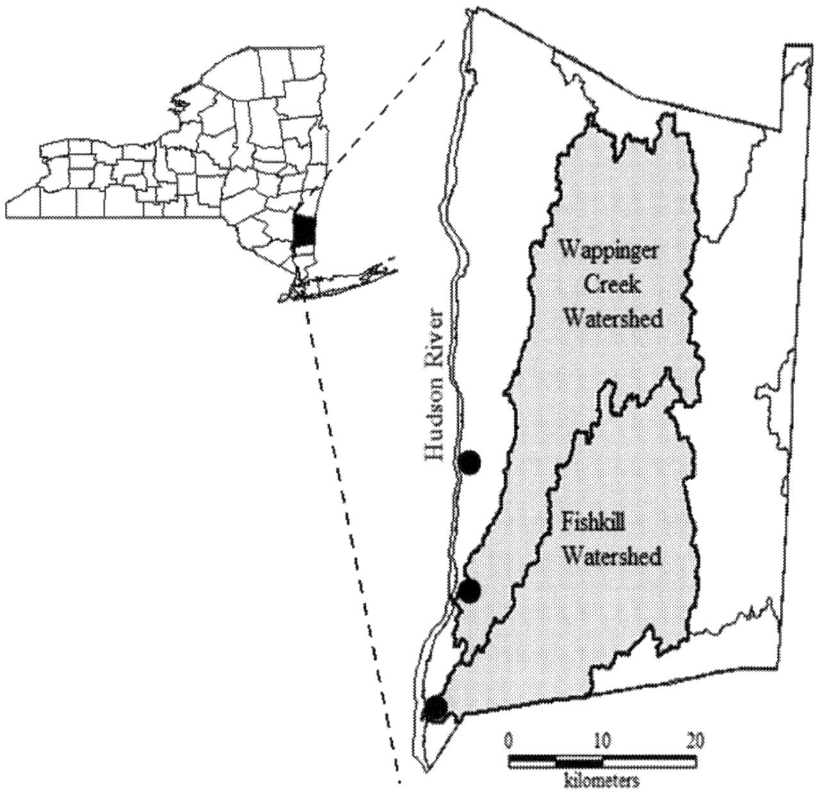

Fig. 3. Map Showing New York State Counties; Dutchess County and its Two Major Watersheds Comprise the Focus of Our Study. *Source:* Limburg and Stainbrook (in press).

Fishkill Creek (521 km^2) watersheds compose over half the drainage area. Both the Wappinger and Fishkill creeks arise in eastern highlands and flow southwest toward the Hudson River.

Land use within Dutchess County is a mix of urban, suburban, agriculture (dairy, pasture, row crop, and orchard), and undeveloped woodlands. Historically, agriculture dominated land use before 1950 (Swaney, Limburg, & Stainbrook, 2006), but beginning in the 1940s, job growth increased in industrial sectors; IBM became a major employer in the county (Erickson et al., 2005). The largest urban centers are located in the southwestern part of the county along the Hudson River (Fig. 3); the cities of Wappingers

Falls and Beacon sit at the mouths of the Wappinger and Fishkill creeks, respectively. Urban flight from New York City, 120 km to the south, has also stimulated development in Dutchess County, primarily in its southern half. Today, the northeast is the least-developed part of the county. Because of the more intensive development to the south, we hypothesized that Fishkill Creek would display lower ecological health than the Wappinger Creek watershed.

3. OVERVIEW OF THE APPROACH

In research supported both by the Hudson River Foundation and the National Science Foundation, we asked the following three questions: (1) How does economic activity create the demand for new land? (2) How does new land demand change the spatial pattern of land-use? and (3) How does land-use change affect watershed health? Each of these questions was explored in separate analyses, and resulted in three different models; a socio-economic model developed by Nowosielski (2002), a land-use model developed by Polimeni (2002), and an ecosystem health assessment developed by Stainbrook (2004). Recently, these three approaches were integrated as three "building blocks" or "sub-models" of an assessment tool (Hong et al., under review).

The socio-economic sub-model (Nowosielski & Erickson, in this volume) is based upon a social accounting matrix (SAM) providing an expanded view of economic activity and interconnections between industries, household income characteristics, and social institutions in the area. The socio-economic sub-model constructs the SAM mostly from the IMPLAN (Impact Analysis for Planning) database and uses it to calculate the Leontief inverse, which shows the requirements from each sector of the economy needed to deliver a dollar's worth of product to final consumers. Although the model is a static "snapshot" of the economy, users of the model can specify various economic-impact scenarios, such as what sectors to increase or decrease, impact location, and percent commuters in the region. The model then estimates the total economic impact for each industrial and household sector individually, using the Leontief inverse. The socio-economic sub-model also estimates the number of households required to meet the new demand by the industries resulting from the economic impact. For example, Fig. 4 shows the new household requirements by various industrial sectors in Dutchess County, produced by the socio-economic sub-model under a scenario in which the semi-conductor industry expands with 1,000 new jobs.

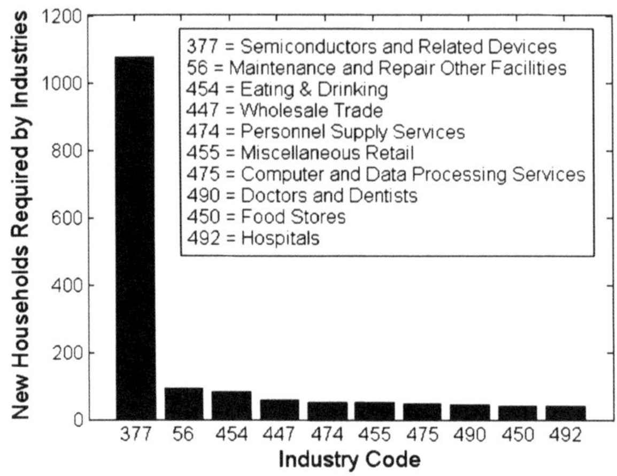

Fig. 4. Bar Graph Produced by Socio-economic Sub-model Showing New Household Requirements Due to a Scenario of an Expanding Semi-conductor Industry.

The land-use sub-model, described in more detail by Polimeni and Erickson in Chapter 9 of this volume, is based upon a binary logit regression model estimating the developmental probabilities of vacant tax parcels in the simulated region. Different sets of independent variables for the logit model (e.g., population variables, income variables, land assessment value, and distance to the central business district, obtained from tax parcel, census, and GIS data available in the region) produce different probabilities of land conversion. The land-use change model uses Monte Carlo simulation to predict the spatial pattern of land development in the near future with or without economic impact, using the estimated developmental probabilities of vacant tax parcels and new household requirement obtained from the socio-economic sub-model. Further refinements to the model include varying assumptions about the employment status of the region (e.g., percent unemployed, socio-economic status), possible restrictions to development (e.g., hydric soils, wetlands, steep area, or otherwise protected lands), other zoning restrictions (e.g., noise pollution), and distribution of new households in relation to the distance to the impact location. The simulation result can be exported to GIS for graphical presentation. Fig. 5 shows an example of land development in 2011 in Dutchess County predicted by the land-use sub-model under the expanding semi-conductor industry scenario.

In order to assess the impacts of land-use change on ecosystem health, we conducted extensive empirical studies of the Fishkill and Wappinger Creek

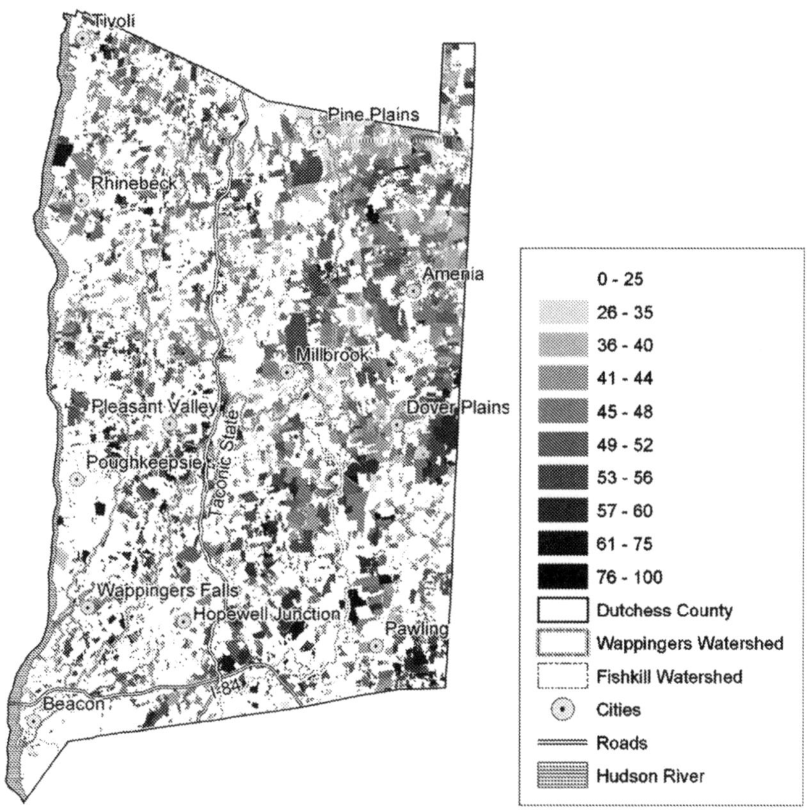

Fig. 5. Relative Probability (%) of Conversion of Vacant Tax Parcels within Dutchess County to Residential use from 2001 to 2011 under Expanding a Scenario of a Semi-conductor Industry, Predicted by the Landuse Sub-model.

watersheds (Stainbrook 2004; Limburg & Stainbrook, in press; Limburg et al., 2005). These studies indicated that both watersheds have been degraded by long-term land-use patterns, but that the press of urbanization is most intense in the Fishkill Creek watershed. Nevertheless, the differences manifested at the whole watershed scale were relatively small, suggesting perhaps some resilience in the systems.

A holistic approach is important in making decisions because different interest groups have different preferences and priorities (Stinner, Stinner, & Martsolf, 1997). For example, although "urban sprawl" is a somewhat value-laden term implying a negative view of increased traffic, decreased water and air quality, and loss of green area and open space, others may see it as a positive sign of increased regional economic activity and more employment opportunity (Steiner, 1994). Decisions from policy makers among different management strategies (e.g., adopting zoning restrictions, enforcing protected lands, etc.) should be based on quantitative evaluation and comparison of possible consequences on the socio-economy, landscape, and environment.

We have been working with a multi-criteria decision assessment methodology (MCDA) (Hermans & Erickson, in this volume). MCDA is a framework transparent to decision-makers, adaptable to many situations across multiple metrics and scales, and amenable to both expert and local stakeholder pools of knowledge. The MCDA process starts with a clear definition of a goal, which is facilitated by some form of participatory process (in this case, aided by the simulation model). This is followed by identification of alternatives to achieve the goal. The future outcome of each of these alternatives is then characterized by a suite of indicators. Criteria are then measured in multiple units (both quantitative and qualitative) and dimensions (both spatial and temporal). Once the MCDA problem is structured, the next step is to elicit the preferences of the stakeholders using one of several methods within the family of MCDA frameworks. For example, PROMETHEE (Preference Ranking Organization Method of Enrichment Evaluation) is a specific sort of outranking method commonly used in MCDA (Brans, Vincke, & Mareschal, 1986). PROMETHEE requires criteria-specific and stakeholder-identified: (1) choice of maximizing or minimizing, (2) weight of importance to the overall decision, (3) preference function that translates quantitative or qualitative metrics to consistent rankings, and (4) various decision threshold parameters for each function (for example, indifference thresholds identify ranges where a decision-maker cannot clearly distinguish their preferences). This exercise is carried out by each stakeholder in a decision problem. During sensitivity analysis, criteria weights, preference functions, and decision thresholds can all be varied to estimate stability intervals for the rankings of alternatives and evaluate both imprecision of criterion measurement and uncertainty of preference. The outcome of PROMETHEE includes both complete and partial rankings (depending on the incomparability of decision alternatives), and both pairwise and global comparisons of decision alternatives. Global comparisons can be illustrated with a GAIA (Graphic Analysis for Interactive Assistance) plane diagrams that represent a complete view of the conflicts between the criteria, of the characteristics of the actions, and of the weighing of the criteria. With multiple stakeholders, MCDA analyses can be used to visualize conflict between stakeholder positions and opportunities for compromise, alliances, and group consensus, or to revisit and redefine the goal, alternatives, and criterion themselves (Macharis, Brans, & Mareschal, 1998). The advent of spatial decision support systems (SDSS) provides an important new opportunity for the evolution of MCDA methods and applications (Malczewski, 1999) including extensions of our work.

4. DISCUSSION

The conceptual frameworks linking human economic activities to ecosystem health have been proposed by many researchers (e.g., McDonnell & Pickett, 1990; Stinner et al., 1997; Alberti et al., 2003; Nilsson et al., 2003; Peterson et al., 2003a). Their conceptual models have similar components (humans, the physical environment, and the ecosystem), and processes interconnecting them. They differ in the levels of detail describing each component (e.g., "one-box" versus detailed food web) and the nature of processes relating them (e.g., unidirectional cause–effect relationships versus feedback loops).

In terms of evaluating the usefulness of each of these frameworks as an assessment tool, one should consider whether all the components in the model and the connections among them are explicitly and quantitatively expressed. After careful consideration, Nilsson et al. (2003) suggested that the uncertainties in the available data and the gaps in our knowledge about complex, non-linear processes are so large that the quantitative description of these systems in an integrated, holistic framework is not yet feasible. Peterson, Cumming, and Carpenter (2003a, 2003b) suggested a way of circumventing overwhelming uncertainties through "scenario planning," in which the responses of future economies, landscapes, and ecosystems to different management strategies are illustrated to the decision-makers as possible outcomes that emerge from quantitative-assessment simulation models. Clark (2002) suggested that the uncertainties in our knowledge should not be used as an overt justification for avoiding the use of quantitative tools in the decision-making process, but rather that the assessment models should deal with the existing uncertainties more rigorously and explicitly. Currently available tools for quantitative assessment of anthropogenic land-use change and resulting stream health (e.g., Costanza et al., 2002) do not have algorithms performing rigorous uncertainty analysis. Ultimately, we intend to develop an assessment tool that will have the full capability of uncertainty analysis, enabling the policy-makers to make decisions based on the quantitative evaluation of possible outcomes, while appreciating the uncertainties in the model predictions at the same time. In addition, a successful decision support system should be truly holistic (Stinner et al., 1997), have multiple endpoints (Santelmann et al., 2004), show explicit linkages among different systems (Young, Lam, Ressel, & Wong, 2000), help the user to select various scenarios and construct decision trees (Djodjic, Montas, Shirmohammadi, Bergstrom, & Ulen, 2002), and have an easy-to-use GUI communicating with the user (Young et al., 2000). We are working toward addressing each of these.

Nilsson et al. (2003, p. 671) state:

> "...environmental forecasting is subject to a variety of technical and resource limitations, many of which will require massive intra and interdisciplinary efforts in the fields of economics, quantitative spatial analysis, hydrology, geomorphology, and ecology to overcome or ameliorate. If researchers can fill – or at least reduce – these gaps, thus improving our ability to forecast environmental change and to advise on the potential effects of different land-use changes on running waters, ecology will play a significant role in formulating land-use policies in the future. This is one of the greatest ecological challenges of our time, yet it is an area where we can reasonably expect to see major breakthroughs."

We echo these sentiments, and feel that we have already come a considerable way toward meeting the goal of linking together these diverse disciplines. We are now poised to continue this exciting work; collectively, we have a rare combination of the transdisciplinary, analytical expertise necessary to take on this challenge. The next step will be to continue to evolve the linkages of the models to include uncertainty analysis, feedback loops, and scale effects. Undoubtedly, such model structures are capable of producing surprising results which may manifest some of the complexity inherent in studying these coupled systems. Not only do we hope to advance the field of environmental assessment to a new level, but in doing so we will address fundamental themes in ecology, economics, geography, hydrology, and geomorphology: namely, the effects of scale, and quantification/ramification of uncertainty.

REFERENCES

Alberti, M., Marzluff, J. M., Shulenberger, E., Bradley, G., Ryan, C., & Zumbrunnen, C. (2003). Integrating humans into ecology: Opportunities and challenges for studying urban ecosystems. *Bioscience*, *53*(12), 1169–1179.

Benitez, J. A., & Fisher, T. R. (2004). Historical land-cover conversion (1665–1820) in the Choptank watershed, Eastern United States. *Ecosystems*, *7*, 219–232.

Brans, J. P., Vincke, P., & Mareschal, B. (1986). How to select and how to rank projects: The PROMETHEE method. *European Journal of Operational Research*, *28*, 228–238.

Brown, L. R., Gray, R. H., Hughes, R. M., & Meador, M. (Eds). (2005). *The Effects of Urbanization on Stream Ecosystems*. American Fisheries Society Symposium 47, American Fisheries Society, Bethesda, MD.

Center for Watershed Protection (2003). *Impacts of impervious cover on aquatic ecosystems*. Watershed Protection Research Monograph No. 1, Center for Watershed Protection, Ellicott City, MD.

Clark, M. J. (2002). Dealing with uncertainty: Adaptive approaches to sustainable river management. *Aquatic Conservation – Marine and Freshwater Ecosystems*, *12*(4), 347–363.

Costanza, R., Voinov, A., Boumans, R., Maxwell, T., Villa, F., Wainger, L., & Voinov, H. (2002). Integrated ecological economic modeling of the Patuxent River watershed, Maryland. *Ecological Monographs*, *72*(2), 203–231.

Djodjic, F., Montas, H., Shirmohammadi, A., Bergstrom, L., & Ulen, B. (2002). A decision support system for phosphorus management at a watershed scale. *Journal of Environmental Quality*, *31*, 937–945.

Erickson, J. D., Limburg, K. E., Gowdy, J. M., Stainbrook, K. M., Nowosielski, A., Hermans, C., & Polimeni, J. (2005). Anticipating change in the Hudson river watershed: An ecological economic model for integrated scenario analysis. In: R. Bruins & M. Heberling (Eds), *Economics and ecological risk assessment: Applications to watershed management* (pp. 341–370). Boca Raton, FL: CRC Press.

Ewing, R. (1994). Characteristics, causes, and effects of sprawl: A literature review. *Environmental and Urban Issues*, *21*(2), 1–15.

Forman, R. T. T., & Alexander, L. E. (1998). Roads and their major ecological effects. *Annual Reviews of Ecology and Systematics*, *29*, 207–231.

Hong, B., Limburg, K. E., Erickson, J. D., Gowdy, J. M., Nowosielski, A. A., Polimeni, J. M., & Stainbrook, K. M. An interdisciplinary modeling framework assessing how human socio-economic activities on the landscape affect stream ecosystem condition. Under review at Ecosystems.

Klein, R. D. (1979). Urbanization and stream water quality impairment. *Water Resources Bulletin*, *15*, 948–963.

Li, L., Sato, Y., & Zhu, H. H. (2003). Simulating spatial urban expansion based on a physical process. *Landscape and Urban Planning*, *64*(1–2), 67–76.

Limburg, K. E., & Stainbrook, K. M. (in press). Assessing ecosystem health in Dutchess County, New York. In: J. D. Erickson & J. Gowdy (Eds), *Frontiers in environmental valuation and policy*. Cheltenham, UK: Edward Elgar.

Limburg, K. E., Stainbrook, K. M., Erickson, J. D., & Gowdy, J. M. (2005). Urbanization consequences: Case studies in the Hudson Valley. In: L. R. Brown, R. H. Gray, R. M. Hughes & M. Meador (Eds), *The effects of urbanization on aquatic ecosystems* (pp. 23–37). Bethesda, MD: American Fisheries Society Symposium 47.

Macharis, C., Brans, J. P., & Mareschal, B. (1998). The GDSS PROMETHEE procedure. *Journal of Decision Systems*, *7*(special issue), 283–307.

Malczewski, J. (1999). *GIS and multicriteria decision analysis*. New York: Wiley.

McDonnell, M. J., & Pickett, S. T. A. (1990). Ecosystem structure and function along urban rural gradients – An unexploited opportunity for ecology. *Ecology*, *71*(4), 1232–1237.

McDonnell, M. J., Pickett, S. T. A., Groffman, P., Bohlen, P., & Pouyat, R. V. (1997). Ecosystem processes along an urban-to-rural gradient. *Urban Ecosystems*, *1*, 21–36.

McKinney, M. L. (2002). Urbanization, biodiversity, and conservation. *Bioscience*, *52*(10), 883–890.

Meador, M. R., & Goldstein, R. M. (2003). Assessing water quality at large geographic scales: Relations among land use, water physicochemistry, riparian condition, and fish community structure. *Environmental Management*, *31*(4), 504–517.

Merriam-Webster Dictionary (2002). Merriam-Webster. On-line at http://www.m-w.com/home.htm

Naiman, R. J., & Turner, M. G. (2000). A future perspective on North America's freshwater ecosystems. *Ecological Applications*, *10*(4), 958–970.

Nilsson, C., Pizzuto, J. E., Moglen, G. E., Palmer, M. A., Stanley, E. H., Bockstael, N. E., & Thompson, L. C. (2003). Ecological forecasting and the urbanization of stream ecosystems: Challenges for economists, hydrologists, geomorphologists, and ecologists. *Ecosystems*, *6*(7), 659–674.

Nowosielski, A. (2002). *Geo-referenced social accounting with application to integrated watershed planning the Hudson River Valley*. Doctoral Dissertation, Department of Economics, Rensselaer Polytechnic Institute, Troy, New York.

Parmenter, A. W., Hansen, A., Kennedy, R. E., Cohen, W., Langner, U., Lawrence, R., Maxwell, B., Gallant, A., & Aspinall, R. (2003). Land use and land cover change in the Greater Yellowstone Ecosystem: 1975–1995. *Ecological Applications*, *13*(3), 687–703.

Paul, M. J., & Meyer, J. L. (2001). Streams in the urban landscape. *Annual Review of Ecology and Systematics*, *32*, 333–365.

Peterson, G. D., Beard, T. D., Beisner, B. E., Bennet, E. M., Carpenter, S. R., Cumming, G. S., Dent, C. L., & Havlicek, T. D. (2003a). Assessing future ecosystem services a case study of the Northern Highlands Lake District, Wisconsin. *Conservation Ecology*, *7*(3), 1–24.

Peterson, G. D., Cumming, G. S., & Carpenter, S. R. (2003b). Scenario planning: A tool for conservation in an uncertain world. *Conservation Biology*, *17*(2), 358–366.

Polimeni, J. (2002). *A dynamic spatial simulation of residential development in the Hudson River Valley, New York State*. Doctoral Dissertation, Ecological Economics, Department of Economics, Rensselaer Polytechnic Institute, Troy, New York.

Postel, S. L. (2000). Entering an era of water scarcity: The challenges ahead. *Ecological Applications*, *10*(4), 941–948.

Rapport, D. J. (1992). Evaluating ecosystem health. *Journal of Aquatic Ecosystem Health*, *1*, 15–24.

Rapport, D. J., Gaudet, C. L., & Calow, P. (Eds). (1995). *Evaluating and monitoring the health of large scale ecosystems*. Berlin: Springer-Verlag.

Santelmann, M. V., White, D., Freemark, K., Nassauer, J. I., Eilers, J. M., Vache, K. B., Danielson, B. J., Corry, R. C., Clark, M. E., Polasky, S., Cruse, R. M., Sifneos, J., Rustigian, H., Coiner, C., Wu, J., & Debinski, D. (2004). Assessing alternative futures for agriculture in Iowa, USA. *Landscape Ecology*, *19*(4), 357–374.

Settle, C., Crocker, T. D., & Shogren, J. F. (2002). On the joint determination of biological and economic systems. *Ecological Economics*, *42*(1), 301–312.

Shrader-Frechette, K. S. (1994). Ecosystem health: A new paradigm for ecological assessment? *Trends in Ecology and Evolution*, *9*, 456–457.

Stainbrook, K. M. (2004). *Using ecological indicators to detect environmental change in urbanizing watersheds: Case study in Dutchess County, NY*. Master's Thesis, SUNY-College of Environmental Science and Forestry, Syracuse, New York.

Steiner, F. (1994). Sprawl can be good. *Planning*, *60*, 14–17.

Stinner, D. H., Stinner, B. R., & Martsolf, E. (1997). Biodiversity as an organizing principle in agroecosystem management: Case studies of holistic resource management practitioners in the USA. *Agriculture Ecosystems & Environment*, *62*(2–3), 199–213.

Swaney, D. P., Limburg, K. E., & Stainbrook, K. M. (2006). Some historical changes in the patterns of population and land use in the Hudson River watershed. *American Fisheries Society Symposium*, *51*, 75–112.

Veldkamp, A., & Verburg, P. H. (2004). Modelling land use change and environmental impact. *Journal of Environmental Management*, *72*(1–2), 1–3.

Vörösmarty, C. J., Green, P., Salisbury, J., & Lammers, R. B. (2000). Global water resources: Vulnerability from climate change acid population growth. *Science*, *289*(5477), 284–288.

Young, W. J., Lam, D. C. L., Ressel, V., & Wong, I. W. (2000). Development of an environmental flows decision support system. *Environmental Modelling & Software*, *15*(3), 257–265.

Zhu, W. X., & Carreiro, M. M. (1999). Chemoautotrophic nitrification in acidic forest soils along an urban-to-rural transect. *Soil Biology & Biochemistry*, *31*(8), 1091–1100.

INTEGRATED AGRICULTURAL AND HYDROLOGICAL MODELING WITHIN AN INTENSIVE LIVESTOCK REGION

Peter Kreins, Horst Gömann, Sylvia Herrmann, Ralf Kunkel and Frank Wendland

ABSTRACT

An interdisciplinary model network consisting of the regional agricultural economic model RAUMIS and the hydro(geo)logical models GROWA/WEKU is used to analyze the effect of different scenarios of maximum agricultural nitrogen balance surplus on water quality. The study area is the federal state of Lower Saxony, Germany, which features heterogeneous natural site conditions as well as agricultural production structures. A focus of the study is the modeling of supra-regional manure transports that, according to the model's results, considerably increase due to a lowering of maximum nitrogen balance surpluses. The assessment of the examined nitrogen reduction measures reveals that adequate indicators have to be applied. In this regard, the model results show that even though the analyzed measure leads to a substantial overall reduction of agricultural nitrogen surpluses, nitrogen discharges into surface and groundwater can regionally increase.

Ecological Economics of Sustainable Watershed Management
Advances in the Economics of Environmental Resources, Volume 7, 113–142

ISSN: 1569-3740/doi:10.1016/S1569-3740(07)07006-X

1. INTRODUCTION AND PROBLEM

In recent decades, purchased feed has allowed regions such as the southwest Netherlands and northwest Germany, Belgium, Denmark, Brittany (France) and Catalonia (Spain) to become sites of concentrated animal production. These concentrations are economically advantageous for producers, as framework conditions in the cluster areas support intensive agricultural animal production.

The farmyard manure produced from this feed import-based form of animal production is generally disposed off in agricultural areas. The utilization rate of manure–nutrients by plants is lower than that of nutrients from mineral fertilizer, so animal production brings about higher unavoidable nutrient surpluses that can be carried over into water bodies. Though some transport of farmyard manure does occur over distances of more than 80 km, most remains in the region because its high water content makes it expensive to transport. Hence, regions with intensive livestock farming typically display the highest nutrient surpluses (Gömann, Kreins, & Møller, 2004b, pp. 81–90).

For example, intensive land use and the application of farm manure close to water bodies pollute river catchments in northwest Germany. The Ems (including tributaries) and the Vechte are considered critically polluted; they are predominantly assigned the biological water quality category II–III (Lower Saxony State Office for Ecology, 2001, p. 31ff), according to a classification system developed by Länderarbeitsgemeinschaft Wasser (LAWA) (1998) ranging from I to IV. With respect to nitrogen, 50% of surface waters were classified in grade level III. High nitrogen concentrations impair the use of groundwater for drinking. Nitrate pollution of groundwater in Lower Saxony is high in comparison to other federal states (Ministry of Environment of Lower Saxony, 2004).

Against the background of these problems, the German Fertilizer Regulations (Düngeverordnung, 1996) were created to put the EU nitrate directive (Council Directive 91/676/EEC of 12 December 1991) into force, and the application of manure was limited to 210 kg total nitrogen per ha of grassland and 170 kg per ha arable land. Farms exceeding these amounts must take steps to comply with the limits. One option is to lease land for the application of the manure, but leasing prices in intensive livestock areas are high due to the high demand for areas for "slurry disposal." In some regions, most land is already at or above the application limit. A further option is the use of nitrogen and phosphorous reduced feed (RAM Feed), which requires a change in feeding. These measures incur different levels of adaptation costs to the affected farms or regions, while the reduction of the animal herd entails the highest adjustment costs for the farms.

A further possibility to comply with the limits is to distribute the farmyard manure in a larger area. The need to dispose off manure from concentration areas is juxtaposed with the growing demand for low-cost fertilizer in arable farm regions. Supra-regional marketing structures for farmyard manure are evolving,[1] with distributions generally administered through contractual supply agreements between farms or farms and haulage contractors. This option has gained significance in recent years and can provide a long-run adaptation possibility for intensive livestock farms/regions, assuming these transports do not violate animal epidemic regulations or meet resistance from residents of importing regions.

From the perspective of water protection, the manure application limits of 210 and 170 kg total nitrogen per ha for grassland and arable land, respectively, are considered too high. In addition, there is no limit on the total amount of fertilizer (Staffel-Schierhoff, 2001, p. 98). In specialized livestock farms, the average additional amount of mineral nitrogen used is between 100–150 kg N/ha/year (Osterburg, Schmidt, & Gay, 2004). To reach the environmental policy goal of reducing nitrogen charges by 50%, a reduction in nitrogen surpluses to 50 kg N/ha is necessary (German Environmental Agency, 2001, p. 50).

However, the environmental and economic implications of lowering the permissible nutrient applications from manure and/or nutrient balances have not yet been assessed. The impacts on water quality are unclear, partly due to an expected increase in supra-regional transport of farmyard manure. While water quality might improve in manure exporting regions, it might deteriorate in manure importing regions. An overall positive effect on water quality is only attainable if the natural site conditions of the manure importing regions are less susceptible to discharges/leaching of nutrients than the exporting regions with the higher nutrient surplus. In addition, farms will incur adjustment costs in order to comply with lower nutrient balance surpluses.

In this study, the effects on water quality as well as the economic impacts of lowering the permissible nitrogen balances are analyzed taking into account supra-regional transports of farmyard manure within the study area of Lower Saxony. The analysis takes place using an interdisciplinary model network made up of the agricultural economic Regional Agricultural and Environmental Information System (RAUMIS) (Henrichsmeyer et al., 1996) and the hydro-geological model GROWA98 (Kunkel & Wendland, 2002) and WEKU (Wendland & Kunkel, 1999). For wide area coverage, the coupled models provide a regionally differentiated, consistent link between the Driving Force Indicator "Nitrogen Balance Surplus," the State Indicator "Nitrogen (Nitrate) Concentration" and Response Indicators (Gömann, Kreins, Kunkel, & Wendland, 2003, 2004a).

The paper proceeds as follows. The following second section provides a theoretical background of the impacts of lowering permissible nutrient balances taking supra-regional transports of manure into account. In the third section, the methodological approach of the interdisciplinary model network is described including details on the module "supra-regional physical transport," which is a central advancement of RAUMIS. The fourth section presents model results concerning the impacts of lowering the permissible nitrogen balances on regional and overall discharges of nitrogen into ground and surface water as well as on agricultural production, potential supra-regional transports of manure, agricultural income and jobs. In a fifth section, the measure is discussed with respect to its efficiency to achieve water quality targets and the section concludes with recommendations.

2. THEORETICAL BACKGROUND

Farmyard manure is an unavoidable product of animal production and must be used by livestock farms, meaning it must be distributed on the field. Farmyard manure has a positive effect on the fertility-promoting humus development in soil and can partially substitute for mineral fertilizers, however, it is not possible to make a complete substitution due to the fixed nutrient proportions in the farmyard manure. The substitution value of an applied unit of farmyard manure may not exceed the manuring costs using mineral fertilizer with similar nutrients.

In general, animal farms distribute farmyard manure on the fields they farm as long as the benefit of an additionally distributed unit (marginal benefit) is greater than the opportunity costs of an alternative use of this unit, such as selling the manure to be distributed on fields of other farms. The optimal farm application level of manure is attained when the marginal benefit of the manure applied on the source farm is in accordance with the profit made when selling it to another farm. With increasing transport expenses, the profit is reduced.

When there are no fertilizer restrictions and no marginal benefit of farmyard manure application to the manure supplying farm, the farmyard manure will be transported as far as the transport costs are in accordance with the substitution value of a mineral fertilizer with the same nutrient value. Even over minimal distances, the transport costs of farmyard manure exceed its substitution value. For this reason inter-farm, or rather supra-regional, farmyard manure transports are economically sensible only in a limited area.

The marginal value of the application of farmyard manure in arable farming declines with increased application levels and shows a negative marginal benefit when it exceeds a plant physiological maximum. This phase begins first at application levels, which are high above an ecologically acceptable level (Oehmichen, 1983, p. 388). The fraction of nutrients that crops can utilize is lower in farmyard manure than it is in mineral fertilizer. Thus, fertilization of arable crops with farmyard manure results in unavoidably higher nutrient surpluses that can leach into water bodies.

In the following, the theoretical impact of the implementation of maximal nutrient balances is derived. This measure represents a more complete and thus more flexible restriction than do fertilizer regulations (DüngeVO), since the application of mineral fertilizers is implicitly included in the nutrient balance. In order to comply with the restrictions, production adjustments are necessary, which take place against the background of complex, partially interdependent alternative actions. Adjustment options include cropping structure and intensity, feeding and size of the animal herd. The expansion of farm area is a less viable option in intensive animal areas, since the availability of non-saturated areas within the region is limited, and as a rule, high demand for agricultural land causes high leasing prices.

Production adjustments to comply with the established restriction on nutrient balances (Nbal) cause a loss in farm profits (abatement costs). In Fig. 1, profits (π^1) are dependent on Nbal without the inclusion of interregional slurry transport. A gradual loosening of the restriction leads to increases in π^1 with declining increments and at a^* reaches a maximum that is no longer bound by the restriction. Taking interregional slurry transports into account, profits (π^2) are higher at low nutrient balance levels, since slurry transports present an option to minimize the production limitations that result in high profit losses.

According to the course of the profit function π^1, the marginal profits (marginal abatement costs mAC^1) decrease with an increase of acceptable nutrient balances. Through the inclusion of interregional transports of farmyard manure, the adjustment range is expanded for agriculture, such that profit losses are lower with a reduction of permissible Nbal. For this reason, the marginal abatement costs including interregional transports (mAC^2) is rotated downward in comparison to mAC^1 with a^* as a pivotal point.

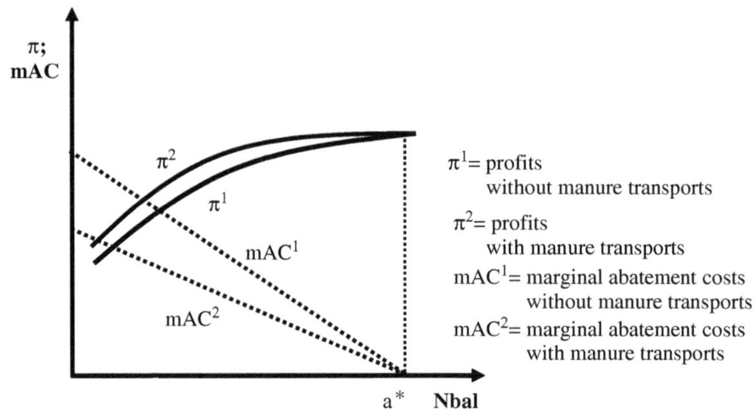

Fig. 1. Abatement Cost and Marginal Abatement Costs of Reducing Balance Surpluses. *Source:* Own Design.

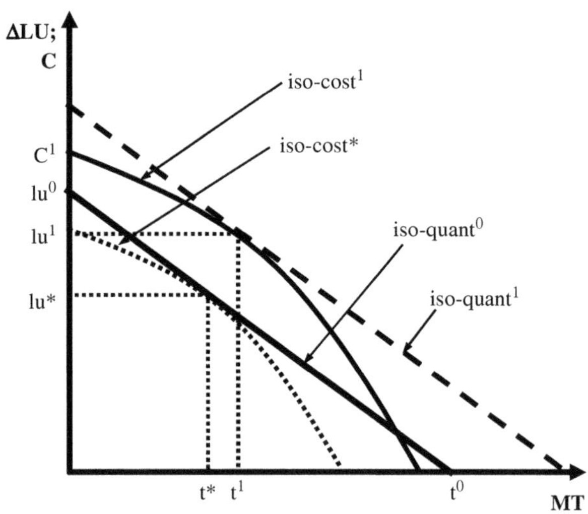

Fig. 2. Optimal Mix of Production Adjustments and Supra-Regional Transports. *Source:* Own Design.

The optimal relationship between production adjustments and interregional transports is based on the relative profitability of the two alternatives. In Fig. 2, production adjustments are represented through reductions in herd sizes (ΔLU) and are compared to interregional slurry transports (MT).

To meet an exogenously determined nutrient balance surplus, it is assumed that either a reduction in herd size at the level of lu^0 or interregional transports at a level of t^0 is necessary. The two options are completely interchangeable with regard to the targeted nutrient reduction. For this reason, combinations of animal herd size reduction and slurry transport with the same nutrient reduction (iso-quant^0) are in a linear relationship to each other.

To derive an iso-cost curve, it is assumed in Fig. 2 that a reduction of herd size of lu^0 causes costs (profit losses) at a level of C^1. From the course of the marginal abatement cost function mAC^1 (see Fig. 1) it follows that the last unit of the animal herd size reduction causes the highest marginal abatement costs. The marginal costs of the first interregional manure transport units are comparably low. Consequently, instead of dismantling the last livestock unit within the lu^0, over-proportionally more units of interregional manure transport can be realized at constant costs (iso-cost^1). If more and more livestock units are kept in production, an increasing amount of manure has to be transported over even longer distances. Hence, marginal transport costs increase while the marginal costs of dismantling the animal herd decrease. The iso-cost^1 is shaped concavely.

The optimal relationship between herd reduction and interregional slurry transports is given if the reverse relationship of marginal costs is equal to the constant marginal rate of technical substitution of both alternatives. For iso-cost^1, this is given in point t^1/lu^1, tangent to the iso-quant^1. However, through iso-quant^1, a higher nutrient reduction is realized as necessary. Through a parallel shifting (scaling) of iso-cost^1, an iso-cost* function belonging to iso-quant^0 can be derived as well as the optimal relationship (t^*/lu^*) between the reduction in herd size and interregional manure transports.

The introduction of a measure such as nutrient balance limits can, due to the possibility of transporting manure, lead to significant changes in the regional nutrient distribution. Through the reduction of nutrient surpluses in problem areas, the total input of diffuse nutrients in water will not necessarily be reduced. This is due to the minimal correlation between the pressure indicators nutrient deficits and the "environmental good" state-indicator water quality. The link between pressure and state indicator is targeted through an integrated illustration of agricultural and hydrological processes in the framework of an interdisciplinary model.

The theoretical issues discussed above are analyzed with an interdisciplinary model network consisting of the RAUMIS and the hydro(geo)logic models GROWA and WEKU.

3. INTEGRATED AGRICULTURAL ECONOMIC AND HYDROLOGICAL MODELING

3.1. Agricultural Economic Modeling

RAUMIS is designed for continuous usage in the scope of medium- and long-term agricultural and environmental policy impact analyses. It comprises more than 50 agricultural products, 40 inputs with exogenously determined prices, and reflects the whole German agricultural sector with its sector linkages. The model consolidates various agricultural data sources and generates base model data with the national agricultural accounts as a framework of consistency.

Due to data availability, the spatial differentiation is currently based on administrative bodies on NUTS III level ("Landkreise"). Some 326 regions are treated as single "region farms" that reach their production decisions autonomously. Hence, adjustments of production at the national level are based on aggregated responses of region farms. Adjustments caused by changes in general conditions (e.g., agricultural policies) are determined using a positive mathematical programming approach (Howitt, 1995) with the following nonlinear objective function for each region:

$$\begin{aligned} \max_x \quad \Pi &= \sum_i z_i(x_i)x_i \\ \text{s.t.} \quad b_i &\geq \sum_i a_i x_i \end{aligned} \tag{1}$$

The objective function is a regional agricultural profit (Π) function maximizing the product of per unit margins z_i between the price and the costs of the *i*th netput[2] and the level of each netput x_i. The objective function is nonlinear since z_i's are functions of their realized netput level x_i. The problem is solved subject to a set of technical, political and economic constraints ($b_i \geq \sum_i a_i x_i$), e.g., land availability, set-aside obligations etc., and proceeds in two stages. In the first stage, optimal variable input coefficients per ha or animal are determined. In the second stage, profit maximizing cropping patterns and animal herds are determined simultaneously with a cost minimizing feed and fertilizer mix.

3.2. Nutrient Balancing

A set of agri-environmental indicators implemented in RAUMIS is linked to agricultural production. Currently, the model comprises indicators such as nutrient surplus (nitrogen, phosphorus and potassium), pesticide expenditures, a biodiversity index and corrosive gas emissions. Regarding diffuse water pollution, the indicator "nitrogen surplus" is of particular importance. The concept of balancing nitrogen follows PARCOM-guidelines (PARCOM, 1993) where the soil surface represents the system border. The long-term regional nitrogen balances (Nbal) averaged over several vegetation periods are calculated as follows:

$$\text{Nbal} = f(x_i, \text{s}, \text{cl}, \text{af}, \text{ad}) \quad (2)$$

The positions of the nitrogen balance are calculated based on the activity framework in RAUMIS. In order to obtain regional input and output (netput) positions, activity-specific coefficients are multiplied with the level of each activity (x_i), e.g., area harvested or livestock units. The primary demand for nitrogen is the nutrient uptake of plants to be harvested. In RAUMIS, crop-specific nitrogen requirements are calculated using linear functions that depend on expected yields as well as on regional soil (s) and climate (cl) dependent requirement functions. Further nitrogen sources are symbiotic and asymbiotic nitrogen fixation (af), as well as atmospheric depositions (ad) that partially originate from agricultural ammonia emissions.

An important source of nitrogen for plant production is manure. RAUMIS differentiates between four processes of manure and its application, i.e., dung and liquid manure from cattle, hogs and poultry. Coefficients representing nutrient contents in manure as well as utilization factors of plants are taken from the literature and are also provided by experts of the German Federal Ministry of Consumer Protection, Food and Agriculture (BMVEL). Following the concept that nitrogen from manure can replace nitrogen from mineral fertilizer, mineral fertilizer equivalents for manure are calculated based on different nitrogen utilization factors of dung and liquid manure from cattle, hogs and poultry. Nitrogen is supplied by mineral fertilizer also. The total national mineral fertilizer supply is consistently broken down to regions and production activities in RAUMIS.

As a rule, regional balances of nitrogen supplies and extractions result in a positive figure. The nitrogen surplus represents a risk potential since it indicates the amount of nitrogen potentially leaching into ground and surface water.

3.3. Specification of a Supra-Regional Manure Transport Module

For most applications of RAUMIS, it can be assumed that region farms reach their production decisions independently from each other, which facilitates computation. However, this assumption cannot be maintained when analyzing the impacts of restricting nutrient balances because supra-regional manure transport might result in interaction between regions. This interaction is implemented into RAUMIS by adding a manure transport module to the input–output matrix of each region farm. The methodology applied is based upon a study that focuses on the implementation of a federal statewide trading system of milk quotas (Kreins & Cypris, 1999). The manure transport module allows region farms to dispose of excess manure while other region farms fulfill nutritional requirements via manure imports.

The general profit maximization problem described in Eq. (1) is expanded to take transport activities between regions into account and optimizes overall profit across all regions, i.e., the regional dimension (index r) is added to the problem. Three transport activities, liquid manure from cattle, hogs and poultry, are implemented into RAUMIS according to the nutrient balancing described in Section 3.2. Activity specific transport costs ($t_{r,i}$) depend on the distance ($d_{r,j}$) between any two regions within the set of (n) regions. Furthermore, a maximum allowable nutrient balance is imposed. The optimization problem becomes

$$\begin{aligned} \max_x \quad \Pi = \quad & \sum_r \sum_i \left(z_{r,i}(x_{r,i}) - \sum_j t_{r,i,j}(d_{r,j}) d_{r,j} \right) x_{r,i} \qquad (j = 1, 2, \ldots, n) \\ \text{s.t.} \quad & b_{r,i} \geq \sum_{r,i} a_{r,i} x_{r,i} \\ & \text{Nbal}_r \geq f_r(x_{r,i}, \text{sc}_r, \text{cl}_r, \text{af}_r, \text{ad}_r) \end{aligned} \tag{3}$$

Activity-specific nutrient utilization factors are taken into account. This implies that exporting liquid manure from a region does not only reduce the physical amount of manure but also consistently lessens the nutrient supplies and balances. In contrast, importing regions do not only absorb the nutrient fraction of manure that plants can exploit but take over the non-accessible nutrients as well, increasing the overall nutrient balance. Since implementing maximum nutrient surpluses is primarily a problem for livestock intensive regions, the manure export activities are charged the total transport costs.

The extent of supra-regional manure transports primarily depends on transport costs that are generally determined by volume rather than by weight. Transport costs for liquid manure or effluent sludge are reported quite differently in the literature. Cost differences are primarily due to the particular transport technology that changes according to distance. While local distances are covered by transport systems that rely mainly on farm equipment/machinery, long-distance systems mostly include heavy goods vehicles (trucks) requiring a transfer to liquid manure spreaders at the destination point.[3]

Transport costs can be cut by 2/3 if "thick" liquid manure is shipped instead of "thin" liquid manure, as increased dry matter fraction increases nutrient content of the manure. However, potential dehydration costs must be compared to transport costs. In addition, the content of dry substance determines the transport technology. Technology differences mainly affect costs that are independent of the distance, such as the freight vehicle, its access route and loading (Laiblin, 1999, p. 44). Transport costs depending on distance are mainly determined by transport time, though the average transport velocity increases with distance and the expansion of the road networks (Arlt, 2003).

Support values for deriving transport costs as a function of the distance d_{r},j were compiled from the literature. Marginal unit transport costs $t_{r,i,j}$ range from 0.56 Euros per metric ton (mt) and kilometer (km) for local distances (Bayerische Landesanstalt für Landwirtschaft, 2002) to 0.13–0.16 Euros per mt and km for larger/regional distances (Betriebshilfsdienste und Maschinenringe, 2004). Based on these support values the following marginal unit cost function is derived:

$$t_{r,i,j}(d_{r,j}) = 57.193 d_{r,j}^{-0.287} \quad (j = 1, 2, \ldots, n) \tag{4}$$

This function is implemented into RAUMIS to calculate total unit transport costs $t_{r,i,j}$ of manure between region farms as follows:

$$t_{r,i,j}(d_{r,j}) = t_{r,i,j}(d_{r,j}) d_{r,j} = (57.193 d_{r,j}^{-0.287}) d_{r,j} = 57.193 d_{r,j}^{0.713} \quad (j = 1, 2, \ldots, n) \tag{5}$$

Both functions are depicted in Fig. 3.

In order to calculate transport costs $t_{r,i,j}$ the distances $d_{r,j}$ need to be determined. Assuming that livestock is homogeneously distributed within regions, actual transport distances are approximated by defining a rectangle around each region and designating its central point as the center of the region. Distances are then calculated between regions' central points. While the overlap of adjacent regions' rectangles results in a systematic overestimation of distances, this effect is presumably outweighed by the calculation of distances as linear "air-line distances," which underestimates distances by disregarding the actual routing and configuration of streets and roads.

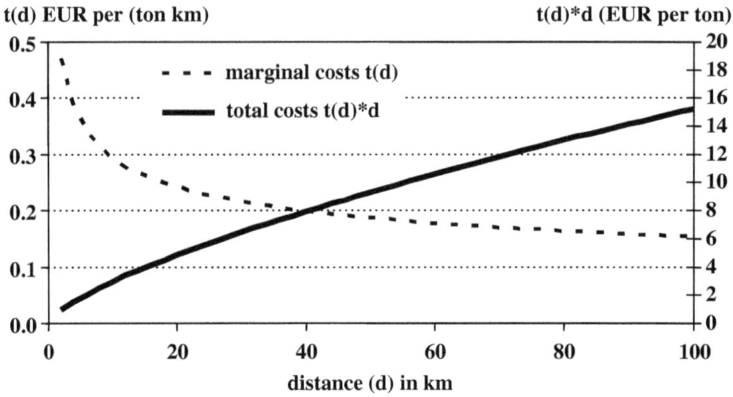

Fig. 3. Transport Cost Functions. *Source:* Own Calculations.

3.4. Description of Hydrological Modeling

In this study, hydrological models were chosen to analyze diffuse nitrate inputs into surface waters. These models correspond to the regionalized agricultural economic approach pursued by RAUMIS and are designed for area-differentiated consideration of nutrient inputs on a supra-regional scale. According to the models' applicability to large river basins, the hydrological, pedological and hydrogeological input parameters needed for modeling are taken from thematic maps. The scale of these maps, ranging from 1:50,000 to 1:200,000, determines the degree of detail of the model input values and defines the validity range of the model results.

Diffuse nitrogen inputs into rivers take place via two runoff pathways: direct runoff and groundwater runoff. Diffuse pollution starts as the nitrogen surplus calculated by RAUMIS, then reduced by the denitrification losses in the soil. This nitrogen surplus is related to the groundwater recharge/total runoff ratio. For example, in areas where groundwater runoff is 90% of the total runoff, it is assumed that 90% of the diffuse nitrogen surpluses are transported to the surface waters via groundwater. During this transport, nitrate degradation may occur. Thus, a calculation of the groundwater borne nitrate inputs into surface waters requires knowledge of the groundwater flow paths, the total residence time of the nitrate and the denitrification kinetics in the upper aquifer. These processes are considered by different models.

The GROWA model is used to carry out area differentiated water balance analyses. The mean long-term total runoff is modeled as a function of the regional interaction of climate, soil, geology, topography and land use conditions. The model separates the total runoff into the direct runoff (interflow and surface runoff) and groundwater runoff (groundwater recharge). The ratio between groundwater recharge and total runoff was taken as a measure for the extent to which diffuse nitrogen surpluses are displaced from soil to groundwater.

Nitrate degradation in soils was calculated according to a Michaelis–Menten kinetics using the approach of Köhne and Wendland (1992). Denitrification losses occur mainly in the effective root zone of the soils, and can be described as a function of the nitrogen surpluses, the average field capacity and the site-specific denitrification conditions.

The WEKU model models the reactive nitrate transport in groundwater. In the first step, groundwater velocities are calculated according to Darcy's law of hydraulic conductivity, effective yield of pore space of the aquifer and the slope of groundwater surface (hydraulic gradient). The residence times of the groundwater runoff are calculated in a second step, which utilizes a digital relief model of the groundwater surface. This is analyzed with respect to the water network, groundwater discharge or transfer areas, lateral flow dynamics and groundwater effective recipients.

The WEKU model has been extended by a module quantifying nitrate degradation in groundwater. According to extensive field studies by Böttcher, Strebel, and Duynisveld (1989) in a catchment area in the north German Lowlands and van Beek (1987) for a site in the Netherlands, a first order denitrification kinetics has been assumed with a reaction constant in the range of 0.17 to 0.56 per year. This corresponds to a half life of nitrogen leached into the groundwater of 1.2 and 4 years. Rather simple indicators, such as the presence of Fe (II) and Mn (II) and the absence of O_2 and NO_3 can be used to decide whether a groundwater province has hydrogeochemical conditions in which denitrification is possible.

The validation of the groundwater borne nitrate inputs into rivers is based on results of the MONERIS model (Behrendt et al., 2000). The MONERIS model distinguishes between point source emissions from wastewater treatment plants and direct industrial discharges and six diffuse pathways, including the inputs via groundwater. It is assumed that the observed nitrogen concentrations in rivers under base flow conditions correspond to groundwater borne nitrate inputs. Thus the modeled nitrogen inputs into surface waters from groundwater were compared to the corresponding values given by the MONERIS model. When the results of the MONERIS model and the WEKU model agree, we conclude that the chosen procedure gives reliable estimations for the groundwater borne nitrate input into the aquifers.

4. RESULTS

4.1. Status quo in the Study Area and Reference Scenario

The study area is the German federal state of Lower Saxony (see its location within Germany in Fig. 4), which features a variety of soil and hydrological site conditions. The north German lowlands, generally consisting of quaternary granular soil (sand and gravel) covered by sandy soil, make up about 75% of the state's area. Fertile loamy soils are prevalent only in the plains in the southern part of the north German lowlands, making intensive market crop farming (wheat and sugar beets) possible in these areas.

In the higher Geest, far from groundwater, fodder cropping is the dominant land use. The thick layers of sand and gravel underlying the Geest result in high groundwater reservoirs, which are significant for water management in Lower Saxony. In this groundwater recharge region, surplus nutrients can

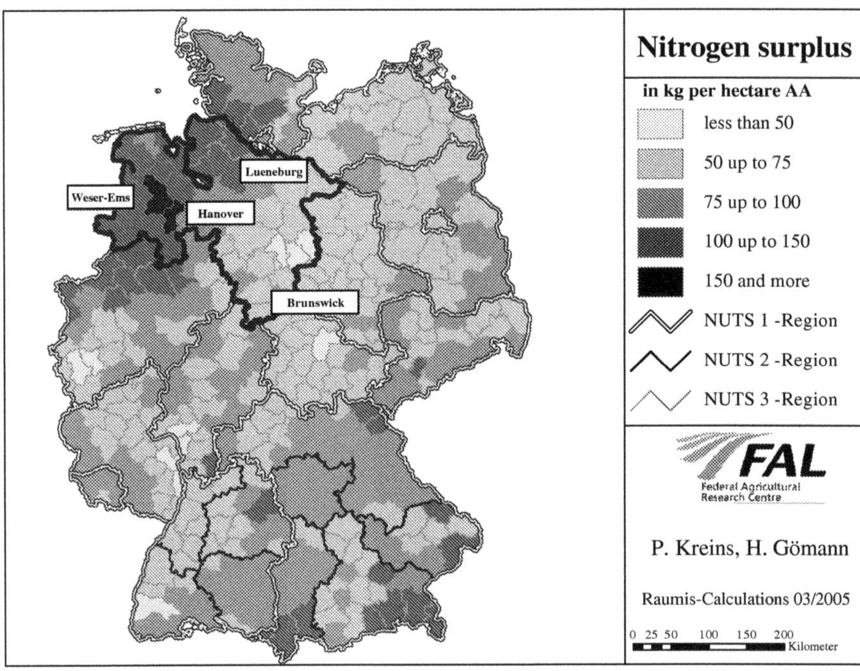

Fig. 4. Nitrogen Surplus in Germany and Lower Saxony in 1999. *Source:* Calculations by FAL Using RAUMIS 03/2005.

be leached from soil and in many places lead to widespread nitrate pollution of the groundwater. In the marsh regions close to the coast, grassland husbandry is predominant. There, as well as in the areas along rivers (lower Geest), artificial drainages result in a high proportion of direct runoff.

In the mountainous and hilly landscape of south Lower Saxony, the geographical relief, climatic conditions and soil thickness are the most significant factors for land use. In the loamy valley and basins, widespread agriculture takes place, while the elevated areas are mostly forested. Here, direct runoff increases the risk of N and P pollution of surface waters.

The state of Lower Saxony covers an area of about 4.8 million ha (see Table 1). It is partitioned in four administrative districts (NUTS II level) ("Regierungsbezirke") that include 47 counties and district towns (NUTS III level) ("Landkreise and kreisfreie Städte"). Some counties and

Table 1. Key Features of the Study Area in the Status Quo Situation 1999.

	Unit[a]	Lower Saxony	Administrative Districts			
			Brunswick	Hanover	Lueneburg	Weser-Ems
Total Area	Mill. hectare	4.8	0.8	0.9	1.6	1.5
Population	Mill. Inh.	8.0	1.7	2.2	1.7	2.5
Population density	Inh./100 ha	167.9	205.3	239.6	109.5	164.7
Share of agricultural GDP	% of total	2.2	1.1	1.3	2.9	3.5
Agricultural area	Mill. hectare	2.7	0.4	0.5	0.8	0.9
Share of arable land	% of AA	67.4	86.6	82.7	59.2	58.4
Share of field crop farms	% of total farms	30.2	68.6	50.3	25.3	15.6
Share of grazing livestock farms	% of total farms	49.0	25.3	32.1	57.7	56.5
Share of pigs and poultry farms	% of total farms	13.3	3.0	10.8	6.7	20.8
Agricultural labor force	1,000 LFU	86.9	6.2	13.3	30.3	37.0
Livestock density	LU/ha AA	1.3	0.4	0.8	1.1	2.2
Nitrogen surplus	Kg/ha AA	95.9	54.5	72.5	88.6	132.0
Nitrogen in surface water	Kg/ha AA	14.3	18.5	13.4	12.9	14.1

Source: Calculations by FAL using RAUMIS 03/2005.

[a]Inh.: Inhabitants; AA: agricultural area; LFU: labor force unit; LU: livestock unit.

district towns are aggregated in RAUMIS so that the state is divided into 39 RAUMIS region farms.

Lower Saxony has been selected as a study area chiefly for the reasons mentioned in Table 1.

Firstly, heterogeneous natural site conditions as described above for Lower Saxony are necessary to adequately consider the different environmental impacts of an increase in supra-regional manure transports as a consequence of agri-environmental policies.

Secondly, Lower Saxony features a very diverse spatial structure and specialization of agricultural production. On the one hand, as mentioned above, one of the most intensive livestock areas in Europe is in the "Weser-Ems region," which is one of the administrative districts of Lower Saxony. This region's significantly above-average share of specialized pigs and poultry farms (see Table 1) as well as an above-average share of specialized grazing livestock farms result in excess manure supply that needs to be exported. On the other hand, the administrative districts Brunswick and Hanover are predominated by field crop farms and below-average livestock densities (see Table 1), meaning they have high potential for importing farmyard manure.

Thirdly, the study is restricted to a single federal state because water management falls within the jurisdiction of federal states. Therefore, it is assumed that farmyard manure must not cross federal state borders.

A *scenario of reference* is necessary for assessing impacts of an implementation of maximum N-balances taking supra-regional manure transports into account. In this study the reference scenario is a projection to the target year 2010 under the Common Agricultural Policy (CAP) regime "Agenda 2000" (assuming no major CAP-reforms following the Mid-Term-Review). A variety of exogenous variables such as implicit costs resulting from positive mathematical programming, input–output coefficients, yields, capacities and prices are forecast. Updates are partially based on trend and yield dependent regression analyses as well as on estimates provided by experts particularly relating to prices and the development of farm structures. It is important to note that the fertilizer regulation is not implemented in the reference scenario.

A comparison between the base year 1999 and the projected reference scenario does not reveal any significant changes in regional N-balance surpluses (see Figs. 4 and 5). Anyway, regional N-surpluses are only of limited significance for diffuse water pollution because hydro(geo)logical conditions determine degradation and denitrification processes. Based on the hydro(geo)logical conditions, regional shares of N-surpluses that are discharged into surface water are calculated and displayed in Fig. 5. These

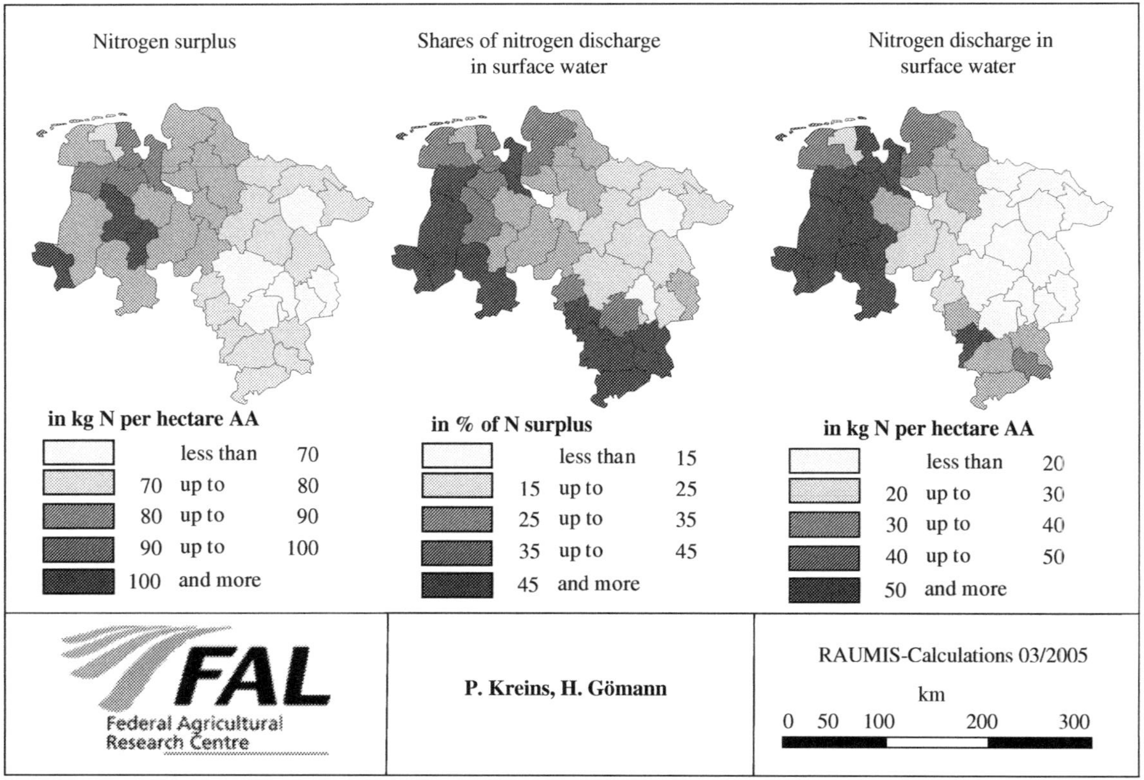

Fig. 5. Regional Nitrogen Surpluses, Shares of Nitrogen Discharges in Surface Water and Nitrogen Discharges in Surface Water in Lower Saxony in 2010. *Source:* Calculations by FZJ Using GROWA/WEKU 03/2005.

regional shares display considerable variation that renders N-surpluses insufficient for measuring diffuse nutrient pollution.

Above-average shares of N-surplus leach into surface water in western and southern regions. In western regions, this is chiefly due to artificial drainages that result in fast runoff with very short residence time. Hence, a large fraction of the N-load in the total runoff is discharged directly into surface water. Only a smaller fraction of the N-load is discharged into deeper soils and aquifers with the groundwater recharge. Thus, even though high denitrification capacities prevail in the groundwater in these regions, the degradation of high N-surpluses is relatively limited. Southern regions of Lower Saxony fall within the highlands of the rivers Weser and Leine with high total area runoff levels, dominated by fast (direct) runoff components. Here, large fractions of the N-surplus are discharged directly into surface water, too. The lowest share of 11% is calculated for the fertile soil region of Peine.

In western regions of Lower Saxony, high N-surpluses coincide with high shares of N-discharges into surface water, an unfavorable overlap. Changing this allocation via supra-regional manure transports might ease water pollution problems in these so-called "hot spot areas" at the expense of impairments in other regions.

4.2. Implementation of Maximum Nitrogen Surpluses

Different mitigation strategies are likely to have different effects on the regional reduction of the N- and P-pollution of surface waters. In order to demonstrate the impacts of the implementation of maximum permissible nutrient surpluses in the context of supra-regional manure transports, three different variants, all of which aim to reduce the total nitrogen discharges throughout Lower Saxony by the same amount, are analyzed and compared to the reference scenario for the target year 2010. In the first variant, a uniform (area wide) maximum N-surplus level of 100 kg N per ha AA is introduced into the RAUMIS model keeping all other parameters and constraints constant. Supra-regional manure transports are not accounted for (Acronym: UniNbalNoTrans). The implications of supra-regional manure transports are analyzed in a second variant that examines a N-surplus restriction at the level of 80 kg N per ha AA (Acronym: UniNbalWith Trans). Adjustments within and among region farms of Lower Saxony are simultaneously optimized in RAUMIS. In the third variant, area differentiated N-balance surplus are derived on the basis of the shares of N-surpluses that are discharged into surface water. Supra-regional manure transports are taken into account (Acronym: DiffNBalWithTrans). It is important to notice that the measure is implemented on the farm level, meaning that not all of the N-balance positions described in Section 3.2 are considered. Atmospheric depositions are an example of excluded N-sources because farmers cannot be expected to have reliable information on this source. Slight deviations may occur because the scenario simulations with RAUMIS apply the complete N-balancing.

The figures given in Table 2 provide an overview about the scenario impacts and the potential magnitude of supra-regional manure transports

Table 2. Impacts of Reductions of Permissible Nitrogen Balance Surplus on Agriculture of Lower Saxony Taking Supra-Regional Manure Transports into Account.

		Reference Scenario	UniNbal-NoTrans 100 kg N/ha	UniNbal-WithTrans 80 kg N/ha	DiffNbal-WithTrans 80 kg N/ha
Nitrogen surplus	Tons	241,155	201,760	210,951	222,320
	Kg per ha AA	93	78	81	85
Nitrogen discharges in surface water	Tons	89,045	73,262	74,152	73,871
	Kg per ha AA	34	28	28	28
Agricultural labor force	1,000 LFU	73.6	65.7	68.4	70.6
Livestock density	LU per ha AA	1.4	1.1	1.2	1.3
Supra-regional manure transport	Million m^3			10.3	14.3
Average transport costs	Euro per m^3			–10.5	–9.8
Total transport costs	Million Euro			–109	–141
Net agricultural value added	Million Euro	2,606	2,393	2,463	2,469
Cost of N discharge reduction in surface water	Euro per kg N		13.5	9.6	9.0

Source: Calculations by FAL using RAUMIS 03/2005.

in Lower Saxony. In the reference scenario, the total supply of farmyard manure in Lower Saxony amounts to about 35 million cubic meters, which contain approximately 217,000 tons of nitrogen. Overall N-surplus from agriculture is calculated at about 241,000 tons, which corresponds to 93 kg per ha AA. Supra-regional manure transports are not accounted for. However, a strict obedience to the provisions of the fertilizer regulation (DüngeVO) would require about 5 million cubic meters of supra-regional farmyard manure transports or an equivalent reduction of livestock in the concerned regions. The total N-discharge into surface water amounts to about 89,000 tons, approximately 37% of the total N-surplus mentioned above. Agricultural income is represented by the agricultural net value added (NVA) and amounts to 2.6 billion EUR.

As intended, the overall N-discharge into surface water is reduced by approximately the same amount in all three variants of N-surplus restriction: 15,000 to 16,000 tons, or about 17% compared to the reference scenario. Because of the almost equivalent reduction level, the measures can be compared with regard to the adjustments in agriculture and the reduction costs (expressed in terms of producer profit changes scaled to the level of net value added).

The consequences, particularly on livestock adjustments, of restricting N-surpluses without allowing for supra-regional manure transports are taken up in the scenario UniNbalNoTrans. Even though a higher N-surplus level is permitted compared to the other variants, total agricultural N-surplus in Lower Saxony is decreased by approximately 22% to about 202,000 tons (78 kg N per ha AA). The regionalized impacts on N-discharge levels into surface water are presented in Fig. 6. The implementation of a uniform maximum N-balance surplus (UniNbalNoTrans) considerably reduces N-discharges in the problem areas in western regions of Lower Saxony. However, some regions, particularly those with high N-discharge fractions, still show elevated N-discharges.

The share of N-surplus that leaches into surface water does not change in comparison to the reference scenario. The reduction is almost exclusively due to a cut back of livestock by about 860,000 livestock units (LU) resulting in a livestock density of 1.1 LUs per ha AA. Abatement costs in terms of agricultural income losses amount to 13.5 Euros per kg reduced N-discharge into surface water. Average cost figures do not display the regional distribution of income losses; in intensive livestock regions of the Weser-Ems administrative district (see Table 1), income losses reach 28% (see Fig. 7).

Allowing for supra-regional manure transports opens up an additional option for adjustment. The implementation of a uniform N-surplus limit of

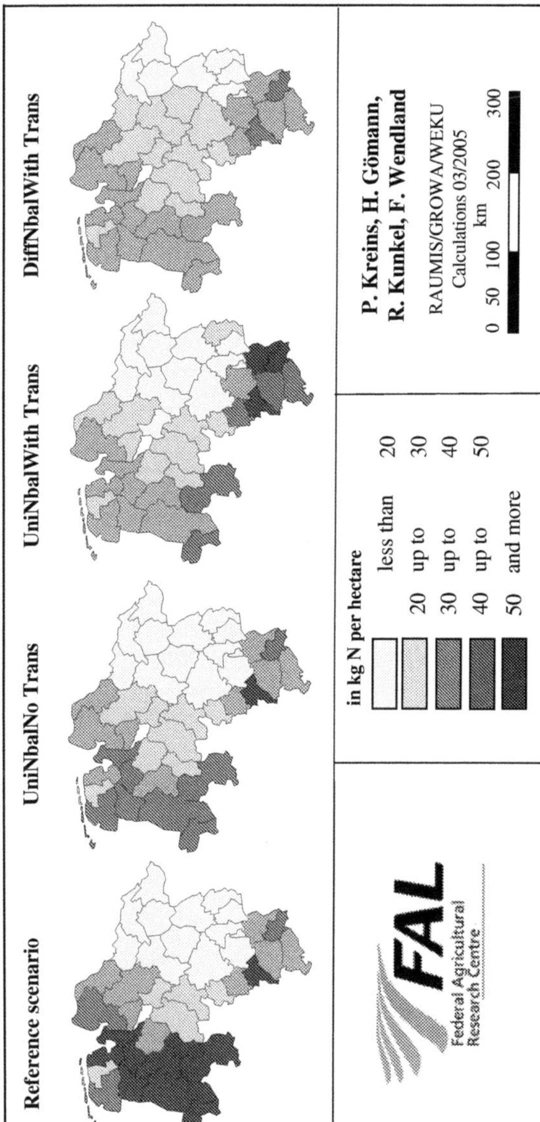

Fig. 6. Regional N-Discharges into Surface Water in the Reference Scenario and Variants of Nitrogen Surplus Limitations in Lower Saxony. *Source:* Calculations by FAL Using RAUMIS 03/2005.

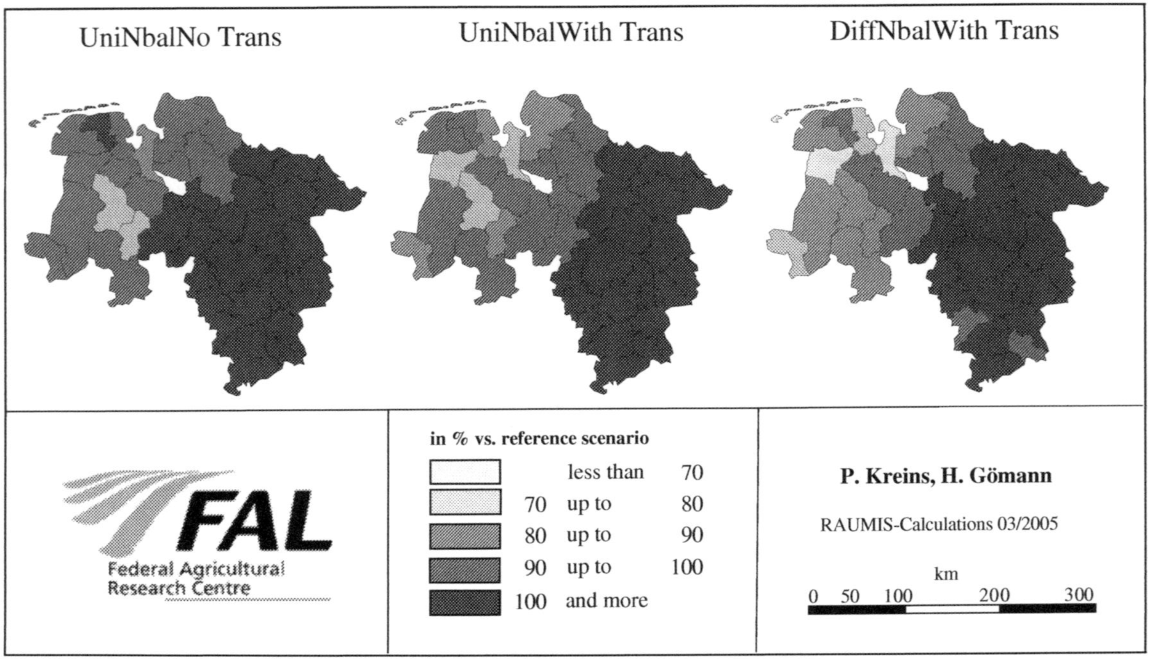

Fig. 7. Regional Agricultural Income Changes vs. a Reference Scenario due to Different Variants of Nitrogen Surplus Limitations in Lower Saxony (2010). *Source:* Calculations by FAL Using RAUMIS 03/2005.

80 kg per ha AA (UniNbalWithTrans) causes a volume of manure transports of about 10 million cubic meters, which costs about 109 million Euros (see Table 2). Since more livestock can be maintained in the problem regions, total N-surplus is higher than in UniNbalNoTrans. However, the N-discharges as a percentage of N-surplus are lower and amount to 35%. Total abatement costs, including both transport costs and agricultural income losses, average 9.6 Euros per kg reduced kg N-discharge. Manure transports increase N-discharges in manure importing regions which poses problems in regions where high shares of N-surplus leach into surface water such as in the southern regions of lower Saxony (see Fig. 6).

With regard to regional income changes it is important to notice that manure exporting regions are charged the total transport cost in the simulations. Hence, the regions that incur a significant income loss under UniNbalNoTrans experience an above-average loss under UniNbalWith Trans as well. However, because of supra-regional manure transports the permissible N-surplus has to be lower to achieve an overall equivalent N-discharge reduction. This leads to considerable income losses in regions that are not affected under UniNbalNoTrans. Manure importing regions generally benefit from manure transports since they receive the nutrients in the manure for free. These regions save expenditures for mineral fertilization that amounts to 2% of the income in the reference situation.

The implementation of area differentiated N-surplus restrictions clearly removes hot-spot areas of diffuse N-pollution and results in a more even spatial distribution of N-discharges (see Fig. 6). This is chiefly due to "targeted" farmyard manure transports that amount to about 14 million cubic meters at a cost of about 141 million Euros. Total N-surplus is only reduced by 8% so that the share of N-discharges decreases to 33%. Taking adjustments of agricultural production into account as well, N-discharge abatement costs are calculated at 9 Euros per kg (see Table 2).

Regional income changes in DiffNbalWithTrans deviate from both uniform N-surplus scenarios because the N-surplus restriction is based on regional shares of N-discharges and may significantly deviate from a uniform maximum N-surplus. While the N-surplus restriction is relaxed in some regions, it is tightened in others. This results in different regional income changes in comparison to the other scenarios.

The analyzed measures explicitly aim at reducing N-discharges into surface water. Of course, groundwater is affected as well; Fig. 8 presents those impacts.

Without allowing for supra-regional manure transports, N-discharges into groundwater are reduced in problem regions without changing the

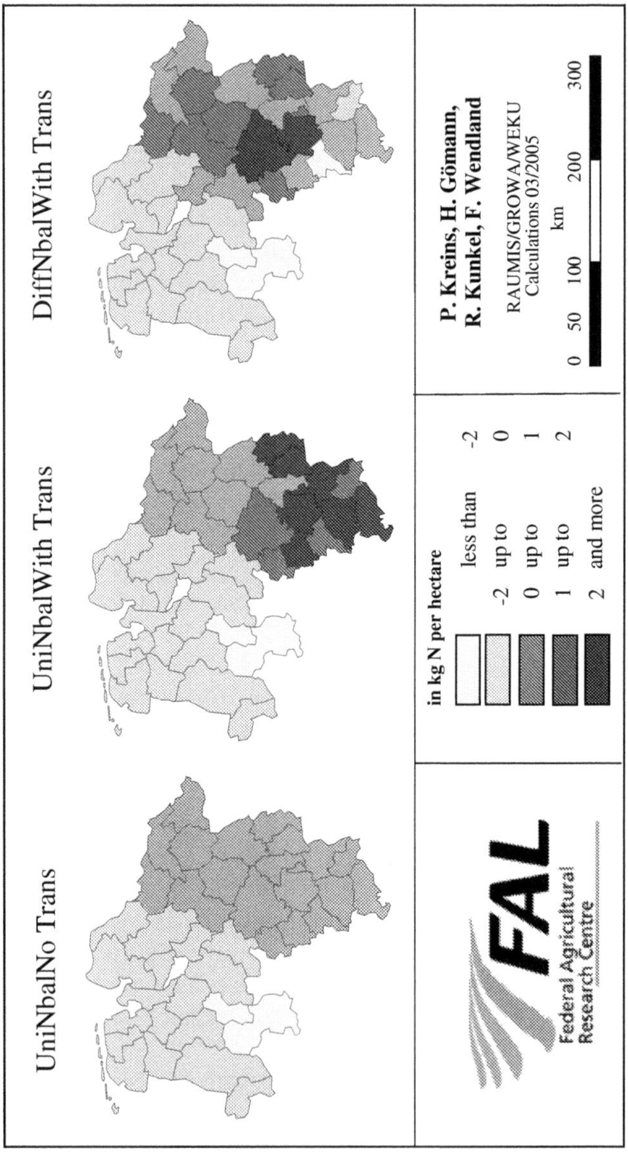

Fig. 8. Change of Regional N-Discharges into Ground Water vs. a Reference Scenario due to Different Variants of Nitrogen Surplus Limitations in Lower Saxony. *Source:* Calculations by FAL and FZJ Using RAUMIS and GROWA/WEKU 03/2005.

situation in other regions. However, when farmyard manure is transported to other regions (as is already the case) then groundwater problems are alleviated in exporting regions at the expense of groundwater impairments in importing regions. The magnitude of the ground and surface water impairments in manure importing regions is chiefly determined by their hydro(geo)logical conditions.

5. CONCLUSIONS AND OUTLOOK

From the results of the study the following conclusions can be drawn:

(1) Impact assessments of agricultural diffuse nutrient reduction measures should not be based on agricultural nutrient surpluses alone but take into account nutrient discharges into ground and surface water as well.

Nutrient discharges from agriculture into ground and surface water have been discussed within the public political debate for several years. The indicator nutrient balance surplus has been gradually established to identify problem regions. However, science, OECD and EU require that the identification of problem regions should be based on indicators that are close to the environmental good to be protected. The study results show only a limited correlation between regional agricultural nitrogen surpluses and regional nitrogen discharges into ground and surface water: while hot-spot areas identified on the basis of nitrogen surplus do not necessarily display the highest nitrogen discharges into surface water, regions with relatively low nitrogen surplus levels have significantly above-average nitrogen discharges. This is due to spatially varying hydro(geo)logical conditions. Hence, indicators to assess reduction measures for diffuse pollution should go beyond nutrient surplus.

(2) Supra-regional manure transports must be considered in evaluations of nutrient reduction measures.

The implementation of measures aimed at reducing agricultural diffuse nitrogen pollution in hot-spot areas can lead to a substantial increase of supra-regional farmyard manure transports. While the value of nutrients in the manure only covers the costs for short local transports, long-distance manure transports are chiefly triggered by high adjustment costs that can be partially avoided by exporting manure to other regions. Calculation of the marginal abatement costs arising from complying with the nutrient surplus constraint is only possible within an agricultural economic model framework such as RAUMIS.

For the economic assessment of nitrogen surplus constraints, abatement costs are derived from changes of the RAUMIS objective function as opposed to a reference scenario. In order to measure agricultural income effects the nonlinear objective function, which has been used primarily to calibrate the model, required some scaling. For future applications a recalibration of the objective function is envisaged such that income effects are represented directly. Anyhow, the derived cost figure allows for a comparison between scenarios and regions.

A reallocation of manure is not captured in the calculation of regional nitrogen balance surpluses following the standard approach based chiefly on regional statistics. Due to a lack of information, supra-regional transports of manure are typically not accounted for. According to the study results, supra-regional manure transports may lead to a considerable relocation of nutrient surpluses with related consequences for the regional discharges into water bodies. In the case of Lower Saxony, nutrients are shipped from hot-spot areas within the Ems catchment to the Weser or even to the Elbe basin after an implementation of nitrogen surplus constraints. Besides the consequences for the water bodies, other environmental effects of these transports should be considered, such as increasing traffic on highways and increasing energy consumption and greenhouse gas emissions. Against this background, the impacts of already established institutions (e.g., "manure trading platforms") should be analyzed.

(3) Diffuse nutrient reduction measures must be target specific.

The measures analyzed in this study focused on reducing nitrogen leaching into surface water. In this regard, the implementation of area differentiated maximum nitrogen surpluses appear to be a cost-effective measure to mitigate hot-spot areas if supra-regional manure transports are allowed. While this is an appropriate measure for achieving environmental targets in surface water, the results show that quality standards for groundwater might not be achieved the same way. Hence, groundwater specific measures are advisable.

Furthermore, the study results raise some more fundamental issues in regard to the valuation of natural resource depletion, burden sharing, cost-effective measures, and compliance with the provisions of the EU Water Framework Directive (WFD) (Directive 2000/60/EC of the European Parliament and of the Council of 23 October 2000). These issues are related to supra-regional farmyard manure transports in three ways. First, a major part of diffuse nitrogen reduction can be attributed to denitrification processes in the aquifer. Area differentiated N-surplus constraints result in "targeted" manure transports into regions that possess high denitrification potentials. However, the denitrification capacity of the aquifer is limited and largely irreversible such that this targeting represents a deliberate depletion of a natural resource. Secondly, supra-regional manure transports alleviate problems in hot-spot areas at the expense of environmental impairments in other regions. Against the background of the polluter pays principle or a reasonable burden sharing, such transports may not be a viable solution for diffuse pollution problems in hot-spot areas. Thirdly, the above-mentioned issues have to be considered in order to derive cost-effective measures.

A fourth major aspect that has to be taken into account in an assessment of diffuse nutrient reduction measures is the WFD, the most substantial instrument of the present EC water legislation. WFD requires that surface waters meet ecological and chemical standards and that groundwater meets chemical and quantitative standards by 2015. In the case of water resources already in good status, the status has to be preserved; in the event that good status is failed or there is a risk that it is failed, measures must be undertaken to mitigate the problem (WFD Article 1). As a result, it is essential that the implementation of nutrient reduction measures to improve water quality in hot-spot areas does not impose a risk of failure on regions that previously met the required targets. This, however, might be an effect of supra-regional manure transport. In order to avoid the undesired impacts, regional input pathways for nitrate into groundwater and surface waters as well as the nitrate retention capacity in the manure importing region have to be considered.

The coupling of RAUMIS, GROWA and WEKU established an integrated agricultural hydro(geo)logical model network that goes beyond the driving force indicator (e.g., nitrogen balance surplus) and calculates a metric (diffuse nitrogen discharges into surface water) closer to the environmental good ("water"). Detection, classification and monitoring of areas with nitrogen problems are more specific on the basis of this improved indicator because natural conditions are taken into account. The model network makes possible an integrated assessment of agricultural nutrient reduction measures, accounting for supra-regional manure transports. Based on the achieved level of model integration, further developments of the model network should address the following issues:

- *Spatial differentiation according to natural sites.* The regional adjustment behavior of agriculture is modeled in RAUMIS on a county ("Landkreis") level that represents region farms. It is assumed that region farms are homogeneous units. This assumption causes an aggregation error that can be reduced through further spatial differentiation. Depending on the availability of data, in particular land use information from remote sensing, a first step is to extend the model to a lower municipality level that is closer to the spatial grid cell resolution of the hydrological models. The next step is to assign "homogeneous natural sites" below the community level and take information about spatial heterogeneity into the simulation of the adjustment behavior of region farms. The disaggregation will enable policy impact analyses on a catchment scale as required by WFD.

- *Consideration of further nutrients.* Regarding the problem of eutrophication, phosphorus is the limiting factor rather than nitrogen. Hence, phosphorus leaching has to be analyzed, too, taking the degree of phosphorus saturation of soils into account.
- *Consideration of all nutrient input pathways.* The RAUMIS/GROWA/WEKU model network focuses on diffuse pollution by agriculture. In addition to diffuse pollution, point sources still play an important role in nutrient discharges into water bodies, though their relevance has decreased in recent years. In this regard, the model network will be complemented by the MONERIS model (see Section 3.4) that considers both nutrient inputs from diffuse and point sources. MONERIS will serve as a framework of consistency for diffuse pollution by agriculture that is modeled by GROWA/WEKU with a high spatial resolution. Having accomplished full model integration, a calibration of the model network will be possible. In this validation process, modeled nitrogen inputs into surface waters as well as pathway simulations will be compared with observed values, e.g., nitrogen concentrations from monitoring stations.

NOTES

1. In north Rhine Westphalia farm aid services, machine rings and private contractors generally organize the creation of nutritional exchanges in order to assure a regulated, transparent supra-farm use of nutrients. The placement guarantee for the nutritional exchange serves as a recognized documentation of land in the permit processes of state agencies (Eisele, 2004, p. 12). In Lower Saxony the voluntary agreement on the supra-farm utilization of organic nutrient carriers serves as an appropriate basis.

2. The notation netput stands for "net output" where positive elements of *xi* denote outputs while negative elements denote inputs or intermediate inputs such as farmyard manure.

3. Due to the low speed of tractors and pump tank vehicles, in contrast to trucks, in transporting farmyard manure, the economic viability of this approach sinks with

increasing distances. Comparative calculations have shown that with trucks, the additional loading and unloading causes higher non-distance-related costs in comparison to transport with a tractor. At a distance of about 15 km, the lower distance-related costs of truck transport compensate for this cost disadvantage in comparison to tractor transport over short distances. For this reason it is generally assumed that the interregional farmyard manure export is carried out with trucks and not tractors.

REFERENCES

Arlt, A. (2003). *Systemanalytischer Vergleich zur Herstellung von Ersatzbrennstoffen aus biogenen Abfällen am Beispiel von kommunalem Klärschlamm, Bioabfall und Grünabfall.* Wissenschaftliche Berichte FZKA 6949, Forschungszentrum Karlsruhe. Gleichzeitig Diss. Universität Stuttgart.

Bayerische Landesanstalt für Landwirtschaft (2002). Ausgleich für Landwirte und Waldbesitzer in Wasser- und Heilquellenschutzgebieten. Bekanntmachung des Staatsministeriums für Landesentwicklung und Umweltfragen vom 12.04.2002. http://www.stmlf.bayern.de

Behrendt, H., Huber, P., Kornmilch, M., Opitz, D., Schmoll, O., Scholz, G., & Uebe, R. (2000). *Nutrient balances of German river basins.* Berlin: Umweltbundesamt, Texte 23/2000.

Betriebshilfsdienste und Maschinenringe (2004). News release. 26.07.2004. http://www.bhd-mr-westfalen.de

Böttcher, P., Strebel, O., & Duynisveld, W. H. M. (1989). Kinetik und Modellierung gekoppelter Stoffumsetzungen im Grundwasser eines Lockergesteinsaquifer. *Geologische Jahrbücher, C51*, 3–40.

Council Directive 91/676/EEC of 12 December 1991 concerning the protection of waters against pollution caused by nitrates from agricultural sources ("EU nitrate directive").

Directive 2000/60/EC of the European Parliament and of the Council of 23 October 2000 establishing a framework for Community action in the field of water policy ("EU water framework directive").

DÜNGEVERORDNUNG (DÜVO) (1996). Verordnung über die Grundsätze der guten fachlichen Praxis beim Düngen (Regulation on the basic principles of best management practices in fertilization), vom 26.01.1996, Bonn. Bundesgesetzblatt Jahrgang 1996, Teil I, Nr. 6.

Eisele, J. (2004). Nutrients between surplus and demand. *Landwirtschaftliches Wochenblatt Westfalen Lippe.* Münster, 11. März 2004.

German Environmental Agency (2001). *Environmental Policy. Water Management in Germany. Part 2 – The Water Quality of Inland Surface Waters.* Berlin.

Gömann, H., Kreins, P., Kunkel, R., & Wendland, F. (2003). The coupling of agricultural economic and hydrological models. *Agrarwirtschaft, 53*(4), 195–203.

Gömann, H., Kreins, P., Kunkel, R., & Wendland, F. (2004a). Model based impact analysis of policy options aiming at reducing diffuse pollution by agriculture – a case study for the river Ems and a sub-catchment of the Rhine. In: R. A. Letcher & C. Giupponi (Eds), *Environmental Modelling and Software. Special issue: Policies and Tools for Sustainable Water Management in the European Union, 20*(2), 261–271.

Gömann, H., Kreins, P., & Møller, Ch. (2004b). Impact of nitrogen reduction measures on nitrogen surplus, income and production of German agriculture. *Water Science and Technology, 49*(3), 81–90.

Henrichsmeyer, W., Cypris, C., Löhe, W., Meudt, M., Sander, R., von Sothen, F., Isermeyer, F., Schefski, A., Schleef, K.-H., Neander, E., Fasterding, F., Helmcke, B., Neumann, M., Nieberg, H., Manegold, D., & Meier, T. (1996). Entwicklung eines gesamtdeutschen Agrarsektormodells RAUMIS96. Endbericht zum Kooperationsprojekt. Forschungsbericht für das BML (94 HS 021), vervielfältigtes Manuskript, Bonn/Braunschweig.

Howitt, R. E. (1995). Positive mathematical programming. *American Journal of Agricultural Economics*, *77*, 329–342.

Köhne, C., & Wendland, F. (1992). *Modellgestützte Berechnung des mikrobiellen Nitratabbaus im Boden*. Internal Report, KFASTE-IB-1/92, Jülich.

Kreins, P., & Cypris, C. (1999). Entwicklung der regionalen Wettbewerbsfähigkeit im Bereich der Milchproduktion und Folgen für die Landnutzung. Vervielfältigtes Manuskript für die 39. Jahrestagung der Gesellschaft für Wirtschafts- und Sozialwissenschaften des Landbaues e. V. in Kiel, 4.–6. Oktober 1999.

Kunkel, R., & Wendland, F. (2002). The GROWA98 model for water balance analysis in large river basins – the river Elbe case study. *Journal of Hydrology*, *259*, 152–162.

Laiblin, S. (1999). Kostenstruktur der landwirtschaftlichen Klärschlammverwertung am Beispiel Brandenburg. Diplomarbeit am Institut für Betriebswirtschaftslehre der Technischen Universität Berlin.

Länderarbeitsgemeinschaft Wasser (LAWA) (1998). *Beurteilung der Wasserbeschaffenheit von Fließgewässern in der Bundesrepublik Deutschland – Chemische Gewässergüteklassifikation*. Berlin: Kulturbuchverlag.

Lower Saxony State Office for Ecology (2001). Water Quality Report 2000. Surface water bodies 13/2001. 1. Edition (2), Hildesheim.

Ministry of Environment of Lower Saxony (2004). http://www.umwelt.niedersachsen.de

Oehmichen, J. (1983). Pflanzenproduktion. Bd. 1: Grundlagen. Berlin und Hamburg. Paul Parey.

Osterburg, B., Schmidt, Th., & Gay, H. (2004). *Evaluation of farm data to calculate the use of nitrogen mineral fertiliser*. Final report for a research project for the Lower Saxony Environmental Ministry. FAL Braun-schweig, June 2004.

PARCOM (Paris-Konvention zur Verhütung der Meeresverschmutzung) (1993). Dritte Sitzung der Ad-hoc-Arbeitsgruppe zur Reduzierung der Nährstoffeinträge aus der Landwirtschaft – Anlage 1: PARCOM-Richtlinien für die Berechnung von Mineralbilanzen.

Staffel-Schierhoff, U. (2001). Schadstofforientierte Aktionskonzepte am Beispiel der Nährstoffreduzierung. In: K.-U. Rudolph (Ed.), *Der Wassersektor in Deutschland – Methoden und Erfahrungen*. Berlin: Federal Ministry for the Environment, Nature Conservation and Nuclear Safety, & German Environmental Agency.

van Beek, C. G. E. M. (Ed.). (1987). Landbouw en Drinkwatervoorziening, orientierend Onderzoek naar de Beinvloeding van de Grondwaterkwaliteit door Bemesting en het Gebruik van Bestrijdingsmiddelen; Onderzoek 1982–1987. Report Meded. 99, Keuringsinstituut voor Waterleidingsartikelen KIWA N.V., 99, Nieuwegein, The Netherlands.

Wendland, F., & Kunkel, R. (1999). *The nitrate decomposition ability in ground water in the Elbe area*. Schriften des Forschungszentrum Jülich, Environmental Series, 13, Jülich.

INTEGRATION OF AGRO-ECONOMIC ANALYSIS AND ECOLOGICAL MODELLING

Thomas G. Schmidt

ABSTRACT

This chapter describes a method to analyse agricultural land use in terms of net value added and employment (working time requirement) in the agricultural sector as well as a corresponding ecological indicator: the nitrogen-leaching-rate. Watershed management demands a basic approach, which deals with common statistics and spatial information from digital maps. This causes a range of uncertainties, which are calculated in relation to the data input. A metamodel derived from a process model calculates the most probable value of the ecological indicator, whereas the economic indicators are estimated by the cumulative numbers of primary production. The uncertainties are expressed as the standard deviation of all impacts as percentages. The method described is applied to a rural district in the Elbe river basin.

1. INTRODUCTION

Agricultural land use covers 54% of the German land surface (29% is forestry and 12% are urban areas) at present. It dominates the landscape and,

Ecological Economics of Sustainable Watershed Management
Advances in the Economics of Environmental Resources, Volume 7, 143–166

ISSN: 1569-3740/doi:10.1016/S1569-3740(07)07007-1

in terms of pollution, the diffuse emissions of nitrogen (N) into rivers. 72% of the diffuse nitrogen emissions into the Elbe river originate from agriculture (German Federal Govermnent, 2000). The Elbe river transports the nitrogen loadings into the North Sea which causes eutrophication (EEA, 2001). Sustainable watershed management requires a significant reduction of diffuse nitrogen emissions from agriculture, and the ecological effect of the land-use intensity is an essential factor to be considered. In this context we deal with two competing aims: On the one hand, farmers aim to maximise economic benefit, which depends on crop yield as well as livestock productivity. Reduced nitrogen emissions usually mean reduced productivity and therefore fewer economic benefits. On the other hand, there are negative ecological impacts of agricultural land management on water quality in terms of nutrient load and pesticides. Therefore it is important to investigate the interrelationship between agricultural land use and its impact on water quality. The design of agri-environmental measures (AEM) creates a link between the two competing aims. The agricultural sector heavily depends on subsidies and financial support, much more than other economic sectors. The design of effective and efficient AEM as part of financial support programmes can contribute to sustainable watershed management. The support programmes mostly have multiple objectives, and in this context, water protection contributes in this area.

How could the agricultural land be managed to reach a certain ecological standard while guaranteeing a certain economic benefit in terms of productivity and employment? Which environmental measures should be developed and what is their economic effect on the agricultural sector in a certain region? How exact or certain are the results?

Existing methods to analyze agricultural land use consider either ecological and economic indicators or uncertainties, or they are not transferable to the Elbe river basin because of a lack of data. Mostly, uncertainties are used in ecological modelling in combination with a Monte-Carlo-Simulation to detect a range of possible scenario effects (e.g., Haberlandt, Krysanova, & Bardossy, 2002). A basic approach in the United Kingdom deals with metamodels,[1] particularly with regard to ecological impacts (Lord & Anthony, 2000). An approach that integrates economic indicators, ecological values and related uncertainties has already been developed in the context of sustainable land use and watershed management in the Elbe river basin (Horsch, Ring, & Herzog, 2001). This approach deals with soil process simulations and the calculation of gross margins to characterise the economic effects of land management applied to a spatially restricted area of about 700 km^2 (Franko, Schmidt, & Volk, 2001). This procedure requires a lot of derivations and estimations of soil and management parameters which are not feasible for larger catchment areas.

The aim of this chapter is to present an approach that is capable of integrating ecological and economic parameters including related uncertainties at a watershed scale. The basic data and additional information come from agricultural statistics, maps and expert knowledge. There is no geo-referenced information about land use except the official statistics, aggregated on a regional level (rural district).

Indicators are used representing the three dimensions of sustainability and reflect the linkage between agricultural land use and water quality. Economics effects are represented by the 'net value added' (NVA) and employment. In terms of watershed management, the net value added of agriculture is a comprehensive value which can be compared with alternative land uses like forestry. Employment (expressed as working time requirement (WTR)) gives an idea of the social component of agricultural land use. The agri-environmental indicator 'N-leaching-rate' represents the impact of agricultural land use on the environment, i.e., groundwater quality. Nitrogen is one of the most important pollutants in waters. These three indicators are just examples to illustrate a first application of this method. Other economic or ecological parameters like productivity of capital and soil erosion could be used to solve different problems.

2. MATERIAL, METHODS AND STUDY AREA

The economic analysis is based on a range of data input which comes from official statistics as well as from sources of standard calculation data for farm management. The official agricultural statistics include specification about crop shares within a region and livestock (Federal Statistical Office Germany, 2002). Standard calculation data about crop production and animal breeding represents an average outlay (KTBL, 2001). Product prices from, and yields for, each year of evaluation are needed.

The ecological approach uses simulation results of complex computer models on agricultural land use and applies statistical methods to deduce a simplified model which can deal with less data input than the computer models.

Models: The model REPRO (Hülsbergen, 2003) calculates the material and energy fluxes of farming units. It analyses the nutrient balance of input and uptake in crop production and in livestock as well as the interaction between both. The model CANDY (Franko, Oelschlägel, & Schenk, 1995) is used to simulate the carbon and nitrogen dynamics in the soil. This application focuses on 'nitrogen leaching towards the groundwater'.

Statistics: A multiple regression analysis shows the linear coherence between a dependent variable (here: N-leaching) and independent variables, which describe the land-use management in statistic parameters. A variance analysis is used to distinguish significantly different areas with samples of simulation objects. These samples have neither the same number of observations nor a normal distribution. Therefore an *H*-test is used to solve this statistical problem. The samples of pairs of areas were compared to summarise these areas in case that the two samples are not significantly different.

A *database* is used to store initial data coming from soil maps, geo-referenced information about land use and a raster map of rainfall data as well as agricultural statistics. The database is a tool for integrating the economic and the ecological analysis.

The study area is a rural district (Kyffhäuserkreis) in Germany of about 1,040 km^2. It is part of the Saale river basin which is a sub-catchment of the Elbe river (Fig. 1). It has been selected because of its heterogeneity in soil and climate. There are five soil types and a range of rainfall between 500 and 700 mm per year. The share of arable land use is 68%, grassland is only 3% (urban land 5%, forests 24%) which is typical for the Elbe river basin. In the

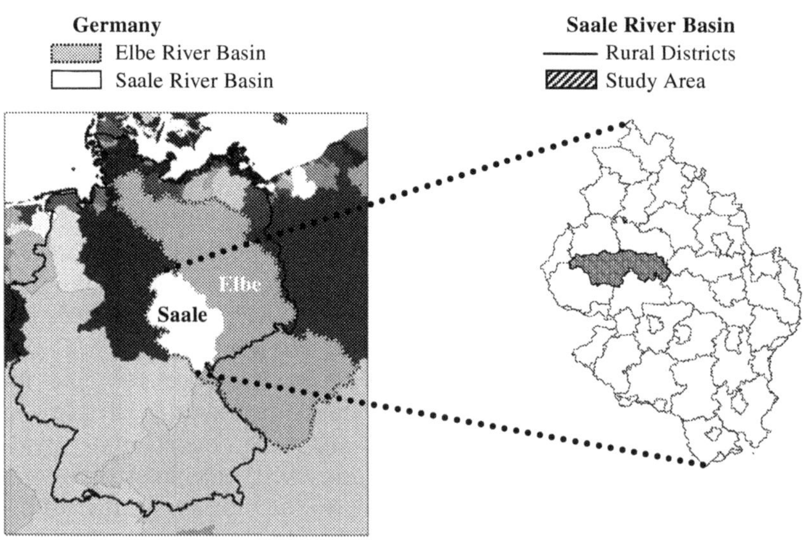

Fig. 1. Study Area as Part of the Wider Saale and Elbe River Basins.

study area there is a livestock density of 0.5 livestock units per hectare on average and 62% of arable land is planted with cereals, others are root crops such as sugar beet and potato.

The effects of changes of agricultural production on nitrogen leaching, farm income and agricultural employment as selected indicators will be demonstrated in this study area. The transfer of this approach to the whole river basin can be realised by aggregating the calculations for each rural district.

3. ECOLOGICAL APPROACH AND ECONOMIC ANALYSIS

The method presented in this chapter includes a land-use analysis in 12 steps which contains the calculation of the most-probable value and related uncertainties of nitrogen leaching, NVA and required working hours in the agricultural sector (Fig. 2).

3.1. Step 1: Organisation of Available Data

There are two categories of available data: geo-referenced information on the natural characteristics of the location, and agricultural statistics providing a fuzzy data set because of their aggregation level. The latter contains a range of parameters of crops and livestock such as yields, cultivated area for each plant species or number of animals (heads) for each year. The aggregation level of this information is the rural district because of data privacy. A rural district (Landkreis) in Germany is an administrative unit of about 520 km^2 on average (34–3,032 km^2). Further references about the share of different farming systems in terms of conventional, integrated or organic farming (in % of the surface area) are available as part of evaluation reports on agri-environmental programmes or the Agenda 2000 that are to be found in the national agricultural report and those of the federal states (BMVEL, 2003; TLL, 2002).

The natural conditions of the region are characterised by the soil type and the annual rainfall in combination with a land-use map (Fig. 3). This information is time independent and usable for each analysis.

The information on soils is derived from a digital soil map and a corresponding data base which contains parameters of the soil types (BGR,

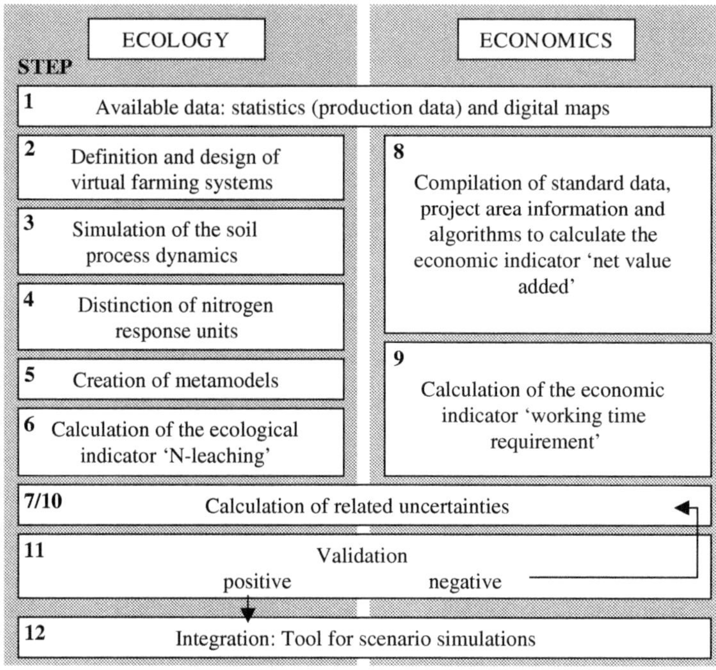

Fig. 2. Scheme of the Method.

1995). The weather conditions are measured in numerous weather stations in different locations of the project area (DWD, 2001). Interpolations between these locations provide a raster map with km^2-grids of annual rainfall. An additional map available distinguishes land use in categories of arable land, grassland, forestry, urban land and others (StBA, 1996).

In this chapter, we only focus on agricultural land use (arable land and grassland).

All three digital maps, together with a map of the rural districts' administrative borders, are the basis of a 'shapefile' which is produced by means of a GIS[2] tool. It unifies the georeferenced information and creates polygons on a new multilayer outline map with a corresponding table of its attributes (annual rainfall, soil type, land use and rural district). A combination of all georeferenced classifications results in 18 different locations (five soil types on arable land and one soil type on grassland times three rainfall categories) in the study area. Only 13 locations with shares of more than 1% of the total area of the district were considered in this approach.

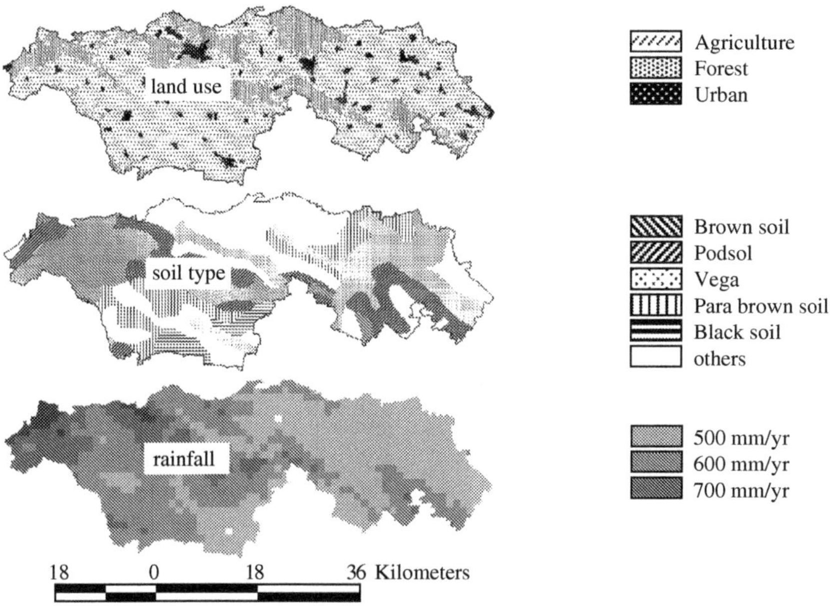

Fig. 3. Geo-Referenced Information on Land Use, Soil Type and Rain Fall.

3.2. Step 2: Definition and Design of Virtual Farming Units

Step 2 needs an input of expert knowledge to create location-adapted farming units with the help of computer models. The model REPRO (Hülsbergen, 2003) is a nutrient and energy balance tool which calculates the nutrient flows of farming units and produces parameter files which are used in the following process simulation. The range of virtual farming units should show the variety of the land use in the project area. The farming units are independent of the locations analysed in Step 1. A total of 125 different crop rotations (single farming unit) were created to describe the range of possible land use by agricultural activities on arable land and five types on grassland. Crop rotations with or without manure from livestock represent the general distinction between cash crop farms and mixed farms. Conventional, integrated and organic farming systems are characterised by the amount of fertiliser input and the diverse crop rotations. The integrated system has 20% less fertiliser input than the conventional system and uses cover crops.[3] In the organic farming system, synthetic fertilisers are prohibited as well as pesticides. In this study, the virtual organic farms always produce livestock and use manure for fertilisation. Grassland is described as meadow with two different levels of fertiliser use or as pasture with one, two or three cows per hectare (1 hectare [ha] = 10,000 m^2).

3.3. Step 3: Simulation of the Soil Process Dynamics

The combination of 130 different virtual crop rotations[4] with five different soil types and three different precipitation levels (annual rainfall 500, 600 and 700 mm) creates 1,950 simulation objects. The soil process model[5] CANDY (Franko et al., 1995) simulates the carbon and nitrogen dynamics in the root zone and creates output data with nitrogen-leaching-rates according to the objects. The model runs (in daily timesteps) for hundred years until the soil-cultivation-system reaches a steady state. This procedure eliminates temporary influences of initial conditions and ensures the only description of the specific farming units. The simulation results are summarised and combined with the initial data in a matrix that describes all objects comprehensively in preparation for the statistical analysis. This aggregation allows both a statistical distinction of nitrogen response units and the creation of metamodels which is described in the next two steps.

3.4. Step 4: Distinction of Nitrogen Response Units

The matrix sorts the objects in groups (land use systems) according to the descriptive characteristics 'land use', 'soil type', 'annual precipitation' and 'farming unit'. A variance analysis (*H*-test) examines the independent variable 'land-use system' with the help of the dependent variable 'N-leaching-rate' on a level of significance of 5%. This statistical analysis allows the grouping of areas without loss of precision and/or shows which groups have significant differences. These united features possess, therefore, a discriminating effect with regard to the area specific N-leaching-rate. Similarly, they mark off retaining areas or, respectively, single 'Nitrogen Response Units' (NRU). Table 1 shows the results of this analysis.

For the study area the statistical analysis generates four NRUs of arable land and one NRU of grassland. Due to the effect of rainfall on nitrogen dynamics, the arable land is subclassified into units of low, middle and high precipitation as well as one unit of soil type BS (black soil). Grassland is represented by only one location which is a separate NRU.

Table 1. Grouping of Locations Without Significant Differences in Terms of N-Leaching-Rate – Nitrogen Response Units.

No	NRU	N-Leaching-Rate ($kg\,ha^{-1}\,a^{-1}$)	Area (%)		Land Use	Soil	Level of Rainfall ($mm\,a^{-1}$)
1	A	40–75	21	42	Arable land	PBS	500
2			4		Arable land	SBS	500
3			11		Arable land	VE	500
4			6		Arable land	PS	500
5	B	0–83	2	26	Arable land	PBS	600
6			3		Arable land	SBS	600
7			10		Arable land	VE	600
8			11		Arable land	PS	600
9	C	2–82	2	2	Arable land	PS	700
10	D	0–103	2	25	Arable land	BS	500
11			13		Arable land	BS	600
12			10		Arable land	BS	700
13	E	2–127	3	3	Grassland	PBS	500

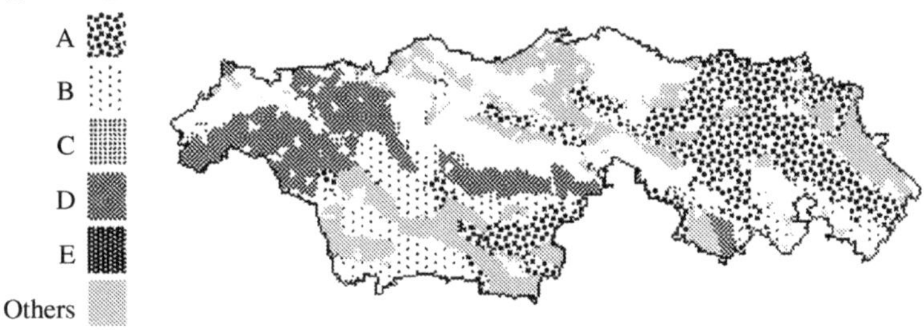

3.5. Step 5: Creation of Metamodels

A metamodel is the abstraction of a very detailed and validated process model. It reflects its system description in a simplified manner and delivers a similar result with less data input and accordingly with larger uncertainties. The metamodels of this study describe the N-leaching-rate [kg N ha^{-1} a^{-1}] of the NRUs based on sensitive parameters.

A multiple regression is used to measure the relationship between one interval-dependent variable, the N-leaching-rate and several independent variables (management parameters of the cropping system and livestock production). The independent variables selected should have strong correlation with the dependent variable but none or only weak correlation with other independent variables. The general structure of the equation is:

$$N_{\text{leach}} = a_0 + \sum_{i=1}^{n}(a_i \cdot \text{SP}_i) \qquad (1)$$

where N_{leach} = N-leaching-rate; a_0 = intercept; a_i = regression coefficients; SP_i = sensitive parameters (independent variables).

All simulation objects refer to one of the five NRUs (Table 1). But only objects with a leaching rate larger than zero can be considered. After sorting the objects, the following regression analysis shows two magnitudes of influence (sensitive parameters) in the simulation of the study area: 'percentage of cereals in the crop rotation' [%] and 'livestock'. The livestock or number of animals is specified as 'Livestock Units' (LU) per hectare (for instance: one cow or three swine are 1 LU) [LU ha^{-1}]. Table 2 shows the parameters of this analysis.

The NRUs are subdivided into farming systems with their specific parameters. An exception is NRU A, which contains only 14 objects and only one set of parameters. The reason for the small number of objects in relation

Table 2. Results of the Regression Analysis.

NRU	Farming System	n	Intercept a_0	Regression Coefficient a_1*	Regression Coefficient a_2*	R^2	p*	Standard Error
A	con	14	49.99	−0.50	21.58	0.78	<0.0002	10.92
	int							
	org							
B	con	183	−0.98	0.07	26.28	0.59	<0.0001	13.71
	int	167	−5.33	0.10	24.30	0.58	<0.0001	11.56
	org	144	26.82	−0.34	14.96	0.52	<0.0001	9.44
C	con	46	4.77	0.10	23.19	0.59	<0.0001	12.56
	int	42	−0.32	0.12	22.26	0.59	<0.0001	10.83
	org	37	31.15	−0.36	17.41	0.69	<0.0001	7.83
D	con	135	1.06	0.06	32.50	0.51	<0.0001	20.25
	int	121	−2.62	0.06	30.69	0.48	<0.0001	17.73
	org	103	27.00	−0.35	16.44	0.53	<0.0001	10.08
E	con	4	30.93		22.89	0.93	0.0335	9.61
	int	2	1.63		41.73	1		
	org	2	2.25		41.11	1		
								∅ 10.35

*a_1 = regression coefficient of the variable 'percentage of cereals in the crop rotation'; a_2 = regression coefficient of the variable 'livestock'; p = probability value.

to the other NRUs on arable land is the low precipitation rate and, as a consequence, no groundwater recharge/no nitrate leaching. The parameters of the NRU E (grassland) are derived from a simple linear regression analysis with only two, or respectively four, virtual farming units because of the small variation of land-use management on grassland. All other NRUs have a high potential variety of land-use systems. The coefficients of determination (R^2) indicate a reasonable relationship between the N-leaching-rate and independent values. The p-levels are highly significant, except for conventional grassland because of the small heterogeneity of the sample size.

3.6. Step 6: Calculation of Nitrogen Leaching

The use of specified metamodels (by means of two sensitive parameters) allows the calculation of the average N-leaching-rate of the study area. Because of missing data about the real cultivation within a rural district, all theoretically possible distribution patterns must be considered in order to determine the interval. The quantity of the theoretically possible results depends on the number of NRUs, the number of sensitive parameters and their intervals. The intervals are a gradation of these parameters ($P_1 = 20$, 50, 80% of cereals in crop rotation; $P_2 = 0$, 1, 2 LU ha^{-1}):

$$K_{\text{num}} = [(Pn_1 + 1) \cdot (Pn_2 + 1)]^{\text{NRU}} = (3 \times 3)^5 = 59,049 \qquad (2)$$

where K_{num} = number of combinations; $Pn_{1/2}$ = number of intervals (Parameter 1 and 2); NRU = number of NRUs.

The calculations were realised by a software application. A total of 59,049 possibilities are tested (Eq. (2)) and a selection of 408 solutions corresponds to the agricultural statistics of the study area. This means that 408 different combinations of the percentage of cereals in the crop rotation and the amount of livestock in combination with 5 NRUs result in an average of 0.5 LU ha^{-1} and about 62% cereals. This conforms to the current land use in the considered rural district. The application of the metamodels allows an estimation of the most-probable N-leaching-rate and their range (Fig. 4).

The N-leaching-rate is between 33 and 94 kg ha^{-1} a^{-1}, at an average of 62.8 kg ha^{-1} a^{-1} and a standard deviation of 12.1 kg ha^{-1} a^{-1}. Low N-leaching-rates indicate a good agricultural management which is adapted to natural conditions. Unfavourable allocation of livestock and cereals in the rural district produce high levels of N-leaching (for a more detailed technical description, please see Schmidt, 2004).

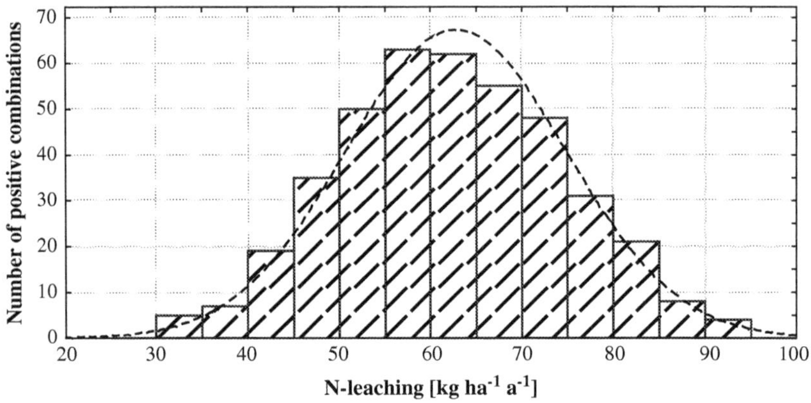

Fig. 4. Range of Possible N-Leaching.

3.7. Step 7: Calculation of the Uncertainties of Ecological Modelling

This section describes the treatment of different independent error influences and the calculation of the total error. First of all there are uncertainties of the calculated N-leaching-rate which result from the non-georeferenced information about land use. The distance between the minimum of 33 kg ha^{-1}, the maximum of 94 kg ha^{-1} and the medium value (63 kg ha^{-1}) cause a possible deviation of $\pm 48\%$ (comp. Fig. 4).

The uncertainty of the soil heterogeneity is estimated at $\pm 15\%$ by using the theoretical range of silt and clay of each soil profile in the project area and additional simulations of all objects.

The influence of the rainfall term on the uncertainties is $\pm 12\%$. This number is derived from the error which accrues by using rainfall grades (500, 600 and 700 mm) instead of real values.

The process model itself has uncertainties in terms of N-leaching. An evaluation by Beblik et al. (2001) showed a variation of $\pm 14\%$.

The derivation of a metamodel causes uncertainties of 34%, measured by the average standard error between the metamodel (regression equation) and the process model.

The *total uncertainty* (ε) is a combination of all single uncertainties by using the standard deviation. This calculation is based on the assumption that all variables are independent:

$$\varepsilon = \sqrt{\sum_{i=1}^{n} (\varepsilon_i)^2} = \sqrt{48^2 + 15^2 + 12^2 + 14^2 + 34^2} = 64\% \tag{3}$$

The maximum deviation from the most-probable value is ±64% which is mainly influenced by the limited knowledge of actual land use and a large variance between the process- and metamodel.

3.8. Step 8: Calculation of the Economic Indicator 'Net Value Added'

NVA is used as an indicator to show the economic effect of agricultural activities. It represents the total value of the farm sector's production of goods and services, less payments to other (non-farm) sectors of the economy. It reflects the contribution of the agricultural sector to the national economic product. It also represents the sum of economic returns to all providers of production factors: farm employees, lenders, landlords and farm operators (ERS, 2003). The calculation is based on common statistics about cultivated area, yield of crops and livestock. The production value is equal to the market price of all crops produced in the area and the value of produced animals per year. On the other hand, there are overheads according to the farming system and variable costs depending on production practices as well as on quantities and prices (before tax and inclusive VAT) of inputs. These include inputs such as seed, fertiliser, feed, chemicals and hired labour. In addition there are general subsidies depending on cultivated surfaces and crops. The payment of AEM is a financial compensation for extraordinary expenditures such as planting cover crops or it countervails a loss of yield caused by restrictions in terms of fertiliser input or use of pesticides.

The calculation in Table 3 does not consider services like the processing of goods or trade. It only focuses on the primary agricultural production. The production value, representing the market performance of crop and livestock products, is calculated by the sum of the production (area × yield respectively heads × annual production) of each species of crops, or respectively, animals multiplied by its price (Table 4). The quotation of prices for ex-post-analysis is derived from annual reports of the ZMP (this company collects and disseminates information on agricultural, food, forestry and timber markets). Ex-ante-analyses need assumptions for the rate of price increase. The differences of yield, product prices and the shares of the farming systems (conventional: 92%, integrated: 5.2%, organic: 2.2%) need to be considered (Table 4).

This compilation of the most-important agricultural products represents 91% of the total crop production and 90% of animal products. Some less-important goods are not considered. Because of this, the total production value has to be multiplied by a corresponding factor or converted into units per hectare[6] which represent an average value. This procedure allows only a

Table 3. Calculation of 'Net Value Added'.

$$\text{Production value (PV)} = \sum(\text{area} \times \text{yield} \times \text{price}) + \sum(\text{annual animal production} \times \text{price}) \quad (4)$$

$$\text{Overheads (OH)} = \sum(\text{farming system} \times \text{area}) \quad (5)$$

$$\text{Variable costs (VC)} = \sum(\text{yield} \times \text{factor variable costs} \times \text{area}) + \sum(\text{number of animals} \times \text{variable costs}) \quad (6)$$

$$\text{Subsidy (SUB)} = \sum(\text{crop} \times \text{area}) + \sum(\text{livestock} \times \text{premium}) \quad (7)$$

$$\text{Taxes (TAX)} = \sum(\text{product} \times \text{tax}) \quad (8)$$

$$\text{Agri-environment measures (AEM)} = \sum(\text{payment} \times \text{area}) \quad (9)$$

$$\text{Net value added (NVA)} = \sum \text{income} - \sum \text{expenditure} = \text{PV} - \text{OH} - \text{VC} + \text{SUB} - \text{TAX} + \text{AEM} \quad (10)$$

rough estimation of the most-significant parameters in calculating the 'net value added', but it exemplifies the methodology of this approach which could be modified to comply with further requirements if necessary.

The expenditures can be divided into fix and variable costs. The total overheads (fix costs) depend on the farming system (conventional, integrated, organic) and its share of the project area (Table 5). All overheads are expressed as € per hectare:

Overheads include depreciation, insurance and rent of buildings and machines. The differences between farming systems come from a more extensive use of equipment in organic farming.

Variable costs include all production factors such as fertilisers, pesticides, seeds and others. They depend on crops, or respectively animals, and yield. They are multiplied by cultivated area or annual production and included in the total costs. In the organic farming system no sugar beet or rape is cultivated (Table 6).

The costs of animals and perennial plants must be converted into annual amounts in order to get unified and comparable values.

Subsidies are described in detail in the AGENDA 2000 publication of the German government (BMVEL, 2002). There are payments depending on the

Table 4. Production Values of Crop and Livestock (Year: 1999, German Average).

Crops	Area (ha)			Yield ($100\,kg\,ha^{-1}$)			Price (€ $100\,kg^{-1}$)			Production Value [1,000 €]		
	con	int	org	con	int	org	con	int	org	con	int	org
W-wheat	19,277	1,083	557	77	69	46	11.0	11.0	16.5	16,328	822	423
S-wheat	1,234	69	36	58	52	35	11.0	11.0	16.5	787	39	21
Rye	1,342	75	39	70	63	42	9.5	9.5	14.3	892	45	23
W-barley	6,204	348	179	73	66	44	10.0	10.0	15.0	4,529	230	118
S-barley	5,200	292	150	57	52	34	10.0	10.0	15.0	2,964	152	77
Oat	371	21	11	58	53	35	11.0	11.0	16.5	237	12	6
Triticale	1,061	60	31	66	60	40	9.0	9.0	13.5	630	32	17
Potato	219	12	6	421	379	253	10.5	10.5	15.8	968	48	24
Sugar beet	1,781	100		507	457		5.2	5.2		4,695	238	
Rape	8,390	471		39	35		21.0	21.0		6,871	346	
Livestock	**Heads (units)**			**Production (Annually)**			**Price (€ $unit^{-1}$)**			**Production Value [1,000 €]**		
	con	int	org	con	int	org	con	int	org	con	int	org
Dairy cows	4,318	242	103	0.3	0.2	0.1	760.0	760.0	1,140.0	985	37	12
Milk				4,355.0	3,920.0	2,352.0	0.3	0.3	0.5	5,641	285	121
Calf				0.9	0.8	0.5	192.0	192.0	288.0	746	37	15
Breeding pigs	3,973	223	94	0.5	0.5	0.3	370.0	370.0	555.0	735	41	16
Piglet				18.0	16.2	9.7	43.0	43.0	64.5	3,075	155	59
Fattening pigs	22,111	1,242	525	105.0	95.0	63.0	1.2	1.2	1.8	2,786	142	60
Sheep	20,665	1,160	491	2.0	1.8	1.1	12.2	12.2	18.3	504	25	10
Lamb				32.0	28.8	17.3	1.5	1.5	2.3	992	50	20
Wool				3.0	2.7	1.6	2.1	2.1	3.1	130	7	2
Hens	82,429	4,629	1,958	0.8	0.8	0.5	0.5	0.5	0.7	33	2	1
Eggs				260.0	234.0	140.0	0.1	0.1	0.1	2,143	108	27
Total											60,576	

Sources: Federal Statistical Office Germany (2002); ZMP (2001a, 2001b, 2001c).

Table 5. Overheads.

Farming System	Area (ha)	Overhead Per Unit (€ ha^{-1})	Total Overheads (1,000 €)
Conventional	62,942	293	18,442
Integrated	3,535	293	1,036
Organic	1,495	352	526
Total			20,004
Average			294 € ha^{-1}

Sources: Federal Statistical Office Germany (2002); BMVEL (2002).

Table 6. Variable Costs.

Crop	Area (ha)			Variable Costs (€ ha^{-1})			Total (1,000 €)		
	con	int	org	con	int	org	con	int	org
W-wheat	19,277	1,083	557	431	388	517	8,308	420	288
S-wheat	1,234	69	36	398	358	477	491	25	17
Rye	1,342	75	39	500	450	600	671	34	23
W-barley	6,204	348	179	449	404	539	2,786	141	96
S-barley	5,200	292	150	399	359	479	2,074	105	72
Oat	371	21	11	348	313	417	129	7	5
Triticale	1,061	60	31	425	382	510	451	23	16
Potato	219	12	6	1,611	1,450	1,933	353	17	12
Sugar beet	1,781	100	–	726	653	–	1,293	65	
Rape	8,390	471	–	550	494	–	4,615	233	

Animal	Heads (units)			Variable Costs (€ unit^{-1})			Total (1,000 €)		
	con	int	org	con	int	org	con	int	org
Dairy cows	4,318	242	103	790	711	948	3,411	172	98
Breeding pigs	3,973	223	94	269	242	323	1,069	54	30
Fattening pigs	22,111	1,242	525	269	242	323	5,948	301	169
Sheep	20,665	1,160	491	28	25	34	579	29	17
Hens	82,429	4,629	1,958	16	14	19	1,319	67	38
Total								33,967	

Source: KTBL (2002).

cultivated area and animal production. The subsidies for animals are paid per head and vary according to the species. The specific payment for land use is conditional on cultivated crop and its surface expansion. Altogether there are 21.78 Mio. € of subsidies in the study area with an average payment of 447 € per hectare. Taxes are paid for milk and non-food products which are not considered in this calculation.

AEM are payments for extensively used and ecological farming systems. This study considers the organic farming system and one extensively (integrated) cropping system which deals with less fertiliser input (80% of conventional agriculture) and restrictions on pesticides. Just a small percentage of farmers participate in the agri-environmental programme in the study area: 5.2% of integrated farming systems and 2.2% organic farming which results in an insignificant number of 705 T€ or 14.5 € ha^{-1} on average.

The NVA, compare Eq. (10), represents the total value of the farm sector's production. It can be expressed as a total figure to compare it with other economies or sectors. Because of a simplified approach, this analysis only considers 91% of agricultural land use and 90% of livestock; this means that the intermediate result of 29.09 Mio. € has to be multiplied by the related factor ($\approx$1.1) to get the total NVA to correspond to the whole district. This generates a final figure of a

$$\text{Net value added} = 31.99 \text{ Mio. €.}$$

A conversion of this amount into a relative figure per area unit (hectare) is useful for the comparison of different districts (598 € ha^{-1}).

3.9. Step 9: Calculation of the Economic Indicator 'Working Time Requirement'

The second economic indicator WTR describes the employment in primary agricultural production. It is derived from crop and animal production depending on area and livestock numbers as well as on working time in farm management. The amount of all individual activities in relation to the annual working time indicates the employment in the sector:

$$\begin{aligned} \text{working time requirement (WTR)} = & \sum(\text{crop} \times \text{working hours per ha}) \\ & + (\text{animals} \times \text{working hours per head}) \\ & + (\text{management in hours per ha} \times \text{area}) \end{aligned} \quad (11)$$

The standard values of working time spent on (KTBL, 2001) livestock production (which are usually expressed in units per head or daily time exposure per stable unit) need to be converted into annual units (Table 7).

The calculated number of the overall WTR of 1.79 Mio. hours has to be multiplied by the factor 1.1 to extrapolate from the considered 90% of the

Table 7. Working Time Requirement.

Crop	Area (ha)			Working Time Per Area (h ha^{-1})			Total Working Time (h)			
	con	int	org	con	int	org	con	int	org	sum
W-wheat	19,277	1,083	557	7.48	6.22	11.93	144,192	6,736	6,645	157,573
S-wheat	1,234	69	36	6.59	6.25	11.11	8,132	431	400	8,963
Rye	1,342	75	39	7.48	6.22	11.93	10,038	467	465	10,970
W-barley	6,204	348	179	7.48	6.22	11.93	46,406	2,165	2,135	50,706
S-barley	5,200	292	150	6.59	6.25	11.11	34,268	1,825	1,667	37,760
Oat	371	21	11	6.59	6.25	11.11	2,445	131	122	2,698
Triticale	1,061	60	31	7.48	6.22	11.93	7,936	373	370	8,679
Potato	219	12	6	23.34	23.16	21.41[a]	5,111	278	128	5,518
Sugar beet	1,781	100		7.53	7.41	[b]	13,411	741	0	14,152
Rape	8,390	471		7.50	7.01	[b]	62,925	3,302	0	66,227
										363,246

Animal	Livestock (Heads)			Working Time (h head^{-1})			Total (h)			
	con	int	org	con	int	org	con	int	org	sum
Dairy cows	4,318	242	103	50.49	50.49	60.59	218,016	12,219	6,241	236,475
Breeding pigs	3,973	223	94	15.82	15.82	18.98	62,853	3,528	1,784	68,165
Fattening pigs	22,111	1,242	525	8.88	8.88	10.66	196,346	11,029	5,597	212,971
Sheep	20,665	1,160	491	11.04	11.04	13.25	228,142	12,806	6,506	247,454
Hens	82,429	4,629	1,958	0.26	0.26	0.31	21,432	1,204	607	23,242
										788,307

Management	Area (ha)	Management (h ha^{-1})	Total (h)
General	48,619	13.10	636,909
			Total: 1.79 Mio. hours

Source: KTBL (2001).
[a]Less working time requirement in organic farming because of less yield and consequently less harvest time per hectare is anticipated.
[b]No organic farming production of sugar beet and rape in the study area.

land surface and livestock to the whole agricultural production. This produces a final result of 1.967 Mio. hours. This corresponds to 1.7 labourers per 100 hectare (at an annual working time of 2,100 hours work force unit) or 938 labourers in the study area. This result reflects a static approach with a given annual working time on average. As a consequence, the number of labourers rises or falls according to WTR, but in actuality, the annual working time of the labourers is variable to a certain limit. Beyond this limit, the number of labourers will change.

3.10. Step 10: Uncertainties of the Economic Approach

The uncertainties in economic data result from a transfer of single data, or respectively averages, to a project region. The regional differences, as well as temporal variations, have to be considered in a sensitivity analysis.

The yields and product prices have the strongest influence on the NVA. All other variables (like variable and fix costs, subsidies, etc.) are estimated with uncertainties of $\pm 10\%$. The yields depend on natural conditions and the technical standard. The gap of information regarding agricultural statistics about yields indicates a range of uncertainties of $\pm 2\%$ which influences the NVA by $\pm 2.4\%$. The product prices vary by 7% within one average year (Commodities futures exchange (CFE), 2002) which cause uncertainties of $\pm 7.5\%$ in the calculation of NVA. In extraordinary years (depending on precipitation) the variation could be wider.

The overall uncertainties are summarised to a total number of $\pm 13\%$:

$$\varepsilon = \sqrt{\sum_{i=1}^{n} (\varepsilon_i)^2} = \sqrt{10^2 + 2.4^2 + 7.5^2} = 13\% \qquad (12)$$

This approach does not predict direct interactions and dependencies between variables (covariance = 0 is assumed), but an overlap is considered by using the standard deviation.

The calculation of WTRs is based on data collection by the KTBL-organisation (KTBL, 2001), which provides a large range of farming systems and its WTRs in detail. This makes a specified evaluation possible with regard to a differentiation of farming systems. Therefore, the WTR varies between 2% and 5% according to the farm and plot sizes and techniques which produce uncertainties of $\pm 3\%$ on the average.

3.11. Step 11: Validation

The results of the ecological approach are compared with data about reference farming systems. The result is verified when the reference system is inside the predicted interval. Otherwise there are more uncertainties to look for.

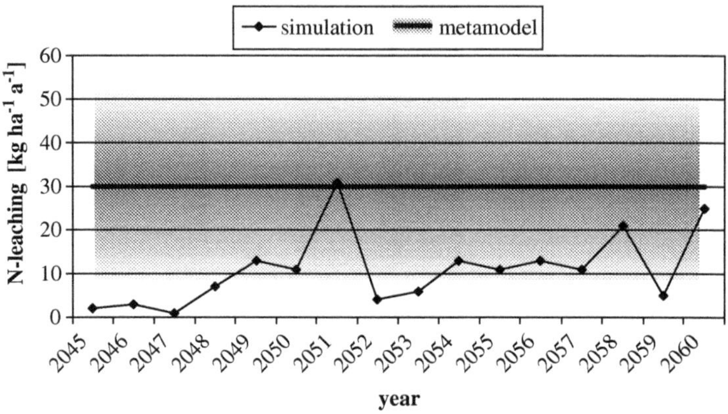

Fig. 5. Comparison of the N-Leaching-Rate Between a Reference System and the Metamodel.

The comparison needs a system in steady-state conditions. Therefore an independent simulation study is used which predicts the N-leaching-rate of a small catchment near the study area (Franko & Schenk, 2000).

Fig. 5 shows the reference system at the lower level of uncertainty. This means that the calculated result by the metamodel is verified, but has a tendency to overestimate the N-leaching-rate.

The calculated values of the economic indicators (Fig. 6) are compared with official data of accounting results in agricultural statistics. The WTR is converted in manpower (MP) per 100 hectare and the NVA [in €] refers to one hectare.

The comparison between data from official sources (BMVEL, 2003) and the calculated results shows the real values of WTR at the top level of the uncertainties. This was expected because of the focus on primary production in the calculation. Official data of the agricultural sector include income and manpower from secondary services like processing of goods and transport. Nevertheless, the NVA fits very well, which indicates that the calculated data are overestimated rather than inverse.

3.12. Step 12: Integration

Integration means the combination of the three main statements (selected indicators of economics, ecology and uncertainty) in relation to each other. It is a linkage between the input data and results of the analysis. The data

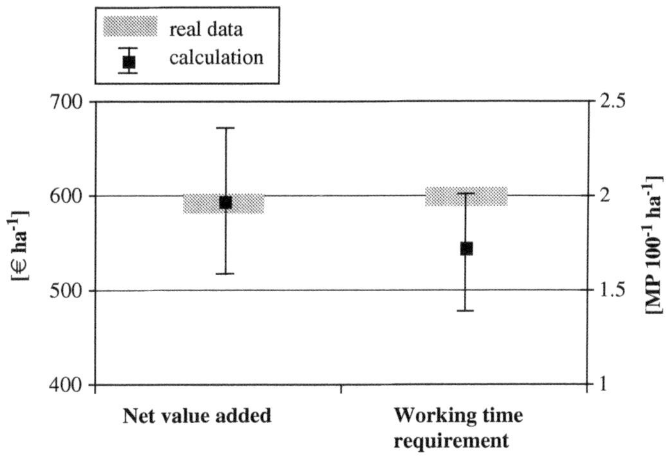

Fig. 6. Uncertainties Related to Economic Indicators.

could be assigned to three different groups:

1. Initial data: agricultural statistics (crops and livestock) and the georeferenced information (soil, rainfall, administration borders).
2. Static parameters: prices, subsidies, working hours per unit, relationship between conventional, integrated and organic parameters.
3. Model- and scenario parameters: the definition of NRUs, parameters of the metamodels, results.

In the first application of this method, the EXCEL software was used to calculate the status quo. Managing large study areas with several rural districts and much more data would require a relational database system. All initial data and the results of Steps 1–11 are stored in the database and are connected by algorithms which calculate the status quo or alternative land-uses. This static approach does not simulate dynamic ecological or economic effects, for instance the trend of prices. It reports the indicators in terms of a given land-use definition.

4. DISCUSSION AND CONCLUSIONS

The aim of this approach is to describe the ecological and economic effects of agricultural land use by using the indicators N-leaching-rate, NVA and employment, or respectively, WTR. The suggested method is based on official data like common statistics and digital maps, which describe the general characteristics of the study area. Because of these fuzzy and raw data there is a large range of uncertainties. The estimation of uncertainties is an important issue besides the calculation of the most probable value.

The ecological approach uses metamodels which are derived from a process model. The possible range of real farming systems is simulated by the process model and the relationship between the dependent variable 'N-leaching-rate' and two independent variables were determined in a linear multiple regression analysis. This procedure transforms the land use description of a detailed process simulation into a simplified calculation basis (metamodel) which deals with spatial information. These spatial data are fuzzy and cause a range of uncertainties of ±64%. The largest uncertainty in terms of N-leaching is the insufficient knowledge about the real agricultural activities. The selected variables (number of livestock and percentage of cereals in the crop rotation) are known as the total number in a rural district. However, the real distribution in the area is unknown and an allocation of the management to the georeferenced locations is impossible. It is approximately feasible on the basis of expert knowledge but the uncertainties of this approach are unknown. Hansen, Thorsen, Pebesma, Kleeschulte, and Svendsen (1999) assessed the uncertainty in simulated nitrate leaching with a Monte Carlo analysis. It was implemented in a case study of different farm types. The results range between 22% for arable farms and 41% for the pig farm rotation. This indicates the high variability within one farm type. Transferred to the whole study area (where the regional distribution of farm types is unknown) the estimated uncertainties of 64% are relatively modest in our case.

The economic results originate from a summarisation of the primary production of agricultural goods.[7] This includes the production values, fix and variable costs, or respectively overheads, as well as subsidies and payments within the framework of AEM. This calculation covers the most-important crops and animals. It could be optimised by including more species of crops and animals in the matrix. But in each case the calculated results have to be multiplied by a related factor in order to expand the values to 100% of agricultural production. This seems to be a pragmatic solution to reduce the research work to an acceptable amount. The considered variables (NVA and labour) allow the interpretation of the agricultural sector in scenario analysis. An extra evaluation is possible in terms of sub-variables, like payments from government. This makes estimation possible which describes the monetary effect of more ecological land use. For instance the establishment of more integrated and organic farming systems affects the water quality positively, but requires more financial compensation from public funds (the federal states, the national government and the European Union).

In general, the economic calculations are incomplete because of data gaps. However, the correct description of relevant scenario parameters such as an upper limit of livestock in the region is more important than a fuzzy approach of all possible influences because of its unpredictable effects.

This study is a first step of an approach to integrate economic and ecological indicators as well as related uncertainties. The same method could be used to predict indicators other than the ones presented but it is not tested yet. Further investigations in a more detailed description of integrated farming systems and the application to, as well as the evaluation of, other regions are suggested.

NOTES

1. A metamodel is a simplification and derived from a more detailed process model.
2. GIS – Geographic Information System.
3. Cover crop: A crop that is primarily planted not to be harvested for food but to prevent soil erosion, control weeds and improve soil quality.
4. 47 crop rotations (CR) of conventional farming, 44 CR of integrated farming and 39 CR of organic farming.
5. A process model is a (simplified) mathematical description of a natural process.
6. 1 hectare = $10{,}000\,m^2$.
7. This approach relates to the national accounts where the agricultural sector is also represented by the primary production.

REFERENCES

Beblik, A., Cepuder, P., Dreyhaupt, J., Frank, J., Feichtinger, F., Franko, U., Kersebaum, K. C., Krysanova, V., & Steinhardt, U. (2001). *Modellanwendungen – Konzepte und Simulationsergebnisse. Stickstoffmodellierung für Lysimeter des Parthegebietes.* UFZ-Bericht *17*(2), UFZ Centre for Environmental Research Leipzig-Halle GmbH, Leipzig, pp. 25–57.

BGR – Bundesanstalt für Geowissenschaften und Rohstoffe (Hrsg.) (1995). *Bodenübersichtskarte der Bundesrepublik Deutschland 1:1000000 (BÜK 1000)*. Hannover.

BMVEL – Bundesministerium für Verbraucherschutz, Ernährung und Landwirtschaft (2002). Agenda 2000, Pflanzlicher Bereich, Agrarumweltmaßnahmen, Tierprämien. Bonn.

BMVEL – Bundesministerium für Verbraucherschutz, Ernährung und Landwirtschaft (2003). Ernährungs- und agrarpolitischer Bericht 2003 der Bundesregierung. Bonn.

CFE – Commodities futures exchange (2002). *Product information.* Hannover

DWD – Deutscher Wetterdienst (2001). *Weather data (unpublished)*. Offenbach.

EEA – European Environment Agency (2001). *Eutrophication in Europe's coastal waters*. Topic report 7/2001, Copenhagen.

ERS – Economic Research Service, U.S. Department of Agriculture (2003). Farm income and costs: glossary. (http://www.ers.usda.gov/Briefing/FarmIncome/Glossary/def_va.htm).

Federal Statistical Office Germany (2002). *Regional Statistics*. Wiesbaden.

Franko, U., Oelschlägel, B., & Schenk, S. (1995). Simulation of temperature-, water- and nitrogen dynamics using the model CANDY. *Ecological Modelling*, *81*, 213–222.

Franko, U., & Schenk, S. (2000). *Einfluss der Bewirtschaftung auf den C-N-Kreislauf im Boden und den N-Austrag in die Umwelt*. UFZ-Report 28/2000, UFZ Centre for Environmental Research Leipzig-Halle GmbH, Leipzig.

Franko, U., Schmidt, T. G., & Volk, M. (2001). Modellierung des Einflusses von Landnutzungsänderungen auf die Nitratkonzentration im Sickerwasser. In: H. Horsch, I. Ring & F. Herzog (Hrsg.), *Nachhaltige Wasserbewirtschaftung und Landnutzung – Methoden und Instrumente der Entscheidungsfindung und Umsetzung* (pp. 165–186). Marburg: Metropolis.

Haberlandt, U., Krysanova, V., & Bardossy, A. (2002). Assessment of nitrogen leaching from arable land in large river basins. Part II: Regionalisation using fuzzy rule based modelling. *Ecological Modelling*, *150*(3), 277–294.

Hansen, S., Thorsen, M., Pebesma, E. J., Kleeschulte, S., & Svendsen, H. (1999). Uncertainty in simulated nitrate leaching due to uncertainty in input data. A case study. *Soil Use and Management*, *15*, 167–175.

Horsch, H., Ring, I., & Herzog, F. (2001). *Nachhaltige Wasserbewirtschaftung und Landnutzung. Methoden und Instrumente der Entscheidungsfindung und -umsetzung*. Marburg: Metropolis.

Hülsbergen, K.-J. (2003). *Entwicklung und Anwendung eines Bilanzierungsmodells zur Bewertung der Nachhaltigkeit landwirtschaftlicher Systeme*. Berichte aus der Agrarwirtschaft. Aachen: Shaker Verlag.

KTBL (2001). *Betriebsplanung Landwirtschaft 2001/2002*. Darmstadt.

KTBL (2002). *Standarddeckungsbeiträge 2001/2002*. Darmstadt.

Lord, E. I., & Anthony, S. G. (2000). MAGPIE: A modelling framework for evaluating nitrate losses at national and catchment scales. *Soil Use and Management*, *16*, 167–174.

Schmidt, T. G. (2004). Integrierte Methodik zur ökologischen und ökonomischen Analyse von Agrarsystemen unter Berücksichtigung von Unsicherheiten. Dissertation. Helmholtz – Centre of Environmental Research - UFZ. Leipzig.

StBA – Statistisches Bundesamt (1996). *CORINE Land Cover*. Wiesbaden.

TLL – Thüringer Landesanstalt für Landwirtschaft (2002). *Thüringer Agrarbericht 2002*. Jena.

ZMP – Zentrale Markt- und Preisberichtstelle (2001a). *ZMP-Marktbilanz Getreide, Ölsaaten, Futtermittel*. Bonn.

ZMP – Zentrale Markt- und Preisberichtstelle (2001b). *ZMP-Marktbilanz Vieh und Fleisch*. Bonn.

ZMP – Zentrale Markt- und Preisberichtstelle (2001c). *ZMP-Marktbilanz Kartoffeln*. Bonn.

REGIONAL ECONOMIC MODELING AND SPATIAL KEY SECTOR ANALYSIS

Audra A. Nowosielski and Jon D. Erickson

ABSTRACT

Direct economic use and changing patterns of human habitation have long been a cause of concern for the ecological health of many rivers and tributaries. Current development trends in many watersheds are driving the conversion of rural, agricultural and forestland to urban or industrial uses. While any single project may not have an adverse effect on the watershed as a whole, the summation of development can rapidly change the character of the landscape and alter the ecosystem functions of a river, its tributaries and an entire watershed. This chapter is a discussion on using available tools to help piece together economic transactions and their relationship to the land.

1. INTRODUCTION

Direct economic use and changing patterns of human habitation have long been a cause of concern for the ecological health of many rivers and tributaries. Today, development in many watersheds often represents a battle

Ecological Economics of Sustainable Watershed Management
Advances in the Economics of Environmental Resources, Volume 7, 167–182

ISSN: 1569-3740/doi:10.1016/S1569-3740(07)07008-3

between economic proponents and citizen groups concerned about the adverse effects development may have on the watershed.

As battles continue to be fought over point sources of pollution in many watersheds, a new generation of debate and policy initiatives has emerged around non-point sources of pollution and the cumulative impact of incremental development. Current development trends in many watersheds are driving the conversion of rural, agricultural and forestland to urban or industrial uses. This conversion is taking place one parcel, one subdivision, one strip mall at a time. While any single project may not have an adverse effect on the watershed as a whole, the summation of sprawling development can rapidly change the character of the landscape and alter the ecosystem functions of a river, tributaries and the entire watershed.

Assessing new development requires tools that local development planners, environmental groups, government officials and concerned citizens can use to evaluate the tradeoffs of development magnitude, scope and patterns. While there are models available that help assess economic or land impacts separately, there are few tools that allow a user to connect how new development not only causes changes in the economy, but also physical changes to the landscape and environment.

This chapter is a discussion on using some of the available tools to help piece together economic transactions and their relationship to the land. By identifying industry relationships and using geographical tools to visualize them, and other characteristics of a watershed, more strategic economic development may occur.

2. LINKAGES BETWEEN ECONOMIC AND LAND USE MODELS

Industry transactions represent the focus of most economic models. Linking these transactions to a specific location can be the first step to creating the link between the economy and land use. By linking economic and land use data, spatial relationships will become part of the model. Although most economic studies ignore them, spatial relationships can be very important, especially when assessing ecosystem functions. According to Bockstael (1996), ecologists put much importance on spatial relationships because "arrangement of land cover, habitat, effluent discharges, etc. can have a 'dramatic effect' on species diversity, natural assimilative capacity, nutrient cycling, etc" (Bockstael, 1996, p. 1171). Also, spatial layout should be important for policy making, as political or economic boundaries may not be adequate divisions when dealing with the environment.

One of the useful goals of linking economic and land use models is furthering understanding of development issues. This assumes participation on the part of planners, citizens and other decision makers or stakeholders.

If a project is to involve significant participation from stakeholders, the institutional capacity of the model is quite important. It also acknowledges the fact that, as stated by Johnson and Campbell (1999, p. 502), "Common to most groups is a concern for healthy communities and healthy ecosystems", however priorities of these concerns are likely to be conflicting for different stakeholders. Communities desire strong economic performance, whether that is through the expansion of existing sectors, or the attraction of new businesses. However, economic development may come with a strong environmental cost, something that may be an undesirable outcome to some or all relevant citizens.

Environmental policy has traditionally been heavily science based, something that may make it difficult for common citizens to participate because of the complex knowledge that is often needed to understand scientific relationships. Linking variables together in a visual representation, such as a map, often provides an easier format for public understanding. In extreme cases, science is chosen as the only rational basis for decisions or action. This leaves other criteria, such as social or cultural, as peripheral (Street, 1997). However, environmental decisions often take on social characteristics, especially when they deal with placement of an undesirable environmental problem. Therefore, environmental planning projects are well suited to combining ecological science with democratic decision making (Johnson & Campbell, 1999). However, it must be noted that including stakeholders does not come without some drawbacks, including upfront time, financial and opportunity costs, as well as the risk of uneven representation of all interested groups (Adams & Rietbergen-McCracken, 1994). Even so, development efforts are more likely to succeed if they include an element of participation, as noted in Adams and Reitbergen-McCracken (1994) and Johnson and Campbell (1999).

3. METHODS

While there are several ways to characterize economic and land use transactions, two are discussed here. Input–output model as an economic framework, and geographical information systems (GISs) as a visualization tool for land use patterns.

3.1. Input–Output Modeling

Input–output modeling is a useful way to represent complicated flows of goods and services between various industries and other sectors of the economy. Wassily Leontief is credited with developing the input–output method, starting with his work in the 1930s. He won the Nobel Prize for his developments in 1973. While there was some similar work done much earlier (such as research by Walras in 1874 and Dmitriev in 1904, as well as François Quesnay's Tableau of Economique in the mid-1700s), none of these quite formed the input–output system (Stone, 1986).

Leontief (1966, p. 14) describes input–output as a "method of analysis that takes advantage of the relatively stable pattern of the flow of goods and services among the elements of our economy".

The method consists of a table representing the sales and purchases of sectors in the economy. Rows show how output of the particular segment is distributed among all other sectors, while columns show the amounts of products that industry absorbs from all other sectors in the economy. There are two conditions that must be met by the flows. First, the "combined inputs of each commodity or service must equal its total output" (Leontief et al., 1953, p. 10), therefore making the model balanced. Second, "there are structural characteristics of all the individual sectors of the economy. These imply the existence of definite relationships between the quantities of all the outputs absorbed by any one particular industry and the level of its total output" (Leontief et al., 1953, p. 10). This implies that there must be a fixed relationship between the level of output and the input requirements for each sector.

In the 1950s, the input–output method gained popularity, not only among academic interests, but also with governmental agencies. The Bureau of Labor Statistics, Bureau of the Mines, Department of Commerce, Bureau of the Budget, the Council of Economic Advisors and the Air Force started to explore the possibilities of putting the method into practice (Leontief, 1966).

Input–output models were also adjusted to reflect regional differences. While some planning requires national estimates, geographic diversity often makes it necessary to revise projections by area (Miernyk, 1968). Input–output has become one of the most accepted methods for accomplishing regional studies. In fact, it has been called the dominant application for regional analysis after World War II (Tiebout, 1957). While Leontief also did early work in creating regional models, some regional input–output methods are credited to Isard. There was a fundamental difference between Leontief's regional models and Isard's, however. Leontief's regional model consisted of disaggregating a national model into different regional components, whereas Isard constructed several different regional tables that could then be aggregated up to a national level (Miernyk, 1982).

Social accounting, another extension of input–output analysis, offers an analytical tool to accomplish the goal of providing a concise view of economic activity and the interconnections between industries, household characteristics, social institutions and even the supporting environmental/natural resource base. A SAM can detail economic flows among sectors of an economy, characteristics of the labor force and income distribution. Stone, Meade and others were the early pioneers in the field of social accounting (see Stone & Croft-Murray, 1959).

By manipulating the values in the SAM, it is possible to analyze the economic effects of changes in employment, output or value added in a particular industry. The full effects of an impact can be found using economic multipliers, which are calculated using the SAM table. These multipliers allow for assessment of the impact of an exogenous change to an economy. An output multiplier for a sector is the total value that all sectors in an economy would need to produce in order to meet an additional dollar's worth of demand of that sector's output. These multipliers may also be computed for employment and value added.

Multipliers may be calculated to model different spending patterns, depending on what the researcher wishes to include. If the aim is only to study industry spending, then Type I multipliers are sufficient as they take only direct and indirect effects into account. Direct effects are those that occur as an initial direct result of the impact. For example, if a new store opens and employs 560 people, the direct effect is 560 employees. Indirect effects go one step further, adding what other industries need to produce in order to meet additional demand. For example, if 1,000 consumers want a new automobile, not only does the automobile industry need to produce 1,000 new cars, the glass industry must supply 1,000 new windshields, the tire/rubber industry needs to supply 4,000 new tires, the steel industry must supply steel for frames, etc. Therefore, not only does the automobile industry experience an increase in activity, but also does each sector that supplies inputs to the automobile industry.

While there are direct and indirect changes in an economy due to an impact, there are also induced effects (including household and/or institution spending) that may be important. Using Type I multipliers will most likely underestimate the total impact, because it does not allow for household spending. Type II and Type SAM multipliers add the induced effect. The induced effect assumes that if 1,000 new people are employed as a result of an impact, they will now spend more money in the economy. Some people may have been unemployed, or in a lower paying job prior to the impact.

These people may go out to dinner more, buy a new car, buy a house, take in more entertainment, etc. Type II multipliers are calculated using the industry columns and rows, as well as the compensation and household values. Type SAM includes household spending, but also the spending of institutions (such as government) that the user wishes to include. Therefore, type SAM would use the industry, household and government columns and rows as well as other institution columns and rows that may be of interest.

3.2. Key Sector Identification

The economic base of a community is largely thought to drive the growth of a region, and therefore is the focus of many regional policies that seek to achieve development. There are several methods to determine a region's economic base. High employment is one, and an in-depth study of input–output linkages is another. Both are highlighted in this chapter.

Much regional theory deals with identifying the industries in a region that produce goods for export, and those that serve the local market. These industries are referred to as the basic (exporting) and non-basic (local) industries Leven (1966, p. 81) states "the region will grow proportionately to the expansion of basic or 'export' industries", therefore movements in the economy should mirror expansions and contractions in the export sectors.

In order to identify basic and non-basic industries it is necessary to examine the concentration of a particular industry in an area. One way to accomplish this is by computing a location quotient (LQ). According to Watkins (1980) and Bogart (1998), the location quotient may be calculated as

$$\mathrm{LQ}_i = (e_i/e_\mathrm{t})/(E_i/E_\mathrm{t}) \tag{1}$$

where e_i = the local employment in industry i; e_t = the total local employment; E_i = national employment in industry i; E_t = total national employment.

If this location quotient is greater than one, the city produces a surplus destined for the export market. Equal to one identifies local self-sufficiency, and a value of less than one indicates the need for imports because of a local deficiency.

The location quotient helps to understand what the impact would be of a local company expanding employment, or of new facilities locating in a region. Bogart (1998, p. 148) offers some cautions about relying heavily on location quotients. Bogart states "if there are imports and exports within the same industry, then the location quotient will underestimate the amount of exports". This is true because the quotient becomes a measure of net exports. This is corrected by disaggregating the data by sector as far as possible. However, when using aggregated data, if an industry is still identified as an exporter with a location quotient greater than one, it still may be said that the industry is an exporter because the bias underestimates the results.

The location quotient may be used as a basis for policy, but the prescriptions in the literature are conflicting. Higgins and Savoie (1995) offer two opposing views on the policy uses of location quotients by stating some feel that industries with high location quotients are areas of strength to an economy, and therefore will help propel the economy when further developed. However, others feel that those industries with a low quotient should be encouraged because it reduces the drain on imports from a region. This is much like the leak-plugging approach to regional growth. Implementing a program of import substitution is also declared as a way to achieve urban growth, although some critics offer the argument that if a good being imported is one that the region does not hold a comparative advantage in, then it is inefficient to produce it instead of another good (Bogart, 1998).

Along with input–output linkages, location quotients can be used to identify keystone sectors. Keystone, a term originally from ecology, may be used to describe those industries whose role is so important in a region, that without the sector, the economy would be fundamentally and detrimentally altered. One way to identify keystone industries is to assess which sectors have a high location quotient, while also possessing strong input–output linkages with other high location quotient industries (Kilkenny & Nalbarte, 1999).

Key sectors in the economy can also be determined by an in-depth study of input–output linkages. The method outlined in Sonis, Hewings, and Guo (2000) involves identifying which sectors have strong forward linkages (that is, they sell a proportionately large amount of inputs to other sectors within the region), as well as those with strong backward linkages (meaning the sector purchases large amounts of inputs from other sectors in the region's economy). An industry where the forward (FL) and backward linkage (BL) indices are both greater than one is said to be a key sector. These calculations and their interpretation are explored with a case study later in this chapter.

3.3. GIS as a Land Use Visualization Framework

There is no comprehensive definition of GIS. DeMers (1997, p. 7) offers the following: "Tools that allow for the processing of spatial data into information, generally information tied explicitly to, and used to make decisions about, some portion of the earth".

GISs began in the early 1960s. The earliest system was built in Canada to automate map information collected by the Canada Land Inventory. During the 1980s, GIS became more prevalent due to the development of affordable desktop computers that made it possible for universities, local planning departments and corporations to invest in and make effective use of GIS (Goodchild, 2000). Today GIS is not just a tool to provide maps, but offers a way to pose and analyze spatial questions (Black, Powers, & Roche, 1994). According to Goodchild (2000), the most common applications of GIS are resources management, utilities management, telecommunications, urban or regional planning, vehicle routing/parcel delivery and in all sciences dealing with the surface of the earth.

One of the appealing features of using GIS is that data from several disciplines are available, or can be constructed from a set of locations. To analyze economic activity geographically, one important dataset is the business point dataset. The business point dataset provides a crucial link between economic activity and land use by coding every business within the United States to a point within a map workspace, as well as allowing the filtering of these businesses by criterion such as primary industry and employee size. By studying where specific types of businesses are located, we can not only learn where different types of businesses have chosen to locate, but assess where new businesses in a specific sector may also choose to locate. Combined with current land use data, vacant areas of the landscapes can be identified, as well as their proximity to a variety of factors including other firm locations.

Industries are not the only part of a social accounting matrix that can be mapped. Households, delineated by characteristics such as income, can also be geographically represented easily. The most useful dataset for household characteristics is U.S. Census Bureau data.

4. CASE STUDY: DUTCHESS COUNTY, NEW YORK, USA

The following section discusses how economic and land use characteristics have been studied for Duchess county in the lower Hudson river valley of New York state. Two 'models' were constructed for the county: an input–output model and a map workspace. The input–output model of the county was based on relationships as defined by the regional modeling software

IMPLAN (see http://www.implan.com). This basis was then refined with local knowledge to arrive at a final social accounting matrix. MapInfo (see http://www.MapInfo.com) was used to create a workspace that mapped all available land and economic geographical data. Several different categories of information in the study area were depicted such as demographic, environmental and economic. This information has been used to characterize the study area as well as answer specific questions regarding where certain activities take place, especially in regards to their position relative to rivers and tributaries. The two were joined together in an interface where a user could access both economic and geographical information. The software used to create the interface was Powersim (see http://www.powersim.com).

4.1. Overview of Dutchess County

Dutchess county is located in downstate New York, between Albany and New York city (see Fig. 3 of Chapter 5 for a map, Hong et al., in this volume). The county lies within the Hudson river watershed, with the Hudson forming its western border. There are over 600 miles of named streams that serve as a source of public water, irrigation, recreational use and waste disposal. The three major streams in the county are the Wappinger creek, Tenmile river and the Fishkill creek. The Fishkill and Wappinger alone drain almost 75% of the county. The Wappinger creek watershed itself drains 210 square miles, about one quarter of Dutchess county. Its drainage basin, the lower portion of which is subject to the tidal influence of the Hudson river, includes the towns of Washington, Pleasant valley, Pine plains, Milan, Stanford, Clinton, Wappinger, Poughkeepsie and LaGrange (Dutchess County Environmental Management Council, 2000).

Current demographic and economic patterns stem from a long history of structural economic change. Before the Erie canal opened, Dutchess county's major economic activity was growing grain, and the village of Poughkeepsie was a shipment center (Griffen & Griffen, 1978). By the 1830s, the county had shifted to wool production. Today, agriculture is no longer the main activity in the county. The major employers in February 2003, according to the Dutchess County Economic Development Corporation, were in the high-tech manufacturing (International Business Machines), health (Health Quest/Vassar Hospital) and government (NYS Fishkill Correctional Institute) sectors. The Dutchess county economy can be characterized by 203 different sectors, and employment of over 120,000.

Dutchess county has a nearly continuous development gradient from northeast to southwest, being mostly rural in the northeast part of the quadrant, becoming suburban in the middle of the county, then residential, urban and industrial in the southwestern area. Land use in the Wappingers creek watershed follows a similar development gradient to the county, which offers an opportunity to study a single tributary of the Hudson at different degrees of development intensity.

Perhaps it is the county's position in the river valley (between New York city and upstate) that has led to this uneven pattern of development between the northern and southern parts of the region. Today, Dutchess county serves as the bridge between New York city and upstate New York. North of Dutchess county is Columbia county, still largely farmland, small towns and a rural landscape. The northern part of Dutchess county has a very similar feel, with large farms, small towns and country roads. Southern Dutchess county is greatly influenced by development pressure as the New York city metropolitan area expands north. People are attracted to Dutchess county to escape the urban feeling and landscape that has taken over counties to the south.

This is a cause for concern for some municipalities in Dutchess. A Red Hook planning report states: "These factors will continue to bring commercial development pressures on any significant highway corridors, as businesses seek to exploit the growing pool of disposable income in Red Hook and Rhinebeck" (Town of Red Hook, 2002). This phenomenon is viewed as a problem by many of the municipalities as they struggle with ways to preserve their rural landscape.

The northeast to southwest development pattern is shown clearly by population density. Dutchess county is one of the most populated counties in New York state (with the exception of the Buffalo and New York city areas). However, that population is not evenly spread throughout the county. Three distinct population levels are shown in Fig. 1, with the highest population density bordering the Hudson in the southwest corner of the county. Once outside the cities of Beacon and Poughkeepsie, there is a drop off of population levels indicating the presence of suburbs. Further to the northeast, there is again another band of population density, this one quite low, indicating the rural area in the county. According to information from the Dutchess County Planning and Development Department, about 75% of the houses in Dutchess county are located in the southern half. This again indicates the differences in development between the northern and southern parts of the county.

4.2. Key Sector Identification in Dutchess County

The identification of key sectors within the economy can help to identify those areas where new development may continue to drive population

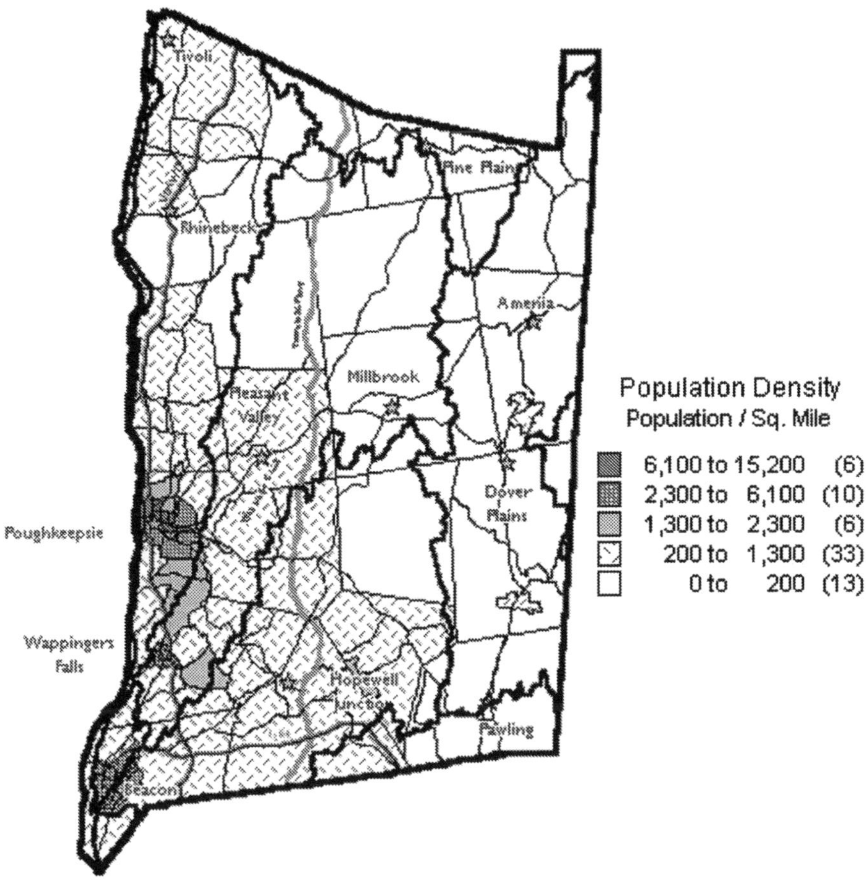

Fig. 1. Population Density in Dutchess County. *Source:* 2000 Census Data.

change and economic growth, not only in that particular sector but also in those sectors that it buys and sells from. The layout of the key sectors also offers implications for policies that may seek to strengthen the economic base of the county, as they may favor one portion of the county over another.

Using the method discussed in the key sector section previously, linkages were calculated for each industry first using both Type I and Type II Leontief inverses. A total of 96 of the 203 industries in Dutchess county are weakly linked sectors, including both mining sectors, all construction industries that involve building new structures, and 7 of the 10 government sectors. These industries are not major suppliers to other industries in the county, nor do they buy large amounts of inputs in the local economy.

Sixty-two industries are backward linked oriented sectors. A change in the final demand of these sectors is expected to cause an above-average increase in the economy. Therefore, an increase in final demand for the sector will cause an increase in the column entries as the industry needs to buy more inputs. The majority of service sectors in Dutchess county are backward linked sectors.

Dutchess county has 21 forward linked sectors. A large forward linkage number indicates that a change in all sectors' final demand would lead to a larger-than-average increase in that particular sector's output. Many of the financial sectors in Dutchess county were linked forward. This may be caused by the fact that new development requires financing; therefore, an increase in any sector will likely impact the banking sector.

Key sectors are those in the economy that have both a forward and backward linkage number greater than one. Therefore, not only will a change in a key sector's final demand give the whole economy a boost, these sectors are more affected by changes in any final demand. Using the Type I Leontief Inverse there are 19 key sectors in Dutchess county. When using the Type II Leontief Inverse, 24 industries are classified as key sectors. Table 1 shows the 11 sectors in Dutchess county that can be classified as key sectors using both a Type I and II Leontief Inverse.

The analysis for key sectors may be taken one step further by assessing location quotients for each industry. Calculating location quotients for every industry in Dutchess county can be done with employment

Table 1. Key Sectors in Dutchess County.

Dutchess County Key Sectors
Electronic computers
Engineering, architectural services
Maintenance and repair, residential
Management and consulting services
Motion pictures
Motor freight transport and warehousing
Newspapers
Other state and local government enterprises
Radio and TV broadcasting
Research, development & testing services
Theatrical producers, bands etc.

Table 2. Dutchess County Industries with a High Location Quotient.

Industry	Location Quotient
Semiconductors and related devices	30.2
Computer peripheral equipment	29.1
Paving mixtures and blocks	24.1
Rubber and plastics hose and belting	20.3
Electronic computers	10.3
Wiring devices	5.3
Woodworking machinery	4.3
Switchgear and switchboard apparatus	4.2
Colleges, universities, schools	4.1
Fertilizers, mixing only	3.7
Miscellaneous livestock	3.4
Elementary and secondary schools	3.3
Sand and gravel	3.2
Curtains and draperies	3.1
Electron tubes	3.1

information for the United States and Dutchess county provided by IMPLAN. There are 72 industries in Dutchess county with a location quotient over 1. Those with a location quotient over 3 are shown in Table 2.

While many of the industries in the county meet the keystone criteria of a high location quotient, only a few possess strong input–output linkages with other high location quotient industries. Three stand out: electronic computers, computer peripheral equipment and semiconductors and related devices. These all have a location quotient over 10. Also, semiconductors and computer peripheral equipment have the highest percentage of national employment out of all 203 industries (about 2%). This is due to the presence of IBM in the county, a major producer of semiconductors and various types of computers. Both electronic computers and computer peripheral equipment make large purchases from all three industries. The semiconductors sector purchases the most input from its own sector.

Based on these results, it appears as if sectors within the computer industry are the keystone sectors of Dutchess county. This certainly supports everyday experience in the county economy. The other industries with high location quotients, while important, do not possess the linkages with other prosperous industries that the computer sector does. In fact, some of the industries with high location quotients have very weak input–output ties to other high location quotient firms, supplying or purchasing very small quantities from other high location quotient industries. For example, residential care and elementary/secondary schools supply nothing to the other 19 industries under consideration. While they do make purchases, they are very small in comparison to the magnitude some of the other industries buy from one another. This phenomenon might be due to the nature of the business of residential care and schools.

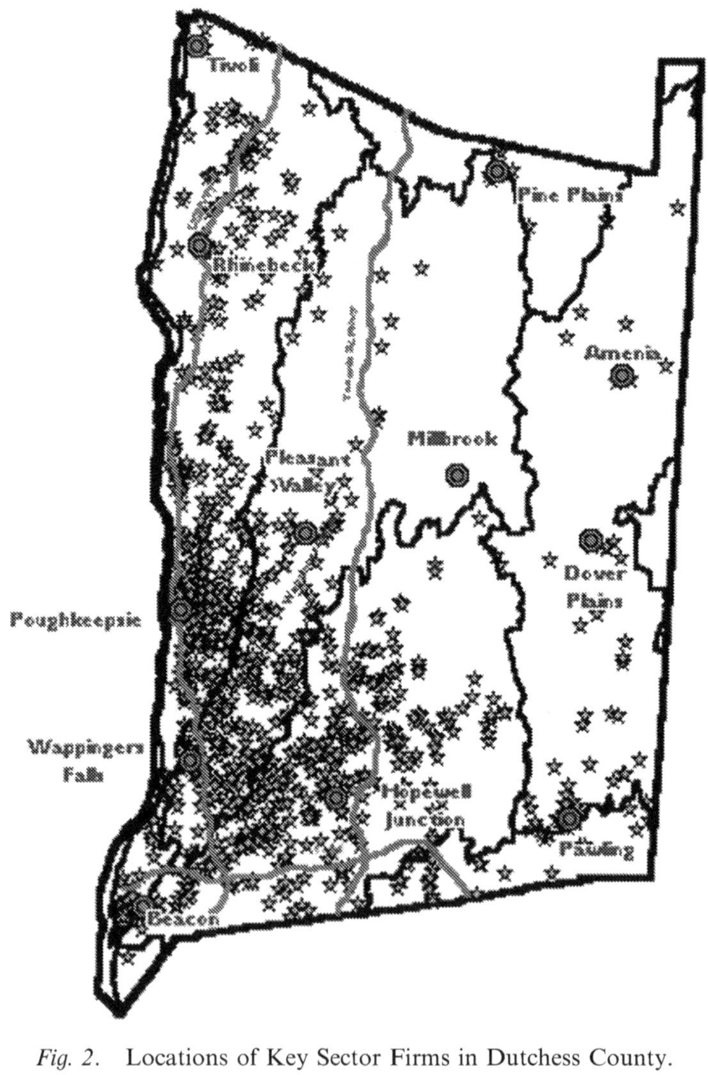

Fig. 2. Locations of Key Sector Firms in Dutchess County.

Fig. 2, created using Axiom's Business Point Data, shows the locations of key sector businesses in Dutchess county. Note the density of businesses in the southwestern part of the county, as well as the location of the majority of businesses between US Route 9 and the Taconic State Parkway, the two major transportation corridors of the county.

Further analysis of key sectors may be done using both GIS and input–output. Input–output is particularly useful for determining the impact of new activity in an economy, and a SAM may offer insights into changes in households. This information, coupled with a GIS representation of industry and household locations, can offer insights as to how much development may take place as a result of a change in the economy and what areas of the region may be affected.

5. DISCUSSION

Economic development has not always been positive or welcomed by residents who often contribute little to the process. While the flow of development may seem unstoppable, this is far from true. With the proper tools and knowledge, concerned people can create an image for what they want their towns to look like in the future and implement policies to achieve that vision.

The Hudson valley has represented an area of increasing development ever since it was discovered and became one of the most important commercial shipping venues in the United States, making it a region experienced with issues regarding new economic development and land use. Dutchess county has already taken a position of caution with regard to development and for protection of what remains of its rural landscape. However, stating this and putting it into practice are two different things. With the right tools, the stakeholders can make a case for their position and help to foster policies that will facilitate the type of community they wish to have.

This chapter has discussed the creation and use of tools that can be used to help these stakeholders visualize the future and plan for it. While it is impossible to predict the exact conditions in the future, it is possible to plan for what you do not want to happen.

REFERENCES

Adams, J., & Rietbergen-McCracken, J. (1994). Participatory development: Getting the key players involved. *Finance & Development*, *31*(3), 36–37.

Black, S., Powers, G., & Roche, M. (1994). Spring. A GIS-based approach to community analysis for targeted marketing. *Economic Development Review*, *12*(2), 12–31.

Bockstael, N. E. (1996). Modeling economics and ecology: The importance of a spatial perspective. *American Journal of Agricultural Economics*, *78*, 1168–1180.

Bogart, W. T. (1998). *The economics of cities and suburbs*. Upper Saddle River: Prentice Hall.

DeMers, M. N. (1997). *Fundamentals of geographic information systems*. New York: Wiley.

Dutchess County Environmental Management Council (2000). Wappingers Creek Watershed. http://www.bestweb.ner/~dcemc/subshed.html

Goodchild, M. F. (2000). The current status of GIS and spatial analysis. *Journal of Geographical Systems*, *2*(1), 5–10.

Griffen, C., & Griffen, S. (1978). *Natives and newcomers*. Cambridge: Harvard University Press.

Higgins, B., & Savoie, D. J. (1995). *Regional development theories and their application*. London: Transaction Publishers.

Johnson, B. R., & Campbell, R. (1999). Ecology and participation in landscape-based planning within the Pacific Northwest. *Policy Studies Journal*, *27*(3), 502–529.

Kilkenny, M., & Nalbarte, L. (1999). *Keystone sector identification. A graph theory-social network analysis approach*. The Web Book of Regional Science. Regional Research Institute, West Virginia University. http://www.rri.wvu.edu/WebBook/Kilkenny/editedkeystone.html

Leontief, W. (1966). *Input–output economics*. New York: Oxford University Press.

Leontief, W., Chenery, H. B., Clark, P. G., Duesenberry, J. S., Ferfuson, A. R., Grosse, P., Grosse, R. N., Holzman, M., Isard, W., & Kistin, H. (1953). *Studies in the Structure of the American Economy*. New York: Oxford University Press.

Leven, C. L. (1966). *The economic base and regional growth. In research and education for regional and area development*. Ames, IA: Iowa State University Press.

Miernyk, W. H. (1968). Long-Range forecasting with a regional input–output model. *The Western Economic Journal*, *6*(3), 165–176.

Miernyk, W. H. (1982). *Regional analysis and Regional Policy*. Cambridge: Oelgeschlager, Gunn & Hain, Publishers, Inc.

Sonis, M., Hewings, G. J. D., & Guo, J. (2000). A new image of classical key sector analysis: Minimum information decomposition of the Leontief Inverse. *Economic Systems Research*, *12*(3), 401–423.

Stone, R. (1986). Where are we now? A short account of the development of input–output studies and their present trends. In: I. Sohn (Ed.), *Readings in input–output analysis: Theory and applications* (pp. 13–31). New York: Oxford University Press.

Stone, R., & Croft-Murray, G. (1959). *Social accounting and economic models*. London: Bowes & Bowes.

Street, P. (1997). Scenario workshops: A participatory approach to sustainable urban living? *Futures*, 29(2), 139–158.

Tiebout, C. M. (1957). Regional and interregional input-output models: An appraisal. *Southern Economic Journal*, *24*, 140–147.

Town of Red Hook (2002). Southern Gateway Small Area Plan, retrieved from: http://www.redhook.org/gateway/Gateway.html

Watkins, A. J. (1980). *The practice of urban economics*. Beverly Hills: Sage Publications.

RESIDENTIAL LOCATION THEORY, MODELING, AND SCENARIO ANALYSIS OF URBAN GROWTH AND PLANNING[☆], [☆☆]

John M. Polimeni and Jon D. Erickson

ABSTRACT

This chapter presents projections of residential development in Wappinger Creek watershed of Dutchess County, New York in the Hudson River Valley. A spatial econometric model is developed based on data from a geographical information system (GIS) of county-level socio-economic trends, tax parcel attributes, town-level zoning restrictions, location variables, and bio-geophysical constraints including slope, soil type, riparian and agricultural zones. Monte Carlo simulation is employed to distribute spatially explicit projections of land-use change under various residential development scenarios. Scenario analysis indicates the likelihood of

[☆] Portions of this chapter have been published as Polimeni (2005) and Polimeni and Polimeni (in press).

[☆☆] Adapted from an article by John Polimeni in the journal *Agriculture and Human Values*, Vol. 22, No. 4, Dec. 2006 entitled "Simulating Agricultural Conversion to Residential use in the Hudson River Valley: Scenario Analyses and Case Studies" with kind permission of Springer Science and Business Media.

Ecological Economics of Sustainable Watershed Management
Advances in the Economics of Environmental Resources, Volume 7, 183–210

ISSN: 1569-3740/doi:10.1016/S1569-3740(07)07009-5

continued residential, decentralized development patterns in formerly agricultural and forested parcels. Policy scenarios demonstrate possible courses of action to direct development and protect watershed health.

1. INTRODUCTION

Undeveloped land provides many benefits to human well-being. Ecosystem services include detoxification and decomposition of wastes, fertile soil, climate regulation, and purification of water and air. For planners to develop effective policy recommendations to safeguard these services, while providing for the myriad of economic services from developed land, land-use change models that are sensitive to local characteristics are needed to develop scenarios for evaluation.

This study projects residential development in the Wappinger Creek Watershed within Dutchess County of the Hudson River Valley of New York State. A binary logit regression model employing spatially explicit socioeconomic and biophysical data layers is used to determine the likelihood that vacant parcels will be converted to residential uses at watershed and sub-watershed development gradients. A combination of tax parcel and census block data is used to calculate and simulate growth trends for land parcels, with emphasis on residential housing growth and the conversion of vacant and agricultural parcels. Tax parcel data includes property class, assessment value, ownership, size, age of infrastructure, and square footage of existing living spaces. Location attributes include analysis of surrounding land uses and distance to central business districts. Census block data supplements tax parcel information to determine likely development probabilities, including detailed household income, educational status, population growth, commuter travel time, and other household characteristics.

The study area is described in more detail in Chapter 5 (Hong et al., in this volume). For the purposes of this chapter it is important to note that Dutchess County is located midway between the state capital of Albany and the burgeoning metropolis of New York City, bordering thirty miles of the Hudson River on the county's western border with Connecticut to the east. The county has been under considerable growth pressure given this proximity to the two major growth poles of New York State, the demand for second-homes in pastoral settings, and expansion of key sector industries such as semi-conductor manufacturing (see Chapter 8, Nowosielski & Erickson, in this volume). Wholly within the county is the Wappinger Creek

Watershed, the principle study area, totaling two hundred-ten square miles, containing ten towns and three villages, and draining approximately 25% of the county's land into the Hudson River. The watershed flows northeast to southwest along a fairly gradually changing development gradient, with the northern part mainly forest and farms, and the southern part transitioning to residential and commercial land use (see Fig. 3 of Chapter 5 for a map of the county and main watersheds).

This chapter begins with a review of the evolving literature on spatial economics, with specific attention to residential location theory and the emergence of research on urban sprawl. The residential development model is then described. Development probability estimates for each tax parcel are then used in scenario analysis. Combined with digitized zoning data and bio-geophysical data such as wetland location, soil type, and land-slope information, probable and allowable development can be projected onto maps of the watershed through a Monte Carlo procedure. A series of scenarios are simulated to demonstrate how different policies and socio-economic trends may affect the landscape of Dutchess County and, more generally, influence the ecosystem health of Hudson River tributaries and the quality of life of human communities.

2. RESIDENTIAL LOCATION THEORY: FROM VON THÜNEN'S RINGS TO URBAN SPRAWL

Research presented in this chapter builds on the rapidly developing area of residential location theory. Residential location theory stems from a long line of study of the effects of geography on economic activities, best described as spatial economics. Space affects economic relationships through market activities in adjacent locations (e.g., neighborhood effects) and in the movement of people and goods (e.g., transportation costs) (Beckmann, 1968). The emphasis of residential location theory is more specifically on the relationship between housing location and values of other correlated economic factors, such as accessibility, land values, and development costs (Ricks, 1970).

The evolution of location theory can be summarized around four main themes: concentric circle models, industrial location, central place theory, and spatial competition. Focusing largely on the location of agricultural activities, Johann Heinrich von Thünen is generally recognized as the founder of location theory. Culminating in *The Isolated State* (1819) von Thünen was primarily concerned with the development of agricultural land use in relation to the distance to a central city (von Thünen, 1966; Beckmann, 1999). He posed the market allocation problem as solving for the optimal amount of agricultural land use given transportation costs to the market through a concentric ring model, with each ring emanating from the central city and composed of agricultural activities dependent on distance to market. William Alonso (1964) some years later extended the von Thünen model to analyze urban land values and uses. Alonso created a model of agricultural rent and land use using bid-rent functions, representing the amount a farmer could pay for rent at different locations with different transportation costs and still remain on the same indifference curve.

With the advent of new spatial datasets, computerized geographical information systems (GISs), and a resurgence of regional economic inquiry, there has been an explosion of research in recent years extending the work of von Thünen and Alonso. For example, Fan, Treyz, and Treyz (2000) used a general economic geography model with multiple industries and regions to study profit-seeking behavior. Hardie, Parks, and Gottleib (2000) performed a land-use analysis for 1,459 counties in the southern United States using both Ricardian and von Thünen land rent models. Block and DuPuis (2001) studied milksheds and dairy policies to explore how von Thünen's model pertains to the fields of geography, agricultural economics, and sociology. de Vries, Nijkamp, and Rietveld (2001) presented a spatial interaction model consistent with Alonso's theory to study spatial origin–destination flows between regions. Knapp, Ding, and Hopkins (2001) expanded on Alonso by developing models of the interrelationship between urban growth and public infrastructure.

While von Thünen and his successors were primarily concerned with patterns of agricultural land use, location theory was also being developed for insight into industrial location. Most notably, Wilhelm Launhardt's *Mathematische Begründung der Volkswirtschaftslehre* (1885) was similarly concerned with minimizing transportation costs. He used a time–distance measure to optimize industrial location such that a consumer can make a journey to and from two market centers in one day, implying a funnel-type pattern of development (Pinto, 1997). Alfred Weber (1909) built on the work of Launhardt allowing for three location types – raw materials, production sites, and a market center. Implied spatial patterns led to early ideas of the positive externalities of agglomeration effects of industrial location.

Industrial location theory received a flurry of attention by the emergence of regional economics in the 50s and 60s, for instance the work of Moses

(1958) and Isard (1969). But it was Paul Krugman (1991) who brought the subject back into vogue in more recent years. Krugman formalized a model of industrial location to investigate why industrial production tends to concentrate in only a few regions. Similar to Weber's earlier conclusions on agglomeration, and expanding on the Dixit and Stiglitz (1977) monopolistic competition model, Krugman argued that firms locate regionally to realize scale economies while also trying to minimize transportation and land costs.

A third line of inquiry under the umbrella of location theory includes work on central place. Walter Christaller (1933) was the first to analyze the spatial dispersion of economic activity in the context of central places. Central places are supply centers, either urban or rural, of different order or rank based on their function and the maximum radius over which a population will travel to buy a product (Bos, 1965). The objective of Christaller's model was to minimize transportation costs by achieving a minimum market size through locational economic rents. He concluded that market region size is exponentially related to order, so that new markets will be created until the smallest and lowest order markets are reached. Building on Christaller's work, Auguste Losch (1944) formalized a central place theory by modeling the minimization of both transportation and production costs to explain the location of centers, distance between centers, amount of goods or services produced, and patterns of trade.

As with recent extensions of agricultural and industrial location theories, recent research has begun to both test and extend the tenets of central place theory. For example, Wang (1999) developed a spatial equilibrium model of city hinterlands, rural areas, and interurban transport costs to investigate the number, size, and location of cities, as well as the wage rates and prices of goods. South and Boots (1999) explored the assumption that consumers patronize businesses at the closest business district by using higher-order Voronoi diagrams. Ishikawa and Toda (2000) constructed a central-place system based on profit maximization to demonstrate that firms tend to concentrate at the largest central place, helping to explain the observed number of central places and the diversity of central places within an urban system.

The notion of spatial competition, with a more explicit focus on market structure, represents yet another body of thought within location theory. Harold Hotelling (1929) was the early pioneer in this work, examining price differences and competition as they relate to location and other factors. He observed that a high price seller will not lose all of their business immediately, but will lose business to rivals gradually. Contrary to assumptions of perfect competition, Hotelling proposed that customers might prefer to buy from the higher-priced seller because of proximity, lower transportation costs, or preference for business style, products, or service. Such circumstances where in conflict with the reigning assumptions of Cournot, Amoroso, and Edgeworth. Rather, Hotelling considered every seller as a quasi-monopolist within a limited class and region, deviating from competitive equilibrium theory.

Arthur Smithies (1941) later built on the findings of Hotelling, investigating various market structures including monopolistic competition, quasi-cooperation, a mix of quasi-cooperation in prices and competition in location, and full competition. Edward Chamberlin (1953) extended this framework into questions of the pace of product innovation, including distinguishing between effects of custom, standards, and profit-maximization behavior. Recent work in this area includes Aoyagi and Okabe's (1993) examination of spatial competition, Tabuchi's (1999) analysis of spatial oligopolies, Rothschild's (2000) study of the relationship between merger activity and choice of location under conditions of spatial price discrimination, Norman and Pepall's (2000) analysis of firm mergers and location coordination, and Collins and Sherstyuk's (2000) work on firm-location decisions within a Nash equilibrium framework.

These pillars of location theory – concentric circle and industrial location models, central place theory, and spatial competition – laid the foundation for the relatively new study of residential location, the main focus of this work, including an emerging literature on urban sprawl. Residential location theory can be roughly divided into two camps – monocentric and polycentric theories. Richard Muth (1969, 1985) formed the basis for the monocentric approach, finding that work and residence location are positively correlated and that wages decline with distance from the central business district. Recent work along these lines includes McMillen and Singell (1992) corroboration of Muth's hypothesis, Bailey's (1999) land-rent model predicting residential location in a metropolitan center with a constraint on suburban land, Bogart and Ferry's (1999) study of the evolution from mono- to polycentric employment centers in the Cleveland area, and Anas, Arnott, and Small (2000) estimate of declining exponential rent and density functions for a monocentric city with commuting costs.

While the monocentric approach helped to develop residential location theory, particularly the incorporation of consumer housing choice, the assumption of a monocentric city has been a major limitation. Shieh's (1987) work is representative of the development of a polycentric approach, estimating bid-rent functions for housing in a city with two commercial centers and one employment center. Shieh's findings are consistent with

Muth's principle that as households move away from employment and commercial centers, the increased travel costs must be counterbalanced with decreased housing costs. Sasaki and Mun (1996) expanded on the polycentric theory modeling the formation of city subcenters, likely when: business districts are close to the city center, commuting costs are high, communication costs are low, larger lot sizes exist, and firm size increases. Henderson and Mitra (1996) expanded residential location theory to a focus on edge cities – complete cities outside of central cities that offer jobs, residences, and shopping with economic output largely from offices rather than manufacturing sites. Along similar lines, Champion (2001) examined the consequences of changing demographics and evolving polycentric urban regions on the size, composition, and distribution of city populations. Buisson, Mignot, and Aguilera-Belanger (2001) examined expanding metropolitan areas and the emergence of intra-urban poles in Lyon, France, finding that while distance to the city center is still relevant to economic activities, the emergence of peripheral employment poles suggest the evolution of a multi-functional, polycentric urban area. McMillen (2001), building on earlier monocentric analysis of Milwaukee, Wisconsin concluded that despite the city's largely monocentric spatial structure, highly dispersed suburban employment is creating a polycentric city.

Moving from monocentric to polycentric residential location models paved the way for the most-recent interest in urban sprawl. Mills (1981) early on defined sprawl as a dispersed residential and commercial development pattern outside of metropolitan centers along major roadways and in rural country-sides, creating a fragmented landscape. He categorized several types of land-use patterns typical of sprawl, including leap-frog, scattered, and mixed development. Recent interest in modeling the process and consequences of sprawl have been motivated by trends in increasing per capita land use. For instance, the U.S. population grew by 92.3% between 1950 and 1990, while land consumption grew by 245.2% (Kahn, 2000). The top 25 metropolitan areas in the United States have almost all decreased their density per acre during the 1980s and early 1990s (Pendall, 1999).

This latest extension to residential location theory includes a number of students relevant to our research in Dutchess County. Lamb (1983) explored the distribution of exurban sprawl around urban areas in the United States for the 1960 to 1970 time period through multiple regression analysis, finding that (1) the older more densely populated centers of the Northeast and the North Central regions and parts of the South hold little promise for absorbing future exurban population increases, (2) no significant relationship exists between distance to urban areas and the success of exurban centers in absorbing population increases, and (3) policies created to control exurban sprawl by encouraging growth of exurban centers would be applicable to controlling growth in outlying rural areas. Building on Lamb's findings, Alig and Healy (1987) examined alternative measures of built-up areas by using six regression models to make long-term national projections under alternative assumptions. McMillen (1989) used a multinomial logit model to predict land use in the urban areas surrounding Chicago incorporating parcel size, distance to Chicago and other nearby localities, and location characteristics as predictors. Walker (2001) more recently explored land-use changes in the Florida Everglades from natural or agricultural cover to suburban uses by combining the models of von Thünen and Alonso, advancing the two-sector model developed by Muth. And finally, Irwin and Bockstael (2002) developed land-use conversion model to explore land fragmentation in the urban fringes of central Maryland between Baltimore and Washington, DC.

3. WAPPINGER CREEK WATERSHED RESIDENTIAL DEVELOPMENT MODEL

The discussion of location theory and its recent application to the phenomena of sprawl helps to lay the foundation for the land-use change model of this study, a sub-model of the larger integrated model described in Chapter 5. Following the collection of land cover, socio-economic, bio-physical, and tax parcel spatial data, a binary logit model was estimated to determine the likelihood of vacant and agricultural parcels in the Wappinger Creek Watershed of Dutchess County converting to residential uses. Land-use conversion probabilities were then combined with digitized zoning and bio-geophysical data to simulate probable development with a Monte Carlo procedure. In the next section, various economic, social, and environmental trend and policy scenarios are then explored.

3.1. Dutchess County Spatial Socio-Economic Database

Tax parcel data was obtained from the New York State GIS Clearinghouse and the Dutchess County Planning Department. Fig. 1 illustrates the residential parcels within the Wappinger Creek Watershed shaded by the age of built structure, with older units more prevalent in the northern agricultural portion and new structures in the south closer to the main cities of the

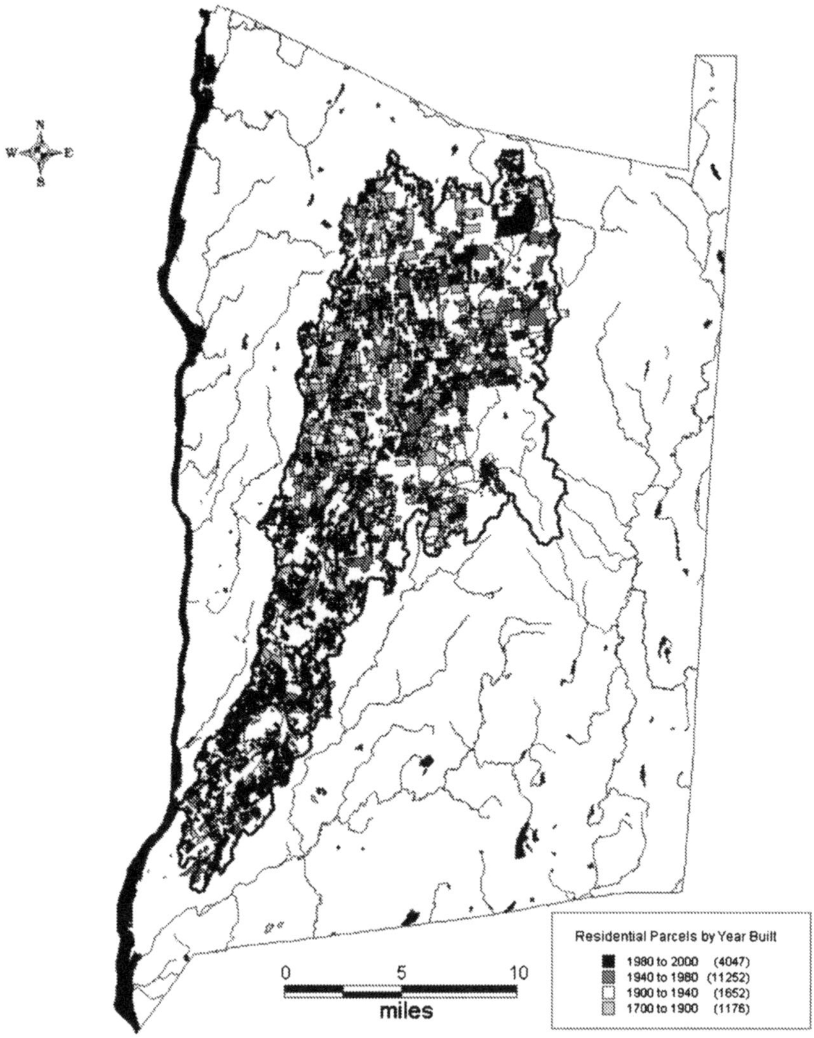

Fig. 1. Residential Tax Parcels in the Wappinger Creek Watershed by Year Built. *Source:* Polimeni (2005, with permission from the Clute Institute for Academic Research).

county. Vacant parcels, including agriculture and forest land, were also identified and served as the inventory for potential new residential development. Tax parcel characteristics included in the database included property class code, assessment value, area in acres, ownership, and age and square footage of existing residential units.

Overlaid on the residential and vacant tax parcels was census data at the neighborhood scale (census blocks), including detailed household income, housing characteristics, population growth, and distances to the nearest business district. Table 1 summarizes the independent variables ultimately explored in the regression analysis for the entire watershed, as well as three sub-watershed scales considered (with the delineation highlighted in Fig. 2). Income metrics are similar throughout the watershed, with the exception of slower income growth in the upper, more rural portion. Population change has been more variable, with the upper watershed growing nearly three times the average of the entire watershed. In fact, the lower watershed had a negative growth rate during the 1990s, indicative of the outmigration from urban to suburban areas of the county. Land-assessment values remain higher in the more urban portions of the watershed, consistent with location theory. Distance to the central business district and the neighborhood index (a ratio of residential to residential plus vacant) follow the urban to suburban to rural gradient, decreasing moving north in the watershed, and population density (people per acre) is the most clear delineation of the three regions. Other independent variables considered but rejected include: distance to New York City, distance to the nearest major road and other public amenities, tax rates, and expenditure per student standardized test scores (a proxy to school quality).

3.2. Binary Logit Model

The final generalized binary logit model took the form:

$$Y = f(I, P, \text{CLAPA}, \text{DCBD}, \text{NI}) \qquad (1)$$

where $Y = 1$ for vacant parcels and 0 for residential parcels; I = income variables; P = population variables; CLAPA = current land assessment per acre; DCBD = distance to central business district; NI = neighborhood index.

Income and population variables were expected to have negative coefficients (i.e. decrease the likelihood of a vacant parcel remaining vacant) assuming housing demand is a function of income and more residential

Table 1. Independent Variable Statistics.

	Entire	Lower	Middle	Upper
Per capita income (US $)				
Mean	30,299	30,263	28,614	31,393
Median	29,128	28,178	28,609	29,272
Standard deviation	8,052.6125	7,394.4468	3,629.3737	10,287.1312
Range	75,730	75,730	11,603	41,484
Minimum	17,700	17,700	21,574	18,747
Maximum	93,430	93,430	33,177	60,231
% Change per capita income				
Mean	0.5181	0.5189	0.5076	0.5179
Median	0.5078	0.5018	0.5060	0.5078
Standard deviation	0.1152	0.1175	0.0648	0.0926
Range	1.6850	0.7820	0.2982	1.4801
Minimum	0.1563	0.1563	0.3239	0.3611
Maximum	1.8412	0.9382	0.6220	1.8412
% Change median household income				
Mean	0.1209	0.1464	0.1200	0.0792
Median	0.1461	0.1624	0.1820	0.0801
Standard deviation	0.0927	0.0831	0.1134	0.0592
Range	0.5716	0.5064	0.3942	0.4645
Minimum	−0.1873	−0.1221	−0.1873	−0.1853
Maximum	0.3843	0.3843	0.2069	0.2792
% Change population				
Mean	0.0410	−0.0384	0.0459	0.1196
Median	−0.0204	−0.0410	−0.0219	0.1602
Standard deviation	0.1460	0.0539	0.1931	0.1235
Range	0.6231	0.2971	0.6163	0.4869
Minimum	−0.1154	−0.1154	−0.1086	−0.0204
Maximum	0.5077	0.1817	0.5077	0.4665
Population density (people/sq. mi.)				
Mean	780	1,401	643	149
Median	455	1,215	455	96
Standard deviation	867.8736	982.8698	353.9139	214.1233
Range	5,619	5,447	1,234	2,882
Minimum	51	223	226	51
Maximum	5,670	5,670	1,459	2,933
Current land assessment per acre (US $)				
Mean	45,943	68,756	54,347	14,433
Median	36,145	61,905	54,038	8,817
Standard deviation	47,120.4783	56,068.5954	31,804.3563	16,592.1706
Range	1,999,982	1,999,870	499,900	271,982
Minimum	18	130	100	18
Maximum	2,000,000	2,000,000	500,000	272,000

Table 1. (*Continued*)

	Entire	Lower	Middle	Upper
Distance to central business district (mi.)				
Mean	2.2698	1.1459	1.7746	3.8727
Median	1.6627	1.1402	1.8382	4.0921
Standard deviation	1.8227	0.5314	0.7205	2.0962
Range	8.4197	3.1057	3.6055	8.4112
Minimum	0.0116	0.0116	0.0355	0.0201
Maximum	8.4313	3.1173	3.6410	8.4313
Neighborhood index				
Mean	0.8015	0.8672	0.8472	0.6973
Median	0.8597	0.9266	0.8691	0.6916
Standard deviation	0.1396	0.1658	0.0598	0.0551
Range	1.0000	1.0000	0.5277	0.2713
Minimum	0.0000	0.0000	0.3846	0.5787
Maximum	1.0000	1.0000	0.9123	0.8500

parcels would likely follow from increases in population. Current land assessment per acre was expected to have a positive coefficient. The sign of the coefficient for distance to the central business district could go either way, with a positive coefficient expected if amenities contained within the central business district are now farther away, and a negative coefficient expected if people have a revealed preference for larger lot sizes, privacy, and other personal amenities. For the neighborhood index, a negative coefficient was expected assuming the more land devoted to residential use, the more likely the remainder of vacant parcels in a neighborhood area would become residential.

The results of seven different specifications of a binary logit model for the entire Wappinger Creek Watershed region are presented in Table 2. For the seven models, the signs are as anticipated, with the exception of the positive coefficients for population density and median household income in Model 7. Four separate goodness-of-fit tests were explored, including the McFadden R^2, Estrella R^2, R_p^2, and sums of fractions correctly predicted. Both the pseudo-R^2 tests (McFadden and Estrella) for the models have values over 0.20, an acceptable value for logit estimates. The pseudo-R^2 tests compare favorably to the 0.223 pseudo-R^2 value achieved by McMillen (1989) in his urban fringe study. R_p^2 as the percentage of correctly predicted parcels as compared against current conditions ranges from 0.7769 to 0.8306, also decent values. The sums of the fractions are the percentage of correctly

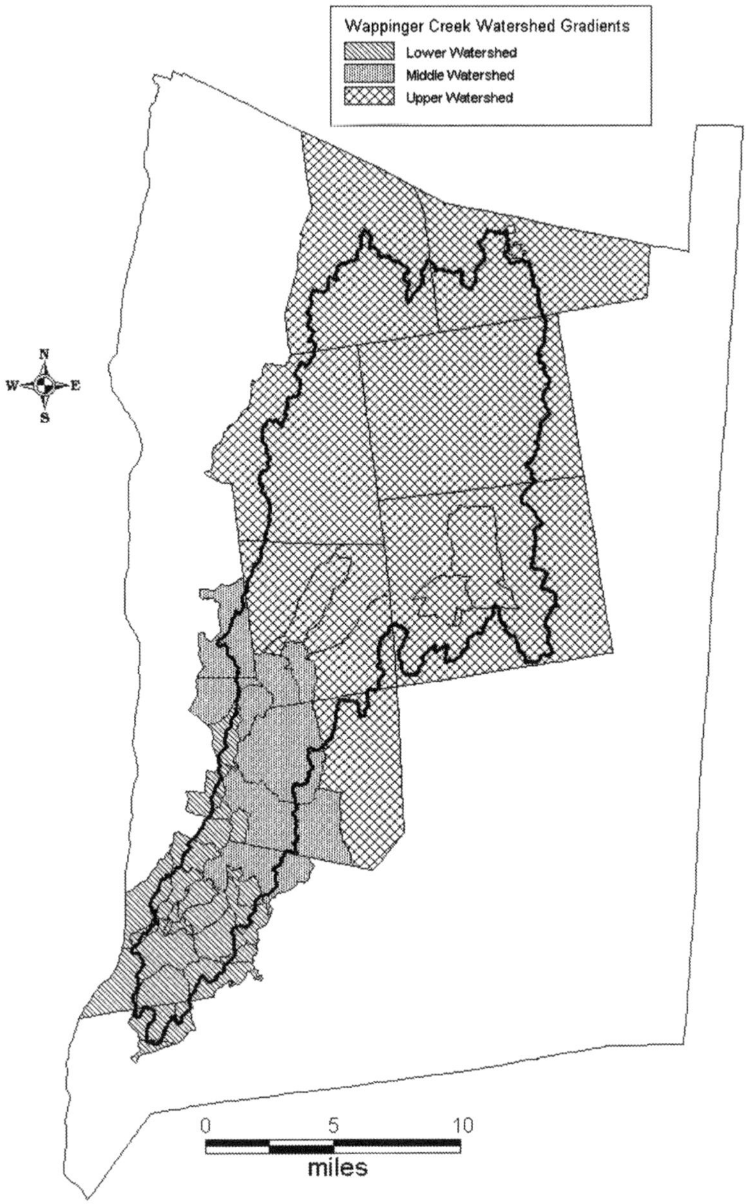

Fig. 2. Wappinger Creek Watershed Gradients. *Source:* Polimeni (2005, with permission from the Clute Institute for Academic Research).

Table 2. Binary Logit Regression Models Tested for Entire Watershed.

Independent Variable	Model						
	(i)	(ii)	(iii)	(iv)	(v)	(vi)	(vii)
Constant	3.7485 (27.93)	4.0062 (24.38)	3.2568 (27.01)	4.095 (22.84)	3.551 (27.92)	4.0849 (29.34)	4.8489 (26.73)
Per capita income (2000)	−0.00001763 (−8.25) [−0.00000147]					−0.00002392 (−11.17) [−0.000002]	−0.000021 (−7.24) [−0.0000017]
% Change in per capita income (1990–2000)		−1.1191 (−6.88) [−0.0946]		−1.1087 (−5.99) [−0.0958]			−1.1367 (−6.29) [−0.0921]
% Change in Median HH income (1990–2000)			−0.3322 (−1.62) [−0.0284]		−1.2882 (−5.85) [−0.1111]		0.755102 (2.45) [0.061187]
Population density (2000)	0.00048111 (14.15) [0.00004]	0.00056908 (16.49) [0.000048]	0.00051132 (15.30) [0.0000437]				0.000478 (13.07) [0.0000387]
% Change in population (1990–2000)				−1.7363 (−11.90) [−0.14998]	−1.8257 (−12.16) [−0.15748]	−1.7088 (−12.05) [−0.1422]	−1.415058 (9.15) [−0.11466]
Current land assessment per acre (2000)	−0.00004925 (−43.04) [−0.0000041]	−0.00004868 (−42.93) [−0.0000041]	−0.00004766 (−42.85) [−0.0000041]	−0.00004672 (−43.14) [−0.000004]	−0.00004657 (−43.05) [−0.000004]	−0.00004897 (−43.34) [−0.000004]	−0.000053 (43.94) [−0.000004295]

Distance to central business district (2000)	−0.09994	−0.0962	−0.09414	−0.15939	−0.16227	−0.15799	−0.114421
	(−9.56)	(−9.24)	(−8.96)	(−15.78)	(−15.90)	(−15.76)	(−10.47)
	[−0.0083245]	[−0.0081]	[−0.0081]	[−0.01377]	[−0.013997]	[−0.01315]	[−0.00927]
Neighborhood index (2000)	−4.2335	−4.6038	−4.3328	−3.9793	−3.8215	−3.7127	−4.568608
	(−28.33)	(−29.90)	(−29.17)	(−27.30)	(−27.04)	(−26.26)	(−28.62)
	[−0.3526]	[−0.3893]	[−0.3705]	[−0.3437]	[−0.3296]	[−0.30899]	[−0.3702]
McFadden R^2	0.2368	0.2360	0.2339	0.2318	0.2314	0.2356	0.2433
Estrella R^2	0.2361	0.2354	0.2332	0.2311	0.2307	0.2349	0.2426
R_p^2	0.7769	0.8306	0.8298	0.8248	0.8281	0.8302	0.6951
Sums of fractions correctly predicted	1.0320	1.2012	1.1976	1.1972	1.2020	1.2175	1.0419
Number of vacant parcels	4,671	4,671	4,671	4,671	4,671	4,671	4,671
Number of residential parcels	18,860	18,860	18,860	18,860	18,860	18,860	18,860
Number of observations:	23,531	23,531	23,531	23,531	23,531	23,531	23,531

Note: Z-statistics (logit uses z instead of t) are reported (in parentheses) below coefficient estimates.
Marginal effects are reported [in brackets] below *Z*-statistics.
Significance tests at the 95% level.
Vacant parcels are those parcels coded as vacant plus agricultural parcels and private forestland parcels.
Vacant parcels are coded as a 1 in the dependent variable and residential parcels are coded as a 0.
R_p^2 is the percentage of total parcels correctly predicted.
Sums of fractions correctly predicted is the sum of the vacant parcels correctly predicted plus the sum of residential parcels correctly predicted.

predicted residential plus the percentage of correctly predicted vacant parcels, and ranges from 1.032 to 1.2175. As Kennedy (1998) states the sum of the fractions should exceed one if the prediction method is worthwhile.

To explore the consistency of the model specifications at the sub-watershed scale, each binary logit model was also estimated separately for each sub-watershed. The results of the lower and middle watershed models were consistent when compared to the entire watershed results, with some insignificance in coefficient estimates turning up at this smaller scale. Overall, the models predicted well in the lower and middle watershed levels, but had inconsistencies in the upper watershed estimates, likely due to the larger and more dispersed parcels in the rural part of the watershed. In addition, the use of dummy variables for the three regions were added to the entire watershed models, but without any significant improvement meriting there inclusion in the model ultimately used for simulation. See Polimeni (2002) and Polimeni and Polimeni (2007) for more detailed analysis and discussion.

3.3. Spatial Autocorrelation

Interdependence of spatial data is commonplace and problematic in location choice analysis (Irwin & Bockstael, 2002). Spatial dependence is most likely positive because factors such as the neighborhood index and distance to the central business district will exhibit positive spatial correlation. As a result, parameter estimates are likely biased in a positive direction, bounding them from above. To test and correct for spatial dependence, a spatial probit model was employed following the work of LeSage (1997, 2000) and LeSage and Smith (2004).

The spatial corrections for the entire watershed model were consistent with the results of the binary logit model. The signs on the coefficients remained the same and the variables remained significant. Results were less than satisfactory at the sub-watershed scale. Most problematic was the use of differing geography scales in the data, with some data collected at the tax parcel level and others at the census block. In the final analysis, the results of the uncorrected binary logit model were considered reliable due to strong goodness-of-fit tests, expected signs, coefficient significance, and accurate prediction of current land-use patterns in repeated Monte Carlo experiments.

3.4. Development Probabilities

Model 4 was chosen for an analysis of development probabilities here, and the scenario analyses described in the section. Vacant parcels were ranked by their probability of residential development, with the top 10%, or 450, chosen for further analysis. Of these parcels with the highest residential development probabilities, 219 were in the lower gradient, 196 in the middle gradient, and 35 in the upper gradient. Only one parcel was classified as agricultural, and 12 border the stream system of the Wappinger Creek. When town-level zoning layers were overlaid, the minimum lot size requirements for these parcels included 247 with lot size requirements of less than one-half acre, 182 in the half-acre to 2 acre range, and 21 with more than two acres required. Average acreage was 0.87 acres, ranging from 0.01 acres to 21.54 acres. In comparison, the mean acreage for all parcels in the watershed was 4.81 acres, ranging from 0.01 acres to 1,953 acres. The mean acreage for just the vacant parcels was 7.5 acres, ranging from 0.01 acres to 697.1 acres. The average land assessment per acre for the 450 parcels was $67,553, nearly identical to the average for all parcels in the lower gradient. The averages of the other independent variables for the 450 parcels were close to the averages of the lower and middle gradients.

4. SCENARIO ANALYSIS

Scenario analysis of future residential location began with an estimate of buildable land within each tax parcel. There were 23,531 parcels in the model with 4,507 vacant, agricultural, or private forest parcels potentially available for new residential development. Maps of wetlands, hydric soils, and slopes of greater than 15% were overlaid on the vacant parcels in order to adjust the total acreage to only buildable land. Of the remaining acreage within each parcel, three different availability scenarios of 100%, 80%, and 50% were applied to simulate the amount of land available for residential construction. The 80% and 50% estimates were suggested by the Dutchess County Environmental Management Council to represent the land available for development after accounting for roads, set-backs, lawns, and other development necessities and amenities.

After purging the database of parcels without buildable land, town-level zoning maps were next overlaid and minimum lot size requirements were divided into each of the available land acreages to determine the maximum number of new residential structures per parcel. For example, a 20 acre parcel with 3.9% of its land as wetlands, steep slopes, or hydric soils has approximately 19.5 acres that could be developed. Multiplied by 80% or 50% scenarios yields only 15.6 and 9.75 acres, respectively, available for development. Assuming a minimum lot size requirement of 5 acres, the maximum number of new residences under the 100%, 80%, and 50% land availability scenarios would be 4, 3, and 2 houses, respectively.

With these restrictions in place, Model 4 of the binary logit regressions was used to estimate development probabilities and undergo a Monte-Carlo simulation for particular trends of independent variables or other model restrictions. The procedure compares development probabilities against random numbers drawn from a uniform distribution between zero and one. If a random number was greater than the development probability of a particular parcel, then that vacant parcel is assumed to switch to residential land use during the simulation period, delineated as a black parcel on map output for a particular simulation. For each simulation, the number of parcels predicted for development and the maximum number of potential homes were calculated for each of the three available land restrictions (100%, 80%, and 50%). One hundred simulations were run for each scenario, with the average reported. In the following sections, unless otherwise noted, Model 4 and the 80% land availability were used for each scenario.

A status quo scenario was developed from which to compare economic, social, and environmental trends and policy scenarios. Dutchess County experienced tremendous growth in the decade of the 1990s, with a 53% increase in per capita income and an 8% increase in population. The status quo scenario projects these same trends from 2000 to 2010. This scenario, highlighted for a particular run in Fig. 3, predicts an average of 1,132 new residential parcels, potentially accommodating an average of 10,370 new houses. Development favored the upper watershed with 677 parcels developed, with an average acreage of 18.39. Development in the middle and lower gradients was more evenly distributed, with the lower watershed containing 228 new residential parcels at an average size of 3.3 acres, and the middle watershed containing 215 new residential parcels at an average size of 8.61 acres.

4.1. Economic Change Scenarios

To illustrate how increasing land values may affect development, low (11.35%), medium (31.35%), and high (51.35%) growth rates were simulated. Averaged across 100 Monte Carlo runs, new parcels developed were 922, 1,011, and 1,010 for low, medium, and high assessment value growth rates, with a maximum of 8,900, 8,915, and 8,946 houses built. The magnitude of the increase in developed parcels is small, and the change from the status quo scenario is negligible. However, the growth in income and

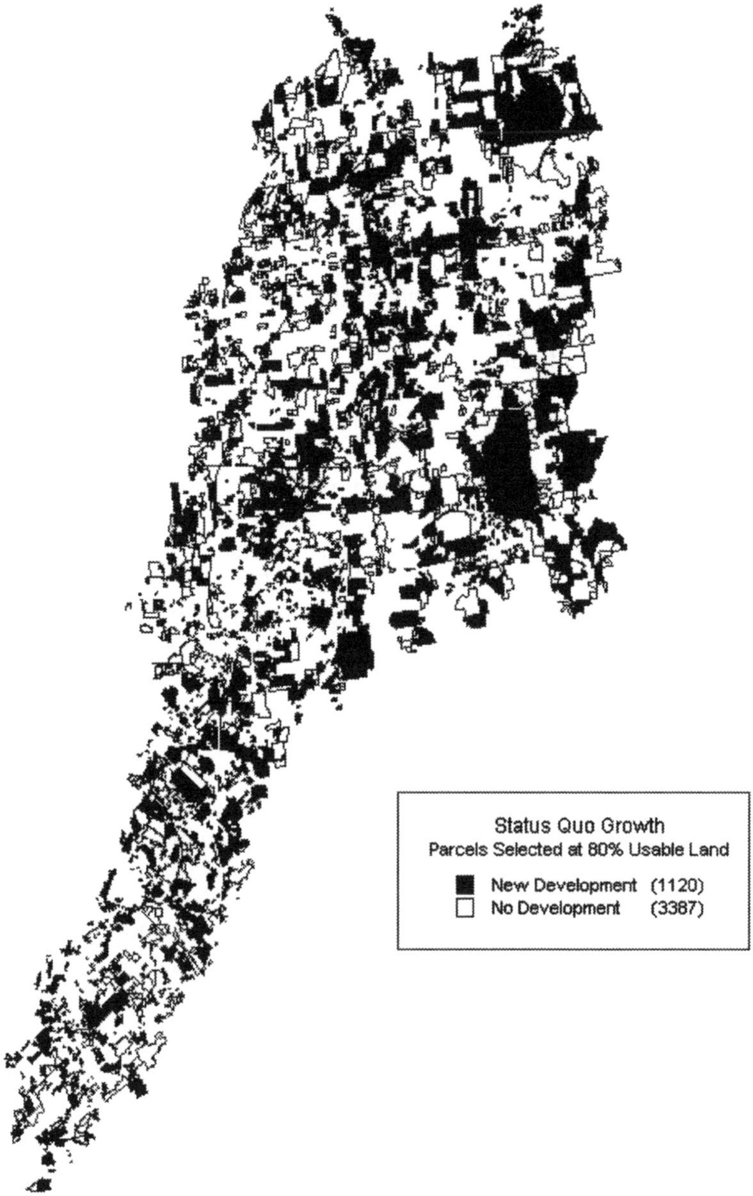

Fig. 3. Status Quo Growth.

population is largely accommodated by a shift from higher-assessed properties to lower-assessed properties, resulting in little change overall.

A second economic scenario explored the development impact of slower (20%) or faster (70%) than status quo growth (53%) in per capita income. Compared to the status quo, the average number of new residential parcels decreased by just 63 parcels (5.8% less) in the low income growth scenario, resulting in 674 fewer potential houses. The high growth scenario over one hundred simulations averaged just four more developed parcels, with five more potential houses. These small changes can be explained by the relatively small marginal effect of income reported in Table 2. When median household income is used instead (as with Model 2), the marginal effect is even less than per capita income.

4.2. Societal Change Scenarios

To examine how land use changes with population growth, low (2%), status quo (8%), and high (14%) growth rates were explored. An average of 982, 987, and 992 new residential parcels were simulated, respectively, with potential new houses ranging from 8,791 to 8,902. This small increase from low to high growth is expected given the small marginal effect of population growth. Compared to models with population density, the population growth variable does have a stronger effect, perhaps accounting for the ability of density increases to be accommodated on fewer parcels.

A second societal scenario explored the influence of distance to the central business district on residential choice. As estimated, the sign for the coefficient on this distance variable was negative, indicating a higher probability of development with greater distance from the nearest central business district (an indication of a sprawl trend in the housing market). Figs. 4(a) and (b) compare simulations of the status quo versus flipping the sign on this one variable, assuming that development probability increases the closer a parcel is to a central business district. Over a hundred simulations, the average number of new residential parcels decreases from 987 in the status quo to 645, a 35% decline. The corresponding number of average potential new houses decreases from 8,791 to 5,961. This might represent a behavioral or demographic change in the housing market, with people drawn more toward town centers, or perhaps the impact of a concerted policy effort to discourage rural housing development or, conversely, encourage urban in-fill.

Scenarios were also explored at the town-level. For instance, Pine Plains in the northern, rural part of the watershed is the only community in Dutchess County without zoning laws. When minimum lot sizes of one-half

(a) (b)

Fig. 4. (a) Development with the Standard Sign on Distance to Central Business District; (b) Development with the Sign Flipped On Distance to the Central Business District.

acre were overlaid on Pine Plains, a dramatic decrease in the number of parcels simulated for development occured. New residential parcels dropped from 27 to 10, with the number of potential new houses dropping from 873 to 82. This can be a powerful illustration of the impact of zoning laws at the town scale.

A final societal scenario examined the impact of focusing new development in urban areas of the watershed. For example, the scenario illustrated by a run in Fig. 5(a) resulted from changing the minimum lot size restrictions in the urban gradient to 0.01 acres and increasing lot size restrictions in the remainder of the watershed to 10 acres. The 0.01 restriction is the smallest for any of the vacant parcels. Under this case, an average of 774 parcels were marked for development, creating a possible 25,046 new

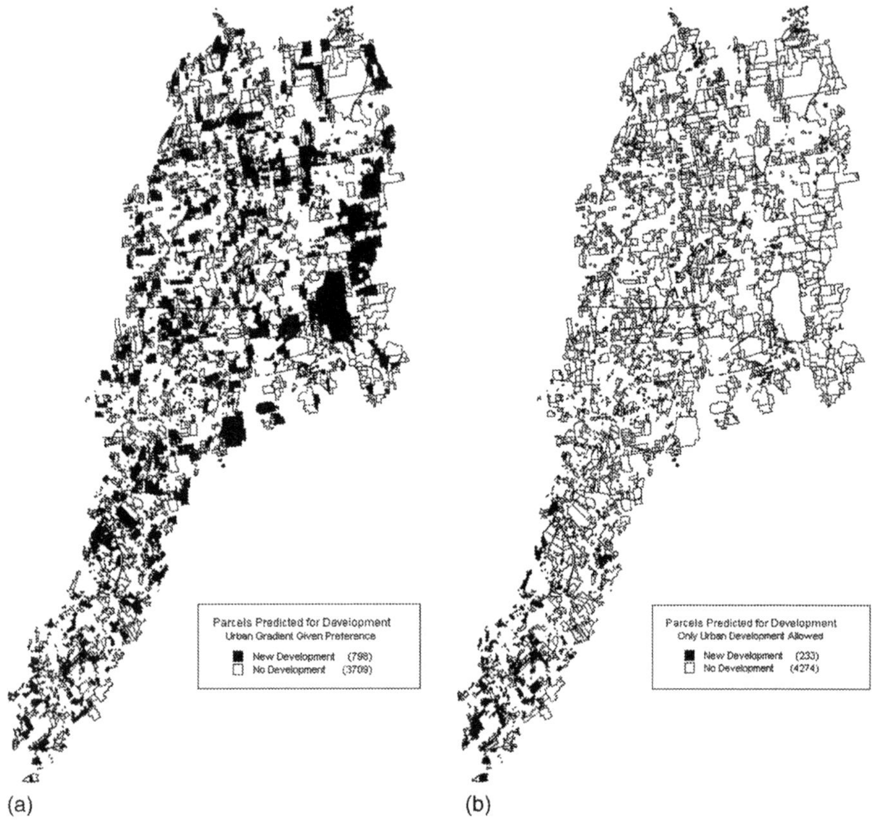

Fig. 5. (a) Urban Gradient Given Preference for Development; (b) Only Urban Development Allowed.

residences. The increase in potential new residences is due to the decrease in lot size requirements and would most likely represent apartment complexes. Fig. 5(b) illustrates the effect if development is restricted only to the urban gradient with current lot size restrictions in place. This is perhaps an unrealistic scenario, but demonstrates one extreme where an average of 230 parcels is predicted for development with a potential 2,251 new residences.

4.3. Land-Use Constraint Scenarios

One of the main concerns of Dutchess County planners and citizens alike is the impact of rapid residential growth on agricultural land and ultimately on the health of the watershed, as outlined in Chapter 5 of this book. To explore the potential impact of possible policy instruments on redirecting development away from farmland and sensitive riparian areas of the watershed, two scenarios were developed.

Fig. 6(a) illustrates the impact of restricting development on agricultural parcels greater than 5 acres. Under status quo growth conditions, this scenario results in an average of 971 parcels predicted for development with a potential for 10,363 new residences. Resulting development shifts from larger sized agricultural parcels to smaller and medium-sized agricultural parcels, moving from the upper gradient to the middle and lower gradients. As with the other scenarios described above, growth is still accommodated, but through more intensive land use closer to central business districts. Even

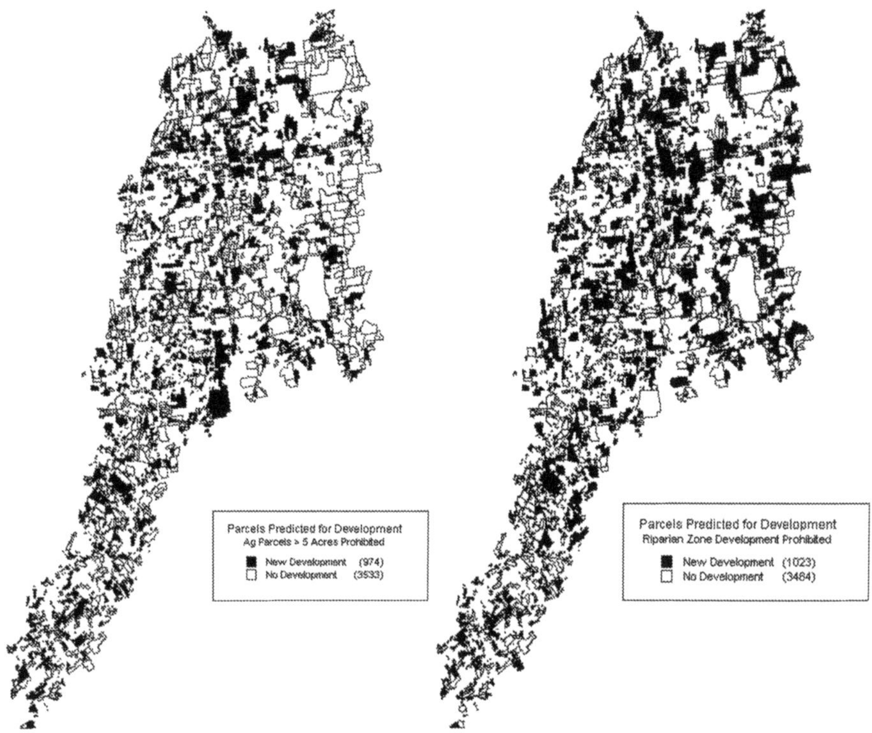

(a) (b)

Fig. 6. (a) Development of Agricultural Parcels of More Than 5 Acres Prohibited; (b) Riparian Zone Development Prohibited.

if development on agricultural land is prohibited all together (preserving over 24,000 acres in current use), the model predicts an average of 920 parcels changed to residential use with a potential of 6,170 new residences.

For watershed protection, the use of buffered riparian zones around tributaries has been discussed extensively in watershed communities. If new development is restricted along waterways then trees, grasslands, and other vegetation can filter impacts from surrounding development. Fig. 6(b) illustrates a simulation with restricted riparian development, resulting in an average of 1,022 parcels predicted for development with a maximum of 8,018 new houses under current zoning. This scenario results in a similar pattern of dispersed development as the status quo growth, but with few large parcels that currently border or encompass waterways taken out of development. Growth is largely accommodated, but with potentially large watershed health benefits, as discussed in Chapter 5.

5. CONCLUDING REMARKS

This chapter discussed the development of a binary logit model to estimate projections of residential development within the Wappinger Creek Watershed of Dutchess County, New York. The model builds on the evolving literature of residential location theory, in particular the newest explorations of urban sprawl, and allows for the exploration of a broad range of economic, social change, and land-use constraints. Development projections can then be the output of models integrated with regional economic models, as described in Chapter 8, and the input to models of ecosystem health, as described in Chapter 5 and in Erickson et al. (2004).

The phenomenon of urban sprawl is often not the future that watershed citizens would chose for their communities as part of a democratic process. However, without the ability to visualize the consequences of incremental development over the long-term, the more immediate, individual economic interests often win out over broader, collective social and environmental goals. Long-term economic health can also be compromised. The development of integrated ecological economic models has increasingly benefited from GIS technology and spatial databases, allowing for the visualization of long-term, landscape-level change.

In this case study of Dutchess County, simulation results reinforce the trend of decentralized development. Scenario analyses indicated that the thrust of new residential development would likely occur in the upper watershed gradient which is primarily agricultural and forest land. The parcels in this part of the watershed are larger with many houses possible under current zoning, and contain many of the rural amenities (away from central business districts) in demand by a growing and wealthier population. The middle watershed provides a transitional area for the ongoing suburbanization trend, also with plenty of large lots available to subdivide.

This work is best characterized as exploratory, with the potential to promote a structured dialogue on the many criteria that make up a desirable future, a process future explored for Dutchess County in Chapter 10 of this book. Limitations to the model include a lack of historical data, particularly given the relatively recent availability of GIS databases. For example, only four years (1996, 1998, 1999, and 2000) of the real property tax data were available. Time series data of land-use change is also complicated by the subdivision of larger parcels to small ownership units each year, and the variability of data quality due to the inconsistent collection of property data by tax assessors of each individual town. Further limitations include the closing of the model to major outside influences (such as vacation home demand from New York City) and the lack of commuting distance and cost data for more accurate central business district effects. The regression procedure itself is somewhat hampered by spatial autocorrelation and heterogeneity, resulting in reduced efficiency of parameter estimates. This is a problem for most spatial regression studies, an area of rapidly evolving research and solutions.

REFERENCES

Alig, R. J., & Healy, R. G. (1987). Urban built-up land area changes in the U.S.: An empirical investigation of determinants. *Land Economics*, *63*, 215–226.

Alonso, W. (1964). *Location and Land Use*. Cambridge: Harvard university Press.

Anas, A., Arnott, R. J., & Small, K. A. (2000). The panexponential monocentric model. *Journal of Urban Economics*, *47*, 165–179.

Aoyagi, M., & Okabe, A. (1993). Spatial competition of firms in a two-dimensional bounded market. *Regional Science and Urban Economics*, *23*, 259–289.

Bailey, T. J. (1999). Modeling the residential sub-market: Breaking the monocentric mould. *Urban Studies*, *7*, 1119–1135.

Beckmann, M. J. (1968). *Location theory*. New York: Random House.

Beckmann, M. J. (1999). *Lectures on location theory*. Berlin: Springer-Verlag.

Block, D., & DuPuis, E. M. (2001). Making the country work for the city: Von Thünen's ideas in geography, agricultural economics and the sociology of agriculture. *American Journal of Economics and Sociology*, *60*, 79–98.

Bogart, W. T., & Ferry, W. C. (1999). Employment centers in greater Cleveland: Evidence of evolution in a formerly monocentric city. *Urban Studies*, *36*, 2099–2110.

Bos, H. C. (1965). *Spatial dispersion of economic activity*. Rotterdam: University Press.

Buisson, M-A., Mignot, D., & Aguilera-Belanger, A. (2001). Metropolization and intra-urban poles: The case of Lyon. *Revue D'Economie Regionale et Urbaine*, *2*, 271–296.
Chamberlin, E. H. (1953). The product as an economic variable. *The Quarterly Journal of Economics*, *1*, 1–29.
Champion, A. G. (2001). A changing demographic regime and evolving polycentric urban regions: Consequences for the size, composition and distribution of city populations. *Urban Studies*, *38*, 657–677.
Christaller, W. (1933). In: Carlisle W. Baskin (Trans.), *Central places in southern Germany*, 1966. Englewood Cliffs: Prentice-Hall.
Collins, R., & Sherstyuk, K. (2000). Spatial competition with three firms: An experimental study. *Economic Inquiry*, *38*, 73–94.
de Vries, J. J., Nijkamp, P., & Rietveld, P. (2001). Alonso's theory of movements: Developments in spatial interaction modeling. *Journal of Geographical Systems*, *3*, 233–256.
Dixit, A. K., & Stiglitz, J. E. (1977). Monopolistic competition and optimum product diversity. *American Economic Review*, *67*, 297–308.
Erickson, J. D., Limburg, K., Gowdy, J., Stainbrook, K., Nowosielski, A., Hermans, C., & Polimeni, J. (2004). Anticipating change in the Hudson river watershed: An ecological economic model for integrated scenario analysis. In: R. Bruins & M. Heberling (Eds), *Economics and ecological risk assessment: Applications to watershed management* (pp. 341–370). Boca Raton, FL: CRC Press.
Fan, W., Treyz, F., & Treyz, G. (2000). An evolutionary new economic geography model. *Journal of Regional Science*, *40*(4), 671–695.
Hardie, I., Parks, P., & Gottleib, P. (2000). Responsiveness of rural and urban land uses to land rent determinants in the U.S. south. *Land Economics*, *76*, 659–673.
Henderson, V., & Mitra, A. (1996). The new urban landscape: Developers and edge cities. *Regional Science and Urban Economics*, *26*, 613–643.
Hotelling, H. (1929). Stability in competition. *The Economic Journal*, *39*, 41–57.
Irwin, E. G., & Bockstael, N. E. (2002). Interacting agents, spatial externalities and the evolution of residential land use patterns. *Journal of Economic Geography*, *2*, 31–54.
Isard, W. (1969). *General theory: Social, political, economic, and regional with particular reference to decision-making analysis*. Cambridge: MIT Press.
Ishikawa, T., & Toda, M. (2000). Some economic extensions of central-place theory involving profit maximisation. *Urban Studies*, *37*, 481–496.
Kahn, M. (2000). The environmental impact of suburbanization. *Journal of Policy Analysis and Management*, *19*(4), 569–586.
Kennedy, P. (1998). *A guide to econometrics*. Cambridge: The MIT Press.
Knapp, G., Ding, C., & Hopkins, L. D. (2001). Managing urban growth for efficient use of public infrastructure: Toward a theory of concurrency. *International Regional Science Review*, *24*, 328–344.
Krugman, P. (1991). Increasing returns and economic geography. *Journal of Political Economy*, *99*, 483–499.
Lamb, R. (1983). The extent and form of exurban sprawl. *Growth and Change*, *14*, 40–47.
Launhardt, W. (1885). *Mathematische Begründung der Volkswirtschaftslehre*. Leipzig: B. G. Teubner.
LeSage, J. (1997). Bayesian estimation of spatial autoregresssive models. *International Regional Science Review*, *20*, 113–129.

LeSage, J. (2000). Bayesian estimation of limited dependent variable spatial autoregressive models. *Geographical Analysis*, *32*, 19–35.

LeSage, J. P., & Smith, T. E. (2004). A Bayesian probit model with spatial dependencies. In: J. P. LeSage, & R. K. Pace (Eds), *Advances in econometrics: Spatial and spatiotemporal econometrics* (Vol. 18, pp. 127–160). Oxford, UK: Elsevier.

Losch, A. (1944). The economics of location. In: W. H. Woglom (Trans., 1954). Yale Connecticut: University Press.

McMillen, D. P. (1989). An empirical model of urban fringe land use. *Land Economics*, *65*, 138–145.

McMillen, D.P. (2001). Polycentric urban structure: The case of Milwaukee. *Economic Perspectives*, (2Q), 15–27.

McMillen, D. P., & Singell, L. D., Jr. (1992). Work location, residence location, and the intraurban wage gradient. *Journal of Urban Economics*, *32*, 195–213.

Mills, D. E. (1981). Growth, speculation and sprawl in a monocentric city. *Journal of Urban Economics*, *10*, 201–226.

Moses, L. (1958). Location and the theory of production. *Quarterly Journal of Economics*, *72*, 259–272.

Muth, R. F. (1969). *Cities and housing: The spatial pattern of urban residential land use*. Chicago: University of Chicago Press.

Muth, R. F. (1985). Models of land-use, housing, and rent: An evaluation. *Journal of Regional Science*, *25*, 593–606.

Norman, G., & Pepall, L. (2000). Spatial competition and location with mergers and product licensing. *Urban Studies*, *37*, 451–471.

Pendall, R. (1999). Do land-use controls cause sprawl? *Environment and Planning B – Planning and Design*, *26*(4), 555–571.

Pinto, J. V. (1997). *English economists' descriptive accounts of location theory*. College of Business Administration, Northern Arizona University, Working Paper Series 97-03.

Polimeni, J. M. (2002). *Dynamic spatial simulation of residential development in the Hudson River Valley, New York State*. Dissertation, Department of Economics, Rensselaer Polytechnic Institute, Troy, NY, USA.

Polimeni, J. M. (2005). Simulating agricultural conversion to residential use in the Hudson river valley: Scenario analysis and case studies. *Agriculture and Human Values*, *22*, 377–393.

Polimeni, J. M., & Polimeni, R. I. (2007). Urban spawl – scenario analysis for public policy makers. *Journal of Business and Economics Research*, *5*(1), 41–58.

Ricks, B. (1970). New town development and the theory of location. *Land Economics*, *46*, 5–11.

Rothschild, R. (2000). Merger under spatial competition. *Urban Studies*, *37*, 443–450.

Sasaki, K., & Mun, S.-I. (1996). A dynamic analysis of multiple-center formation in a city. *Journal of Urban Economics*, *40*, 257–278.

Shieh, Y.-N. (1987). A model of residential location in a two dimension multi-center city. *Papers of the Regional Science Association*, *63*, 71–79.

Smithies, A. (1941). Optimum location in spatial competition. *The Journal of Political Economy*, *49*, 423–439.

South, R., & Boots, B. (1999). Relaxing the nearest centre assumption in central place theory. *Papers in Regional Science*, *78*, 157–177.

Tabuchi, T. (1999). Pricing policy in spatial competition. *Regional Science and Urban Economics*, *29*, 617–631.

von Thünen, J. H. (1966). *The Isolated State*. Carla Wartenberg (trans.), Peter Hall (Ed.). New York: Pergamon Press.
Walker, R. (2001). Urban sprawl and natural areas encroachment: Linking land cover change and economic development in the Florida Everglades. *Ecological Economics*, *37*, 357–369.
Wang, F. (1999). Modeling a central place system with interurban transport costs and complex rural hinterlands. *Regional Science and Urban Economics*, *29*, 381–409.
Weber, A. (1909). Theory of the location of industries. In: C. J. Friedrich (Trans., 1971). New York: Russell & Russell.

PART IV: EVALUATION TOOLS AND PARTICIPATORY WATERSHED MANAGEMENT

MULTICRITERIA DECISION ANALYSIS: OVERVIEW AND IMPLICATIONS FOR ENVIRONMENTAL DECISION MAKING

Caroline M. Hermans and Jon D. Erickson

ABSTRACT

Environmental decision making involving multiple stakeholders can benefit from the use of a formal process to structure stakeholder interactions, leading to more successful outcomes than traditional discursive decision processes. There are many tools available to handle complex decision making. Here we illustrate the use of a multicriteria decision analysis (MCDA) outranking tool (PROMETHEE) to facilitate decision making at the watershed scale, involving multiple stakeholders, multiple criteria, and multiple objectives. We compare various MCDA methods and their theoretical underpinnings, examining methods that most realistically model complex decision problems in ways that are understandable and transparent to stakeholders.

Ecological Economics of Sustainable Watershed Management
Advances in the Economics of Environmental Resources, Volume 7, 213–228

ISSN: 1569-3740/doi:10.1016/S1569-3740(07)07010-1

1. INTRODUCTION

Local decisions are increasingly made on a regional scale in the United States. Emphasis is on regional economies, regional environmental problems, and regional partnerships and collaborations (Sellers, 2002). As local interests, such as environmental issues, expand spatially, they necessarily become regional. Local regulations are no longer effective to deal with the now regional issue and thus regional agreements are forged. This deficiency results in collaborations, partnerships, councils, etc. that operate on an intermunicipal, translocal, or regional level (Lubell, Schneider, Scholz, & Mete, 2002). Now decisions need to be reached that involve not one, but many localities. Additionally, more citizens are becoming involved in local political processes through the creation of citizen groups (Thomson, 2001). The decision-making process is now more complex, involving multiple communities and increased public participation. New ways of arriving at regional decisions are needed as formal governance/control is limited and local governance is insufficient.

Here we look at regional governance at the watershed scale. *Watershed* has become an increasingly common way to define spatial relationships and boundaries, rather than political jurisdiction. In New Zealand, for example, political boundaries have been redrawn to coincide with watershed boundaries.

Watershed-based environmental management and planning decision problems, regardless of place, share several common characteristics. Multiple localities, with diverse local objectives, are involved in reaching a common regional decision. However, local objectives are often in conflict with each other. Often, there is a great deal of uncertainty surrounding the issue. This discrepancy between local objectives and overarching regional decisions needs to be addressed in the decision-making framework used. The challenge is to find a decision method that deals with these regional issues. Many valuation/decision frameworks commonly used (for example, cost–benefit analysis, CBA) result in decisions that have not accounted for the complexity that characterizes the reality of the situation. This often results in a decision that does not have the support of the localities and thus has poor implementation potential.

This article proposes the use of a multicriteria decision analysis (MCDA), although widely used throughout the European Union, has not been widely accepted in the United States. It is proposed that this tool is especially well suited for environmental management and planning at the watershed scale.

For the purposes of this discussion, a simplified environmental decision-making problem is used. It is based on integrated modeling and evaluation research from the Hudson river valley of New York state, summarized in Chapter 5 of this book and in Erickson et al. (2004). A watershed council comprising members from various municipalities in the watershed and various regional organizations and not-for-profit groups has been formed. Their objective is to develop a land management plan that will ensure future water quality and maintain the rural character of the watershed. Current pressures facing the watershed are a high degree of urbanization resulting in sprawl, a corresponding decline in the water quality of rivers and streams, loss of agricultural and rural lands to development, loss of traditional farming, a greater presence of industry, and a declining number of small businesses.

As in most states, local municipalities in New York have political autonomy, and land management power rests at the municipal, not county level. The towns make their own decisions regarding local zoning and development. Historically, there has not been a great deal of regional cooperation between local governments regarding land-use issues. Also, regional and county government has not had a strong role in addressing local development issues. An MCDA is being developed to assist the council in making intermunicipal decisions regarding future watershed development.

Decision makers have identified several alternatives for development: (1) status quo, to do nothing and let current development take its course; (2) to focus on improving water quality through strict business and residential zoning in environmentally sensitive areas adjacent to rivers and streams; (3) to focus on the preservation of agricultural lands and open space by limiting housing/business development in the north–central portion of the watershed; (4) to encourage housing and business development in core urban areas. (Other alternatives consist of a combination of the above options, but for the sake of simplicity, we consider each alternative individually.)

The council must now judge these alternatives against various environmental, social, and economic criteria. Criteria data are qualitative, quantitative, or spatial and are available for water and ecosystem quality, land-use patterns, and various socioeconomic drivers in the watershed. Data are not complete and a great deal of uncertainty exists due to the long development time horizon being considered. As will be discussed below, there are several ways to approach this complex decision problem. Use of this example throughout the text attempts to clarify the distinctions between decision methods.

MCDA is a relatively young family of decision tools, gaining popularity only in the last 20 years. A branch of operations research, MCDA has been primarily used in industrial, corporate, and medical settings. Examples include engineering design, quality assurance, production planning, transportation, strategic planning, medical planning, and proposal analysis (Vincke, 1992). While well established in the domain of private decision making, MCDA has more recently been employed to aid social decisions, including environmental and natural resource problems in areas such as forestry (Kangas, Kangas, Leskinen, & Pykalainen, 2001; Salminen, Hokkanen, & Lahdelma, 1998; Van Elegem, Embo, Muys, & Lust, 2002), water resources (Hyde, Maier, & Colby, 2004; Abu-Taleb & Mareschal, 1995; Joubert, Leiman, De Klerk, Katua, & Aggenbach, 1997; Martin, St. Onge, & Waaub, 1999; Prato, 2000), urban and transportation planning (Roy, 1985; Brand, Mattarelli, Moon, & Wolfler Calvo, 2002), energy policy and planning (Lootsma, 1990; Rozakis, Soldatos, Kallivroussis, & Nicolaou, 2001; Van Groenendaal, 2003; Mirasgedis & Diakoulaki, 1997), pollution (Salt & Dunsmore, 2000; Brigs, Kunsch, & Mareschal, 1990; Joerin, Golay, & Musy, 1998), ecosystem restoration (Qureshi & Harrison, 2001), and in the siting of facilities such as nuclear or thermal power plants and industrial services (Vaillancourt & Waaub, 2002).

Multicriteria decision problems involve making choices from among a number of alternatives in order to achieve or balance a number of objectives. An explicit consideration of multiple objectives or criteria is in contrast to decision frameworks that seek to maximize one objective or reach one optimal solution. Conventional economic tools such as CBA or cost-effective analysis (CEA) seek to reduce complexity to single dimensions, units, and value systems. MCDA, however, strives to *structure this complexity* and *find compromise solutions*.

There are many MCDA methodologies based on several key theoretical foundations. The single purpose of all methods is to be able to evaluate and choose from solutions to a problem based on multiple criteria or objectives. Some techniques rank options, some sort options, some identify a single optimal alternative, some provide an incomplete ranking, and others differentiate between acceptable and non-acceptable alternatives (Vincke, 1992; Iz & Gardiner, 1993; Salminen et al., 1998; Stewart, 1992; Triantaphyllou, 2000). MCDA can include quantitative and qualitative criteria, crisp and fuzzy variables, multiple preferences among decision makers, and a decision environment amenable to scenario analysis, shared learning, and consensus building.

The purpose of this article is to review the diversity of MCDA tools, particularly as they apply to environmental management. In order to benchmark MCDA to a more familiar decision tool in environmental management, the theoretical basis of CBA and MCDA are first compared. The primary schools within MCDA are discussed, and the outranking method and its applicability to resolving environmental problems and conflict are detailed. The example of PROMETHEE (preference ranking organization method of enrichment evaluation) is used to illustrate one particular method amendable to complex environmental management decision problems that are characterized by numerous decision makers, alternatives, and criteria, uncertainty and ignorance, and incommensurable criteria.

2. THEORETICAL BASIS FOR MULTICRITERIA AND SINGLE-CRITERION DECISION FRAMEWORKS

Most decision problems, such as the one described above, involve more than the optimization of a single objective (Guitouni & Martel, 1998). Environmental decision problems, in particular, typically involve multiple objectives, criteria, and decision makers. In addition, ignorance and uncertainty among decision makers is more often the case (especially when there is a high level of public participation), as well as the existence of a high level of imprecision and incompleteness of the data being used (Faber, Manstetten, & Proops, 1992). Yet the dominate tool for U.S. environmental policy analysis has been and remains the mono-criterion, uncertainty-free approach of CBA. CBA, however, is severely limited in situations where change is not marginal, uncertainty is high, criteria are qualitative, markets are imperfect, time horizons are lengthy, and public goods are involved (Joubert et al., 1997; Bouyssou et al., 2000; Prato, 2000; Messner, 2006). CBA reduces the complexity of possible solutions to the sole criterion of economic value, and views decision problems as merely issues of rational utility maximization (Gowdy & Erickson, 2005a).

CBA approaches the watershed development problem by breaking down the problem into costs and benefits to social welfare. To arrive at social costs and benefits, individual's preferences are aggregated into social values. Where no markets exist for "goods" such as beauty, rural character, solitude, sprawl, etc., willingness-to-pay or willingness-to-accept estimates are used (Erickson, 2000). For example, how much will you be willing to pay for a safe water supply might be asked to assign a market value to the non-marketed service of water purification by well-functioning watersheds. In this way, social welfare is measured only in terms of economic efficiency (Sugden & Williams, 1978). The objective of CBA is to choose an alternative that provides the greatest social welfare (utility) measured by the sole criteria of market exchange value (money). The alternative chosen would be a potential pareto improvement, implying that those that are made better off by the solution could compensate those who are made worse (van den Doel & van Velthoven, 1993). Assume, in the watershed example, that the water quality alternative provides the highest social welfare and that this alternative makes small business owners and developers worse off as they lose future monetary benefits due to strict zoning. However, the alternative provides high benefits to current watershed residents. Based on WTP surveys, residents give water quality a high monetary value. As long as the benefit to residents offsets the costs to developers and business owners the alternative provides positive net social welfare. The issue of the equity of the distribution of this welfare is not addressed in CBA.

CBA makes several strong assumptions in order to capture individuals' values as costs and benefits and to translate these into a measure of social welfare. These include that individuals seek to maximize their utility, possess complete knowledge, and have transitive preferences. However, human beings are not perfectly rational, and emotion plays a large role in the decision-making process. We are not able to perfectly articulate our preferences in all circumstances, our preferences are changeable and can be intransitive (Guitouni & Martel, 1998). This means that while we can prefer object A to object B, and object B to object C, we may also prefer object C to object A, under certain circumstances. The notion that decision makers are rational utility maximizers with social preferences captured by market behavior, the underlying assumption of CBA, is often too limiting and unrealistic (Gowdy & Erickson, 2005a, 2005b). For instance, contrary to strict economic assumptions on human behavior, results from behavioral economics indicate the existence of *endowment effects* (people place higher values on things they already possess), *hyperbolic discounting* (people discount the near future at a higher rate than the distant future), *loss aversion* (people are much more averse to taking a loss than to enjoying an equal gain), the *part–whole* problem (people consistently place higher values on the sum of individual components of an object of utility than on the whole thing itself) and many other "anomalies" in consumer choice theory (Gintis, 2000). Decision making also involves more than just "evaluation." The determination of decision makers, modeling of their preferences and objectives, and the development of alternatives are a crucial part of the decision process most often overlooked by CBA (Bouyssou et al., 2000).

Various MCDA techniques can be used to contend with these limitations of CBA. In MCDA, the impacts of a decision are not measured strictly monetarily. Possible solutions are ranked or rated based on multiple criteria that can be quantitative or qualitative. Several MCDA methods are designed to accommodate the existence of uncertainty and/or ignorance in the decision problem (Faber et al., 1992; Kangas et al., 2001). Certain MCDA processes can be adjusted as new and more precise information is obtained during the decision-making process, or as decision makers change and shape their preferences in response to new information (Wilson & Howarth, 2002). Additionally, many stakeholders or decision makers can be involved. To various degrees, MCDA methods preserve the uniqueness of each participant's value systems, points of view, and particular goals in reaching a solution. The solution to the decision problem can consist of many possible alternatives, and MCDA does not limit its search to only pareto-optimal solutions narrowly defined by economic efficiency.

To simplify the classification of MCDA methods, they are often divided into two broad groups: multi-attribute and multi-objective decision-making methods (Malczewski, 1999; Guitouni & Martel, 1998; Joubert et al., 1997). The goal of multi-objective methods, such as goal and compromise programming, is to compute an optimal solution in a continuous decision space with an infinite number of alternatives (Geldermann & Rentz, 2000). Many environmental planning decisions involve evaluating tradeoffs between discrete, feasible alternatives, rather than designing a single optimal, potentially unattainable solution. Therefore, this review concentrates on multi-attribute methods that do not involve an infinite number of alternatives, but where there are a finite number of pre-selected alternatives and the goal is to rank or rate the possible alternatives, not to arrive at an optimal solution.

Multi-attribute methods can be further divided into two schools of thought: American/Anglo-Saxon and European/French (Brans & Mareschal, 1989; Fishburn, 1991; Rogers & Bruen, 1998; Lootsma, 1990). Multi-attribute utility or value theory (MAUT or MAVT) and the analytical hierarchy process (AHP) are the main methods used in the United States. The French school developed the outranking methods of ELECTRE (elimination et choix traduisant la realité) and PROMETHEE and these are used primarily in Europe. Table 1 summarizes the key differences between the two schools of thought, in comparison to CBA. In the following sections, each school is summarized and put in the context of the watershed planning example presented above.

2.1. The American School of MAUT and AHP

Although AHP differs from MAUT and MAVT in some aspects, it shares many of the same theoretical foundations, and thus for the purposes of this review are grouped together. MAUT/MAVT, originating with Keeney and Raiffa's (1976) set of multicriteria decision-making procedures, includes many techniques such as, simple additive weighting (SAW), simple multi-attribute rating technique, weighted product method (SMART), technique for order preference by similarity to ideal solution (TOPSIS), weighted sum, and fuzzy weighted sum (Abi-Zeid, Belanger, Guitouni, Martel, & Jabeur, 1998; Guitouni & Martel, 1998; Yoe, 2002). AHP is a hierarchical decision process developed by Saaty in the 1980s (Saaty, 1980).

The assumptions of the American school are modeled after the tradition of CBA. Decision makers are assumed to be certain about their preferences in regard to the criteria and how they weigh the criteria. Preferences are transitive and not changeable, and the purpose of MCDA is to elicit these well-established preferences (Geldermann, Spengler, & Rentz, 2000). Furthermore, these known preferences can be boiled down to having one attribute: *utility*. The goal of the American methods is optimization. Additionally, these methods allow for compensation of an alternative's poor performance on a criterion by its good performance on another.

To illustrate, consider the AHP method applied to questions of development in a watershed. AHP requires each individual decision maker to compare each pair of criteria on a comparison scale in order to assign a preference index. For example, stakeholders are asked, "In terms of increased economic growth from local businesses (criterion A) and increased preservation of agricultural land (criterion B); is economic growth equally important, moderately more important, strongly more important, very strongly more important, or overwhelmingly more important than agricultural land preservation." This assumes that decision makers have complete understanding of the decision problem and the criteria and can articulate their preferences. They are also asked to compare dissimilar criteria such as water quality and social welfare measures. The preference scale can be inconsistent, violating the axiom of transitivity. Additionally, decision makers are asked to define their criteria preferences without the context of the alternatives.

The next steps are the creation of a pairwise comparison matrix and the establishment of criteria weights. The same process is repeated for each pair of alternatives against each criterion to come up with performance scores. Stakeholders are asked, "In terms of agricultural land (criterion B), how important is Alternative 2 (focus on improving water quality through strict business and residential zoning in environmentally sensitive areas adjacent to rivers and streams) relative to Alternative 4 (focus on encouraging housing and business development in core urban areas)." Again, this assumes that decision makers have known preferences that they can express with certainty. A matrix is created for each criterion and each pair of alternatives. If there are four alternatives and six criteria, six 4×6 matrices are processed. This process becomes very cumbersome with more alternatives and criteria (Mahmoud & Garcia, 2000). Alternatives are then evaluated using a simple linear additive model or a multiplicative model. This will result in a weighted score between 0 and 1 for each alternative. The alternative with the highest score is the preferred alternative. A poor performance on a criterion, such as water quality, can be compensated by a good performance on other criterion, such as potential for economic growth. When making complex environmental decisions, the compensatory nature of AHP (and MAUT) belies the importance of criteria on which alternatives may perform badly.

To illustrate MAUT, the SAW method is used. SAW is one of the best known and widely used MAUT methods. In the watershed case, a score for each alternative is calculated by multiplying the alternative's performance on a criterion by the weight assigned to the criterion. The performance on a criterion is expressed as a rating or utility, regardless of whether the criterion is qualitative or quantitative. These scores are summed for each alternative over all the criteria. The alternative with the highest score, or utility, is the optimal solution. It is important to note that, as with AHP, the assigned weights must be accepted by all decision makers, unlike in the outranking methods. This is not realistic, as one of the key components of environmental decision making is the fact that decision makers and stakeholders value (and thus weight) criteria differently. In MAUT methods, weights are often derived by averaging individual ratings for each criterion. This results in a loss of information about how individual council members feel about the criteria and alternatives.

2.2. The European School of Outranking

There are two primary outranking methods, ELECTRE (Roy, 1985) and PROMETHEE (Brans & Mareschal, 1984), both developed and used extensively in Europe. Outranking attempts to better inform the decision process by incorporating the preferences of the DM (thought process, feelings, and values) into the decision process. Outranking methods do not look for one optimal solution, but aid the decision process by ranking or partially ranking alternatives. These alternatives are ranked based on decision maker preferences for alternatives across criteria. Whereas MAUT decision making is based on normative theory, outranking approaches are based on descriptive or behavioral theory (Fishburn, 1991).

PROMETHEE is used to illustrate the outranking method applied to the watershed problem. Unlike the American school, PROMETHEE evaluates both (1) the degree of advantage or outperformance of one alternative over another over all the criteria, and (2) the degree of disadvantage or underperformance of that same alternative against the other alternative over all the criteria (Vincke, 1992; Brans, Vincke, & Mareschal, 1986). Degrees of preference for one alternative over another are expressed on the interval [0,1], with "0" denoting indifference, "1" denoting strict preference, and numbers between 0 and 1 expressing degrees of relative preference. Preference and indifference thresholds can be elicited from decision makers to further capture the reality of the decision problem. For example, decision makers can be asked, "Based on the criterion of economic growth, is there a point at which you are indifferent between two alternatives?" Additionally, they can be asked, "Based on the criterion of economic growth, is there a point at which you would strictly prefer one alternative over another." Note that decision makers are not asked to express preference for each pair of alternatives; instead these preferences are elicited using preference functions for each criterion. Also, criteria can be weighted by the individual decision maker. For example, a member of the watershed council who is a farmer might weigh the environmental criteria more than the economic criteria, while a small business owner might weigh the economic criteria more.

Alternative 2 (focus on water quality) is said to *outrank* Alternative 3 (focus on agricultural land preservation) if it performs better on most criteria (or, if the criteria are weighted, performs better on the significant criteria) *and* if the water quality alternative is not significantly *outranked* or *outperformed* by preservation of agricultural land alternative on any one criteria. According to Kangas et al. (2001) the key question regarding alternatives is "whether there is enough information to state that one alternative is at least as good as another." The outranked alternative is referred to as *dominated.* An alternative is dominated if there are other alternatives that outperform it on one or more criteria and equal it on the remaining criteria. The first result of these outrankings is to produce a partial ranking of the alternatives based on decision maker preferences and weights. This partial ranking captures any indifference or incomparability between alternatives. Below, Alternatives 3 and 4 are incomparable, meaning that Alternative 4 performs well on some criteria and Alternative 3 performs well on others. The ranking of alternatives can be done for each individual council member as well as for the council as a whole. This is key in that council members will each have their own weights and preferences for each criterion. PROMETHEE allows these individual weights and preferences to be viewed visually and changed, if desired, by the council members. This leads to a better understanding of the perspectives of individual members and can facilitate the decision process. Fig. 1.

Outranking is unique in that it is based on the elicited preference functions (with indifference and preference thresholds) of the decision makers vis-a-vis the established criteria. The result is a partial or complete ranking of the alternatives across the criteria. In outranking, the decision maker can be *indifferent* to the available alternatives, not preferring one to the other. Outranking methods also allow for incomparability between alternatives and partial aggregation. This is not the same as indifference, but results from uncertainty, missing or insufficient data, decision-maker ignorance, or the fact that the alternatives are too different to be compared (Bouyssou et al., 2000; Bender & Simonovic, 2000; Klauer, Drechsler, & Messner, 2006). Incomparability occurs between two alternatives when there is no clear evidence in favor of either, as in the ranking illustrated above. MAUT and AHP models have no mechanism to treat incomparable alternatives and do not distinguish between indifference and incomparability. Unlike MAUT/AHP and CBA, which attempt to find the optimal alternative, the goal in outranking is "compromise" between criteria rather than optimization (Brans & Mareschal, 1984).

In outranking, preferences are not certain, nor are they assumed reducible to a utility criterion. They are allowed to change or evolve with the decision process. In the example above, for example, the small business owner, through a more thorough understanding of the development options, may

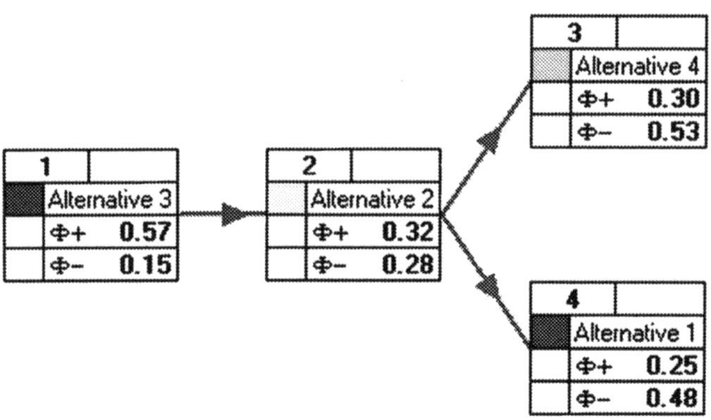

Fig. 1. Illustration of Partial Ranking in PROMETHEE.

decide that social criteria are more important than he previously thought. PROMETHEE can accommodate his changes in weights and/or preferences.

Criteria in outranking, unlike MAUT methods, are non-compensatory. A poor performance of an alternative on a criterion cannot be compensated by a greater performance on another. If Alternative 1, for example, performs well on economic criteria, but badly on ecosystem health criteria, the good performance on the economic criteria cannot compensate for the bad performance on the ecosystem health criteria. This is a key distinction between MAUT and outranking. Additionally, outranking focus on the subjective preferences of the decision maker and a great deal of effort is made to understand and model these preferences.

Proponents of outranking techniques argue that the lack of transitivity and the existence of incomparabilities more realistically represent decision-maker preferences than the restrictive assumptions of MAUT and AHP (Fishburn, 1991; Tversky, 1969). It can be argued that preferences may not even pre-exist the process from which they emerge (Bouyssou & Vincke, 1997). Furthermore, the outranking method is a constructivist approach, allowing for the process inputs to be modified as new information is obtained during the process (Roy, 1993). Additionally, the emphasis in outranking methods is on the *comparison* of individual decision-maker preferences, rather than *aggregation* of them as in MAUT/AHP (Belton & Pictet, 1996).

PROMETHEE is chosen for review over the other dominant outranking method of ELECTRE for several reasons. The PROMETHEE theory and methodology are easier for decision makers to understand (Klauer, Drechsler, & Messner, 2006). PROMETHEE also allows for decision-maker involvement at every stage of a transparent process. Criteria weights, preference functions, and thresholds can all be manipulated at any point in the process allowing for a more dynamic interface than ELECTRE (Brans et al., 1986; Pomeral & Barba-Romero, 2000; Mahmoud & Garcia, 2000). In addition, Brans et al. (1986) found PROMETHEE rankings to be more stable than ELECTRE rankings.

Another key advantage of the PROMETHEE methodology is its ability to handle a large number of decision makers, criteria, and alternatives. This makes it particularly well suited for use in group decision support systems (GDSS). In GDSS PROMETHEE each decision maker can have their own preferences and establish their own criteria weights (Macharis et al., 1998). MCDA as a group process can be understood as a way to develop a shared understanding of the different worldviews and perceived problem situation of the involved decision makers.

3. CONCLUSION

MCDA methods differ primarily in the theoretical assumptions they make about decision-maker's preferences and how they process the data contained in the evaluation matrix, the centerpiece to MCDA procedures used to compare and rank criteria and decision alternatives. In group environmental decision making, it is important to model the decision problem as realistically as possible and to use a method that allows for stakeholder participation and process transparency. Specificities of the decision problem should dictate the method used (e.g., the number of decision makers, alternatives, and criteria, the scale and specific objectives of the problem, and whether data is qualitative, quantitative, or spatial). Outranking is an ideal method in many environmental problems as it is able to realistically capture the complexity of environmental issues, allows for group participation, can accommodate multiple decision makers, criteria, and alternatives in the process, and permits modifications to the process at any stage.

REFERENCES

Abi-Zeid, I., Belanger, M., Guitouni, A., Martel, J. M., Jabeur, K. (1998). *A multicriteria method for evaluating courses of action in Canadian airspace violation situations* (pp. 521–533). Stockholm, Sweden: Department of Defense Command and Control Research and Technology Symposium.

Abu-Taleb, M. F., & Mareschal, B. (1995). Water resources planning in the Middle East: Application of the PROMETHEE V multicriteria method. *European Journal of Operational Research*, *81*, 500–511.

Belton, V., & Pictet, J. (1996). *A framework for group decision using a MCDA model: Sharing, aggregating or comparing individual information?* Working Paper no. 16. Management science: Theory, method and practice series. University of Strathclyde business school.

Bender, M. J., & Simonovic, S. P. (2000). A fuzzy compromise approach to water resource systems planning under uncertainty. *Fuzzy Sets and Systems*, *115*(1), 35–44.

Bouyssou, D., Marchant, T., Pirlot, M., Perny, P., Tsoukias, A., & Vincke, P. (2000). *Evaluation and decision models: A critical perspective*. Boston: Kluwer Academic Publishers.

Bouyssou, D., & Vincke, P. (1997). Ranking alternatives on the basis of preference relations: A progress report with special emphasis on outranking relations. *Journal of Multicriteria Decision Analysis*, *6*, 77–85.

Brand, C., Mattarelli, M., Moon, D., & Wolfler Calvo, R. (2002). STEEDS: A strategic transport-energy-environmental decision support. *European Journal of Operational Research*, *139*(2), 416–435.

Brans, J. P., & Mareschal, B. (1984). PROMETHEE: A new family of outranking methods in multicriteria analysis. *Operational Research*, *84*, 477–490.

Brans, J. P., & Mareschal, B. (1989). *The PROMETHEE method of MCDM*. Brussels: Vrije Universiteit.

Brans, J. P., Vincke, P., & Mareschal, B. (1986). How to select and how to rank projects: The PROMETHEE method. *European Journal of Operational Research*, *28*, 228–238.

Brigs, T., Kunsch, P., & Mareschal, B. (1990). Nuclear waste management: An application of the multicriteria PROMETHEE methods. *European Journal of Operational Research*, *44*, 1–10.

Erickson, J. D. (2000). Endangering the economics of extinction. *Wildlife Society Bulletin*, *28*(1), 34–41.

Erickson, J. D., Limburg, K., Gowdy, J., Stainbrook, K., Nowosielski, A., Hermans, C., & Polimeni, J. (2004). Anticipating change in the Hudson river watershed: An ecological economic model for integrated scenario analysis. In: R. Bruins & M. Heberling (Eds), *Economics and ecological risk assessment: Applications to watershed management* (pp. 341–370). Boca Raton, FL: CRC Press.

Faber, M., Manstetten, R., & Proops, J. L. R. (1992). Humankind and the environment: An anatomy of surprise and ignorance. *Environmental Values*, *1*, 217–242.

Fishburn, P. C. (1991). Nontransitive preferences in decision theory. *Journal of Risk and Uncertainty*, *4*, 113–134.

Geldermann, J., Spengler, T., & Rentz, O. (2000). Fuzzy outranking for environment assessment. Case study: Iron and steel making industry. *Fuzzy Sets and Systems*, *115*, 45–65.

Gintis, H. (2000). Beyond homo economics: Evidence from experimental economics. *Ecological Economics*, *35*, 311–322.

Gowdy, J., & Erickson, J. D. (2005a). The approach of ecological economics. *Cambridge Journal of Economics*, *29*(2), 207–222.

Gowdy, J., & Erickson, J. D. (2005b). Ecological economics at a crossroads. *Ecological Economics*, *53*(1), 17–20.

Guitouni, A., & Martel, J. M. (1998). Tentative guidelines to help choosing an appropriate MCDA method. *European Journal of Operational Research*, *109*, 501–521.

Hyde, K., Maier, H., & Colby, C. (2004). Reliability-based approach to multicriteria decision analysis for water resources. *Journal of Water Resources Planning and Management*, *130*(6), 429–438.

Iz, P. H., & Gardiner, L. R. (1993). A survey of integrated group decision support systems involving multiple criteria. *Group Decision and Negotiation*, *2*(1), 61–79.

Joerin, F., Golay, F., & Musy, A. (1998). GIS and multicriteria analysis for land management. COST C4 Final Conference. Jukkasjärvi, Kiruna, Sweden, September 21–22.

Joubert, A. J., Leiman, A., De Klerk, H. M., Katua, S., & Aggenbach, J. C. (1997). Fynbos (fine bush) vegetation and the supply of water: A comparison of multi-criteria decision analysis and cost–benefit analysis. *Ecological Economics*, *22*, 123–140.

Kangas, J., Kangas, A. ., Leskinen, P., & Pykalainen, J. (2001). MCDM methods in strategic planning of forestry on state-owned lands in Finland: Applications and experiences. *Journal of Multicriteria Decision Analysis*, *10*, 257–271.

Keeney, R., & Raiffa, H. (1976). *Decisions with multiple objectives*. New York: Wiley.

Klauer, B., Drechsler, M., & Messner, F. (2006). Multicriteria analysis under uncertainty with IANUS – method and empirical results. *Environment and Planning C: Government and Policy*, *24*, 235–256.

Lootsma, F. (1990). The French and the American school in multi-criteria decision analysis. *European Journal of Operational Research*, *23*(3), 263–285.

Lubell, M., Schneider, M., Scholz, J. T., & Mete., M. (2002). Watershed partnerships and the emergence of collective action institutions. *American Journal of Political Science*, *46*(1), 148–163.

Macharis, C., Brans, J. P., & Mareschal, B. (1998). The GDSS PROMETHEE procedure. *Journal of Decision Systems*, *7*, 283–307.

Mahmoud, M. R., & Garcia, L. A. (2000). Comparison of different multicriteria evaluation methods of the red bluff diversion dam. *Environmental Modelling and Software*, *15*(5), 471–478.

Malczewski, J. (1999). *GIS and multicriteria decision analysis*. New York: Wiley.

Martin, N. J., St. Onge, B., & Waaub, J. P. (1999). An integrated decision aid system for development of Saint-Charles River's alluvial plain. *International Journal of Environment and Pollution*, *12*(2/3), 264–279.

Messner, F. (2006). Applying participatory multicriteria methods to river basin management: Improving the implementation of the water framework directive. *Environment and Planning C: Government and Policy*, *24*, 159–167.

Mirasgedis, S., & Diakoulaki, D. (1997). Multicriteria analysis vs. externalities assessment for the comparative evaluation of electricity generation systems. *European Journal of Operational Research*, *102*, 364–379.

Pomeral, J.-C., & Barba-Romero, S. (2000). *Decision in management: Principles and practice*. Norwell, MA: Kluwer Academic Publishers.

Prato, T. (2000). Multiple attribute evaluation of landscape management. *Journal of Environmental Management*, *60*, 325–337.

Qureshi, M. E., & Harrison, S. R. (2001). A decision support process to compare riparian revegetation options in Scheu creek catchment in north Queensland. *Journal of Environmental Management*, *62*, 101–112.

Rogers, M., & Bruen, M. (1998). A new system for weighting environment criteria for use within ELECTRE III. *European Journal of Operational Research*, *107*, 552–563.

Roy, B. (1985). *Méthodologie Mulitcritère d'Aide à la Décision*. Paris: Economica.

Roy, B. (1993). Decision science or decision-aid science? *European Journal of Operational Research*, *66*, 184–204.

Rozakis, S., Soldatos, P. G., Kallivroussis, L., & Nicolaou, I. (2001). Multiple criteria analysis of bio-energy projects: Evaluation of bio-electricity production in Farsala plain, Greece. *Journal of Geographic Information and Decision Analysis*, *5*(1), 49–64.

Saaty, T. L. (1980). *The analytic hierarchy process*. New York: McGraw-Hill.

Salminen, P., Hokkanen, J., & Lahdelma, R. (1998). Comparing multicriteria methods in the context of environmental problems. *European Journal of Operational Research*, *104*, 485–496.

Salt, C. A., & Dunsmore, M. C. (2000). Development of a spatial decision support system for post-emergency management of radioactively contaminated land. *Journal of Environmental Management*, *58*, 169–178.

Sellers, J. (2002). *Governing from below: Urban regions and the global economy*. Cambridge: Cambridge University Press.

Stewart, T. J. (1992). A critical survey on the status of multiple criteria decision making theory and practice. *OMEGA*, *20*(5/6), 569–586.

Sugden, R., & Williams, A. (1978). *The principles of practical cost–benefit analysis*. Oxford: Oxford University Press.

Thomson, K. (2001). *From neighborhood to nation: The democratic foundations of civil society*. Boston: Tufts University Press/University Press of New England.

Triantaphyllou, E. (2000). *Multi-criteria decision making methods: A comparative study*. Boston: Kluwer Academic Publishers.

Tversky, A. (1969). Intransitivity of preferences. *Psychological Review*, *76*, 31–48.
Vaillancourt, K., & Waaub, J.-P. (2002). Environmental site evaluation of waste management facilities embedded into EUGENE Model: A multicriteria approach. *European Journal of Operational Research*, *139*, 436–448.
van den Doel, J., & van Velthoven, B. C. J. (1993). *Democracy and welfare economics*. Cambridge: Cambridge University Press.
Van Elegem, B., Embo, T., Muys, B., & Lust, N. (2002). A methodology to select the best locations for new urban forests using multicriteria analysis. *Forestry*, *75*(1), 13–23.
Van Groenendaal, W. J. H. (2003). Group decision support for public policy planning. *Information and Management*, *40*(5), 371–380.
Vincke, P. (1992). *Multicriteria decision aid*. New York: Wiley.
Wilson, M. A., & Howarth, R. B. (2002). Valuation techniques for achieving social fairness in the distribution of ecosystem services. *Ecological Economics*, *41*, 431–443.
Yoe, C. (2002). *Trade-off analysis planning and procedure guidebook*. Planning and Management Consultants, Ltd. A report presented to the U.S. Army Corps of Engineers, Institute for Water Resources.

INTEGRATION OF ECONOMIC EVALUATION INTO WATER MANAGEMENT SIMULATION

Frank Messner, Hagen Koch and Michael Kaltofen

ABSTRACT

In this chapter it is shown how economic evaluation algorithms of water use can be integrated into a long-term water management model such that surface-water availability and economic evaluation of various levels of water availability to different uses can be modeled simultaneously. This approach makes it possible to include essential features of economic analyses of water use into water resource modeling and thus improves the capability of such models to support decision making in water management. This is especially relevant for the implementation of the Water Framework Directive, which requires economic analyses to be included in the decision process about future water management strategies.

The water management simulation model WBalMo is presented and the integration of economic-evaluation algorithms is demonstrated for the examples of surface-water use for fish farming and for filling open-cast mining pits in order to achieve acceptable water-quality levels in the emerging pit lakes. Results of applying this integrated evaluation approach are shown for different water management scenarios under conditions of global change in the East German Spree and Schwarze Elster river basins, where water scarcity is an urgent issue. Among the lessons

Ecological Economics of Sustainable Watershed Management
Advances in the Economics of Environmental Resources, Volume 7, 229–263

ISSN: 1569-3740/doi:10.1016/S1569-3740(07)07011-3

which are drawn by the authors one lesson reads that integrating economic evaluation algorithms into a pre-existing model might bring enormous problems. Therefore, such model approaches should be developed together by water engineers and economists in an interdisciplinary endeavor right from the start.

1. INTRODUCTION

Sustainable river-basin management has the higher goal to identify appropriate water management schemes to meet the needs of present and future generations with respect to the ecological, social, and economic functions of the water cycle. The scientific foundations of river-basin management are hydrological and water management models to estimate water availability in river basins over space and time. However, in the European tradition, the economic, social, and ecological impacts of different levels of water availability have rarely been evaluated explicitly to support water management decisions. Rather, expected water availability for different water users were evaluated implicitly according to political priorities and to water-allocation principles deemed to ensure a "reasonable" water supply for all water users.[1] This practice comes to an end with the implementation of the European Water Framework Directive (EU-WFD, 2000), which requires an economic analysis of water use (see Petry & Dombrowsky, in this volume). In the following we want to present an interdisciplinary approach to integrate economic evaluation into water management modeling in order to create an improved scientific basis for river-basin management decision making.

Certainly, the evaluation of economic impacts of surface water and groundwater use and changes in water availability in river basins or watersheds depends in many cases on hydrological and water management modeling results. However, in most cases economic evaluation of non-market goods related to water – e.g., wetland areas and the existence of marine species – is practiced largely separate from natural science modeling efforts (cf. e.g. Gren, Groth, & Sylvén, 1995; Brouwer, Langford, Bateman, & Turner, 1999). This is reasonable for two reasons. First, if the use or the existence of a natural non-market good must be evaluated, there are usually different kinds of scientific information required to apply one or the other benefit–cost method. For example, to evaluate the benefits of a wetland area before and after the reconstruction of a new dam using willingness to pay (WTP) methods, information is needed about the biological state of the wetland before and after the intervention. Many pieces of scientific information must be brought together to characterize the change in biological quality and its further implications. Since this characterization is a very region-specific task, the development of an integrated interdisciplinary model to estimate the hydrological and biological effects and to evaluate them based on the WTP results would require large efforts and the evaluation would become exceptionally expensive. Second, the evaluation of the benefits of a change in wetland quality does not need very precise results regarding the biological changes in space and time, but only rough estimates of the state before and after the intervention. It is simply not necessary and not possible to inform the interviewed people precisely and comprehensively about all biological consequences due to limited interview time and cognitive reasons. Hence, to estimate rough figures for one evaluation issue, nobody would endeavor to develop highly specific and elaborate interdisciplinary models.

Things are different if water availability is not just a secondary factor in the determination of the value of a natural good, but if the value of water and its various functions itself are the central issues. Water is a unique natural good. It has multiple functions, it provides benefits to many economic and natural processes, and it is a dynamic resource that moves continuously within the water cycle and produces upstream–downstream conflicts time and again (cf. Cech, 2003). Therefore, the determination of the value of water is strongly linked to its distribution in space and time, which is influenced by natural and anthropogenic processes (cf. Messner, 2005). As a consequence, the economic evaluation of the benefits and costs of changes in water availability requires precise estimates of the temporal and spatial distribution of water in order to evaluate the importance of its many services and functions for different anthropogenic and natural systems or agents in the spatial context of river basins. For the evaluation of such a complex natural good the development of integrated interdisciplinary models to estimate water availability and to evaluate its economic consequences is indispensable.

This fact has been recognized by several agricultural economists, who are acquainted with the vital significance of water for agricultural production. They started various attempts to build up interdisciplinary model systems to link agricultural and hydrological models in the spatial context of river basins (cf. Rosegrant et al., 2000; Yang, Khanna, Farnsworth, & Önal, 2003). Yet, these important efforts are still mainly focused on the agricultural sector in order to assist agricultural policy making. Interdisciplinary hydrological–economic model systems to support water management in general are still lacking. This is the case at least in Europe, where the policy strategies for water management in the last decades were basically built on the knowledge of hydrologists and water management engineers.

However, due to significant changes in European water-policy legislation this situation may change in the coming decades. The new Water Framework Directive of the European Union requires explicitly the incorporation of economic analyses into the decision-making process for water management in river basins. Both, important economic driving forces which may influence the availability and/or the quality of water resources must be identified for the next decades. Furthermore, in 2010 the price of water in all EU countries shall reflect all financial, environmental, and resource costs and must be designed according to the polluter pays principle (cf. EU-WFD, 2000, Art. No. 9; Hansjürgens & Messner, 2002; as well as Chapters 2 (Petry/Dombrowsky) and 15 (Unnerstall/Messner), in this volume). The implementation of this new body of European water legislation requires accurate knowledge about the interaction of the socio-economic and water systems. Among others, knowledge about the opportunity cost of water use will become an essential piece of information to design cost-recovery water prices. Developing integrated economic-water-management model systems to estimate and evaluate water availability and water-quality trends under the circumstances of societal, economic, natural, and climate change will be a prerequisite to attain this kind of information.

In this chapter, the research results presented deal with the integration of economic evaluation algorithms into the water management simulation model WBalMo. This integration offers the possibility to model water availability in space and time under different future circumstances and, simultaneously, to evaluate the economic consequences of various water-allocation developments in space and time. The model approach has been applied to the river basins of the German *Spree* and *Schwarze Elster* rivers with their specific water-scarcity situation, which is described in Chapter 4 (Messner, in this volume). The scenarios of global change and alternative policy strategies presented there were modeled and evaluated with the integrated economic-water-management simulation model approach.

This chapter is organized as follows: Section 2 describes the water management simulation model WBalMo in its original structure and presents some results regarding the effects of global change on the river basins of the Spree and Schwarze Elster rivers. Section 3 deals with the integration of evaluation and transfer algorithms into the model, using the economic effects of varying water availability on fish farming and on water quality of new pit lakes as examples. In Section 4 modeling results are described and discussed. Finally, some conclusions are drawn in Section 5, including the lessons learned from this interdisciplinary endeavor.

2. THE WATER MANAGEMENT SIMULATION MODEL WBALMO

2.1. Long-Term Water Resources Management Models

Water resources management models can generally be classified into optimization and simulation models (Yeh, 1985; Cunningham & Amend, 1986). With optimization models the optimal solution regarding one or several objectives with set constraints is searched for. Since the objectives and constraints are often vague and subject to dispute, these optimization techniques are rarely used in project planning (Rogers & Fiering, 1986). Simulation models approximate the behavior of a system and provide the response of this system to certain changes in input variables, e.g., changing decision rules relating to the release of a reservoir. Thereby it enables the decision maker to examine the consequences of various scenarios (Yeh, 1985). For an overview of some of the models developed over the last decades we refer to Wurbs (2005).

Hydro-meteorological processes and the resulting spatial–temporal distribution of runoff generation are uncertain. Therefore the runoff process should be treated as a random process over long periods of time. When considering these uncertainties regarding natural water yield, stochastic models deliver more reliable results than deterministic models (Chow & Kareliotis, 1970; Hirsch, 1981). On the other hand, user demands are deterministic in time and space, but may change depending on socio-economic development (Kaden, Schramm, & Redetzky, 2004).

2.2. Structure of the Simulation Model WBalMo

The simulation model used in this analysis is the result of a development that has taken place since the end of the 1970s. The first large-scale management model called GRM (abbreviation of the German notion "*GrossRaumModell*") has been developed in Berlin by the Institute for Water Resources as a long-term management model. In 1992, a desktop version of the GRM model was developed by WASY Ltd. This PC-GRM was of static type, with conditions of runoff generation being time-invariant.

Subsequently, a PC-GRMDYN was developed from PC-GRM for mining regions with wide-ranging, time-variant groundwater depression cones. Finally, an ArcView desktop implementation of the GRM management model was generated, which can model both stationary (PC-GRM) and dynamic processes (PC-GRMDYN) within river basins. The latest version of this model development is called WBalMo (abbreviation for Water Balance Model).

Based on the knowledge of a river basin's structure, its characteristic natural runoff, the water utilizations, the water users' surface-water demand and the water resources management rules and processes, the quantitative behavior of a river basin's water resource system can be examined with the WBalMo simulation model under various conditions.

The management model, which forms the basis of the simulation model, operates according to the Monte-Carlo-Method. River basin's water utilization processes can be reproduced, covering any time interval in time-steps of one month. However, for special investigations also shorter time-steps can be used. The registration of relevant system states allows a statistical analysis of registered events after completion of the simulation. As a result, approximate probability distributions for values, such as reservoir storage level, water supply deficiency for individual water users or for discharge at selected river profiles are available. Thus, the quality of a selected management strategy can be assessed for the river basin under investigation.

The treatment of stochastic input parameters (i.e., natural water yield) and the deterministic reproduction of water utilization and management processes are strictly separated in the program. The basis for the calculation of the natural water yield is a chronological series of input parameters for the chosen time interval. Usually these input parameters are generated by a stochastic simulation model under consideration of time-dependent conditions for drainage formation.

The simulation of the water utilization processes is based on (see Fig. 1):

- A schematic representation of hydrological processes in a river basin by means of running waters and balance profiles.
- A subdivision of the total basin area: simulation sub areas (SSA) are created and the above-named series of discharge are assigned; this discharge is then distributed among balance profiles within the SSA as natural water yield.
- The integration of reservoirs (barrage systems, lakes) by considering their location, capacity, and release-elements, which describe their demand-oriented operation.

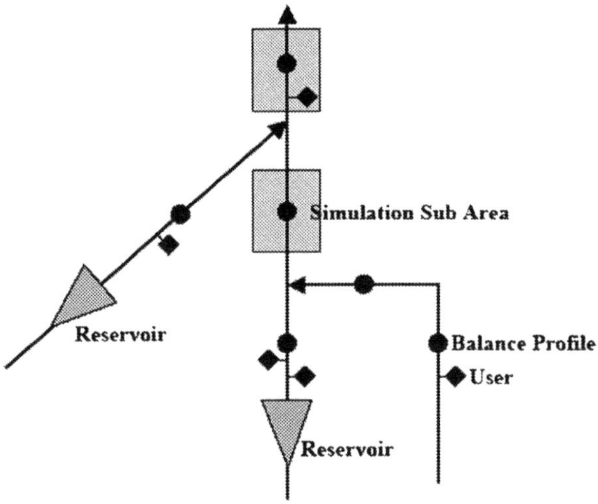

Fig. 1. System Structure of a Water Resources Management Model.

- The consideration of water users: utilizations take place according to their location and size, by withdrawing and/or returning the demanded water from/to balance profiles, respectively. Input data on water demand generated by economic models or expert estimations can be integrated into WBalMo as far as they are fitting to the spatial balance profiles of WBalMo.

Fundamental characteristics of the simulation model are:

- GIS-visualization of the river basins' system sketch within the simulation model,
- balancing of natural water yield and water users' demand in consideration of reservoir releases and water transfers,
- filing of model data in a data base, and
- an interface for external models, which enables specific investigations (e.g., water quality, simulation of daily values, integration of economic evaluation algorithms) on the basis of the WBalMo quantity model.

The WBalMo simulation model provides a useful tool for the analysis of questions concerning the water management of river basins, in which serious water-balance interventions are expected in the future (e.g., intensive mining with connected groundwater draw-down, commissioning of new reservoirs). It allows the reproduction of water yield and water-utilization processes either in a balance period or in a balance year. A balance period is usually a future space of time divided in equally long periods, consisting of several years. The conditions for natural water yield and water demand may change from period to period. Examples are variable flow regimes in mining areas caused by changing catchment area borders or climatic changes. Furthermore, the expected development of water demand of water users or the non-stationary filling phase of new reservoirs can be arranged precisely in time.

An application of the simulation model requires the availability of time series of mean monthly natural yield for those SSA, which constitute the river basin under study. The data of these time series can be observed ones, revised by management effects – if necessary. They also may be stochastically generated on the basis of revised discharges. In the latter case the Monte-Carlo-Method for water management problems is applied. With this method each balance period is sufficiently often simulated with a different, stochastically generated water yield. In this way, precise results concerning the effectiveness of the water management system can be achieved. A sequential (monthly) reproduction of one balance period is called "realization." Usually 100 to 1,000 realizations are used.

Applying the simulation model, water management problems in river basins can be analyzed on the basis of a balance simulation. This is carried out by a monthly location-specific comparison (= balancing) of natural water yield and the user demands over a sufficient number of simulations of the balance period. In this context, water-supply requirements on reservoirs are considered. Balancing includes registration of relevant events (see Section 2.3 for examples). This enables the calculation of water-supply reliabilities by means of statistical analysis (mean, extreme values) after completion of the simulation.

Water-utilization processes are modeled under consideration of water users' demand, reservoir releases, and so-called Dynamic Elements (DYN-elements – programmed in FORTRAN). These model elements receive a rank. Concerning one utilization this rank represents its importance in the system of all utilizations in the river basin. Release-element and DYN-element ranks allow their classification in the users' hierarchy. In this way water management strategies can be incorporated in the simulation process.

The processing of the ranking list gradually transfers the natural discharge of a river basin into a managed final state for each month examined. Registration of relevant state-variables such as discharge at chosen balance profiles, actual water withdrawal by particular users, or current storage levels of individual reservoirs only takes place at the end of the monthly balance.

Certain river-basin management rules, registration requirements, as well as other necessary operations cannot be formulated by the model's standard elements. However, their consideration is possible by the definition of DYN-elements. The performance of the standard algorithm is interrupted by DYN-elements in order to process a given individual user algorithm. This algorithm in general contains relevant values of the system's state-variables and other program variables. The classification of DYN-elements is established by means of their setting in the ranking list of all users and release elements. In this way it is possible to select the moment of interruption of the standard algorithm.

Examples for the application of DYN-elements are:

- Setting state-variables at the beginning of a realization, a balance period or a year.
- Calculating the evaporation loss from reservoirs depending on current reservoir storage level and simulated potential evaporation.
- Calculation of variable water-transfer volumes depending on the discharge rate at the profile of withdrawal.
- Registration of hydrographs of relevant system states in individually defined dry periods.
- Integration of rainfall-runoff models, flood management with time-steps smaller than one month, calculating dependencies between groundwater and reservoir storage level or integration of water-quality criteria in management.
- Integration of evaluation algorithms.

In respect of the specific problems related to water quantity and quality in the Spree and Schwarze Elster river basins (see Messner, Chapter 4, in this volume) the concerned federal states, mainly Brandenburg and Saxony, decided to develop a water resources management plan based on results of a simulation model. Therefore, based on WBalMo a model with spatial reference to the watersheds of the Spree river and the Schwarze Elster river was developed by the water authorities of the concerned federal states and the state-owned company LMBV. This model called *WBalMo Spree/Schwarze Elster* was made available to the GLOWA-Elbe project. Within the project, the model was revised in order to take the specific circumstances of global change into account. For example, the planning period of the model from 1998 to 2032 was adapted to the time horizon of the GLOWA Elbe project, being 2003 to 2052. Specific characteristics of the revised model are the consideration of more than 170 sites (balancing profiles) of the river system, about 400 different water utilizations of water users' with varying water demand throughout each year, water releases of 14 reservoirs, about 50 DYN-elements, and registration of more than 200 indicators (e.g., reservoir releases, throughflow at balancing profiles, deficits for users). Input data for water demand was based on modeling results for the water usage of the energy sector, using the IKARUS data base and KaSIM model system (Vögele, Markewitz, & Martinsen, 2001; Martinsen, Kraft, & Markewitz, 2001). With regard to water demand of other water users expert estimations were executed. Finally, this revised model was called WBalMo GLOWA.

2.3. Forms of Results

Each simulation of WBalMo produces an enormous amount of data sets on frequency distributions, extreme values, and other statistic variables related to surface-water availability and discharge. Further examination of these data requires an elaborate data organization. Therefore, WBalMo output data is organized by three different types of output tables.

The first type of output tables includes percentage values of state variables in form of frequencies of exceedance of pre-determined water users' demand. To put it in another way, these tables comprise the information how secure water provision will be for the respective water utilization.

The second type of output tables enables the consideration of the duration of an event, e.g., the falling short of a determined level in a reservoir. The respective data in these tables are organized in the form of relative frequencies regarding the occurrence of events of fixed duration. In this context events may begin in each calendar month.

In the third type of output tables mean values, standard deviations, minima, and maxima of state variables are registered. These values are calculated separately for each month and calendar year, respectively.

As already mentioned in Section 2.2, the registration of additional events or variables of interest, e.g., absolute numbers of water demand for a specific economic sector, is also possible by applying DYN-elements.

Fig. 2 presents an example for results received by using the data of the first type of WBalMo output tables. The figure shows probabilty values for meeting the pre-determined demand of 8 m^3 per second at the gauge (balance profile) Grosse Traenke near Berlin. The probabilities reflect monthly values of 5-year periods between 2003 and 2052. The figure clearly indicates that the safety of meeting the suface-water demand is distinctly lower in a scenario which takes the incidence of climate change into account.

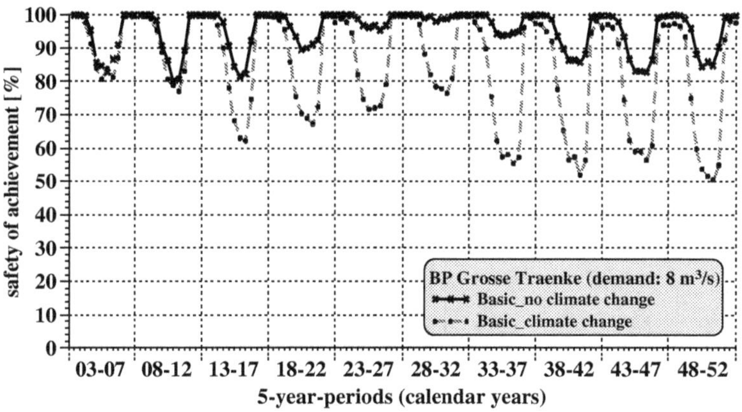

Fig. 2. Probability to Meet the Demand at Gauge Grosse Traenke/Spree River (Scenarios According to Messner, Chapter 4, in this Volume).

Fig. 3 presents an example for specific registrations in output tables programmed by means of a DYN-element in order to compare water availability in scenarios with and without climate change. With regard to the 100 realizations simulated with WBalMo the dashed lines show natural throughflow values with a 20% probability of exceedance. This means that such a value, e.g., 5 m^3 per second in January of the 2003–2007 period of the scenario *Basic_no climate change*, will only be realized once in 5 years. Such a value is not very secure and rather represents the incidence of very wet years with a high natural throughflow. Conversely, the solid lines show a 80% probability of exceedance, meaning that such a value is exceeded in 4 of 5 years. These values are relatively secure. They reflect the hydrological situation of dry years and are often used as a reference point in water resources management and planning. Fig. 3 indicates that the natural throughflow of the scenario *Basic_climate change* is distincly below the values of the scenario without climate change – especially during spring and summertime.

To evaluate the influence of climate change on water availability probabilities of exceedance of 20% ('optimistic') and 80% ('pessimistic') of *natural* throughflow, i.e., without any management, of the scenario *Basic_no climate change* and the scenario *Basic_climate change* are compared. The figure shows the natural throughflow at the gauge (balance profile) Bautzen for 5-year periods from 2003–2052 with annual cycle.

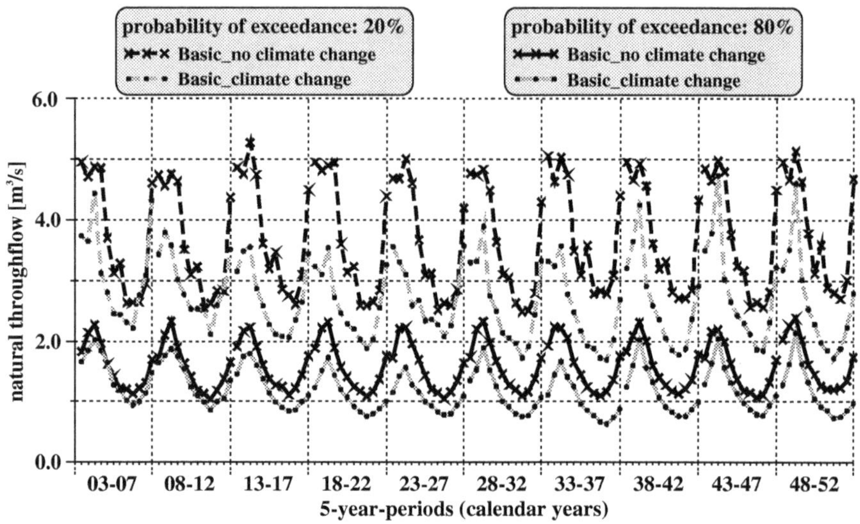

Fig. 3. Natural Throughflow at Gauge Bautzen/Spree River with 20% and 80% Probability of Exceedance (Scenarios According to Messner, Chapter 4, in this Volume).

3. INTEGRATING ECONOMIC EVALUATION INTO WBALMO

How economic evaluation algorithms were directly integrated into the water management simulation model WBalMo in order to simultaneously model and evaluate varying water availabilities of different policy action and global change scenarios is addressed in this chapter.

3.1. Analysis of Economic Impacts

The starting point of the economic impact analysis of changing water availabilities in the Spree and Schwarze Elster river basins was the examination of the model structure of WBalMo and, at an early stage, its initial modeling results regarding the status quo water situation. This analysis clearly indicated that WBalMo considers water utilization in the river basin through a large net of water users. Calculated on the basis of data on all water users' water demand and stochastic data on water availability in space and time, the results of WBalMo give information about the reliabilities on

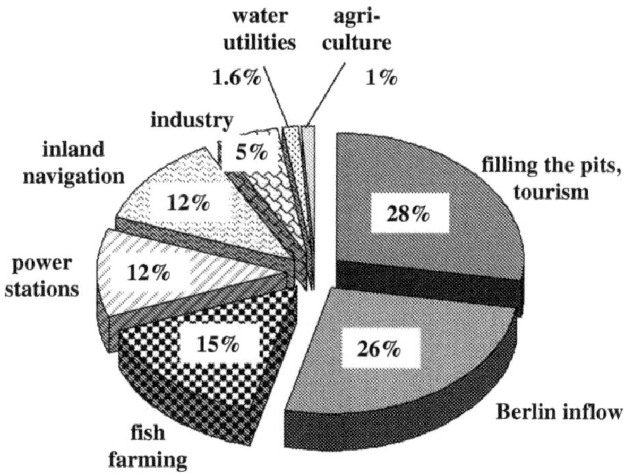

Fig. 4. Share of Various Water Uses in Direct Surface-Water Demand in the Spree River (WBalMo Modeling Results, Mean of Month July in the Period 2003–2007).

covering water users' water demand on a monthly basis. Fig. 4 shows the water demand of major direct water users which withdraw surface water directly from a stream and which are, therefore, included in WBalMo. The highest water demand (28%) is featured by the public rehabilitation firm LMBV, which will rehabilitate the old mining pits in the coming decades by filling them with surface water and thereby preparing the landscape for future tourism. The capital city of Berlin also shows a high demand for drinking water, water for cooling, industry, and tourism (26%).[2] Furthermore, with regard to water withdrawals fish farmers (15%), power stations (12%), inland navigation (12%), and the industry upstream of Berlin (5%) are of prime significance. Less important are surface-water withdrawals of water utilities and agriculture (both below 2%).

The water demand for ecological requirements is not displayed in Fig. 4, because these instream flow requirements do not withdraw water. However, a minimum discharge in a river is needed to sustain the ecological systems. Therefore, *ecologically required minimum discharges*, which are defined by environmental authorities, are considered in WBalMo for specific locations in the watershed.

In a subsequent analysis of all direct water users important information was revealed with regard to political priorities and economic vulnerability. With respect to political priorities it became apparent that power stations in general but also inland navigation on the canal connecting the Spree and

Odra rivers is given high political priority at the national level. Furthermore, the water demand of the power stations is covered by the mining discharges of the active mines, while the canal can also be supplied with water from the Odra river. Therefore, an evaluation of changing water availabilities was not necessary for these users. Additionally, in the examination of the economic vulnerability of water users, it appeared that drinking water provision in the region is mainly based on ground-water sources. Moreover, drinking water demand in general is decreasing due to a declining population. As a result of these two facts the incidence of reduced surface-water availability in the future is not likely to affect regional drinking-water provision.

Eventually, in the analysis of economic vulnerability in the study region the following water users were identified to be the most important ones with medium or high vulnerabilities and thus little possibilities to substitute profitably for surface water: fish farming, industry, rehabilitation of mining pits, tourism at mining lakes, and assisting water management efforts. For these activities evaluation algorithms had to be defined and integrated into the water management simulation model WBalMo. The examples of fish farming and water quality in pit mining lakes were chosen to demonstrate the integration of economic evaluation into WBalMo.

3.2. Transfer and Evaluation Algorithms for Fish Farming

One of the major challenges of integrating economic evaluation algorithms into a water management model consists in finding an appropriate connection between the modeling results for the water users and the accruing costs and benefits of water utilization. In most cases there is no immediate relationship between the discharge at a specific point in the river (in m^3 per second) and the benefits of economic water use (in €). Consequently, an adequate connection which fits to the structure of the water management simulation model has to be identified.

Let us have a closer look at the evaluation of fish-farming activities to make this point clear. Fish farming in the Spree and Schwarze Elster river basins includes small to large-scale aquaculture activities with artificial ponds having existed for centuries with individual pond surface areas of up to several dozen hectares. The pond areas form a semi-natural landscape and also serve as a rich habitat for endangered species. In our case study we only considered the producer side of fish farming, because the fish-market conditions in the Lusatia region are very elastic on the demand side with many substitution possibilities – e.g., there is plenty and cheap fish supply from all countries located at the Baltic Sea. Therefore, benefit reductions on the demand side were assumed to be near zero. The benefit–cost approach chosen to quantify the monetary effects on the producer side was based upon the application of actual economic data from companies and the regional fish-farming sector.

The carp fish-farming activities in the Lusatia study area highly depend on availability of surface water. This water is mainly used to fill the empty fish ponds in spring and to compensate for evaporation and infiltration losses throughout the year. Each autumn the water of ponds containing mature fish is drained to the natural streams and the fish is harvested. The economic optimal mean water level in fish ponds is about 1.3 m. If surface-water availability is low the fish ponds' water levels decrease successively. As a consequence, the ponds' oxygen content is declining and the living condition of the fish population deteriorates. If the water level in any pond falls short of 1 m for more than about eight weeks before the harvest in autumn the emergency case occurs. In order to prevent a large-scale dying of fish the water and the fish of the emergency ponds need to be allotted to other ponds. While the fish population in the remaining ponds increases, profits will fall (on average about 30%), because the fish in general will either not reach their normal mature weight or they will need much more time to grow. These technical pieces of information about fish farming in the study area were obtained from expert interviews (Langner, 2002). They elucidate that profitability depends on pond water levels and pond area – and of course they are closely linked to the water flow of the river from which the surface water for the fish ponds is withdrawn.

The economic first-best approach to evaluate the economic impact of varying surface-water availabilities on fish-farming activities would be:

- first, to translate the WBalMo modeling results on temporal and spatial availability of surface water in the river basin to water levels in each fish pond during summer and autumn;
- second, to identify the marginal ponds with a high probability of emergencies occurring in drought years and;
- third, to estimate the specific losses of every emergency.

Unfortunately, due to the structure of the data contained in the water management model and data availability problems this approach was not realizable. The foremost obstacle to apply the first-best evaluation approach is the fact that in the water management simulation model fish ponds of the study area are aggregated to larger pond units, which may contain ponds of different fish-farming companies. This model approach is reasonable from a water management point of view to study and understand water-scarcity situations in general, but it complicates specific economic evaluations based on its modeling results. Thus, using the model in the present state it was not possible to identify the marginal fish lakes, which will be affected first in the event of low surface-water availability. Moreover, it is not possible to use specific company data to estimate profits and losses either.

After having understood this problem in the interdisciplinary research team, there remained two options: to change WBalMo and rebuild it in order to allow the integration of economic first-best evaluation algorithms or to find a sound second-best evaluation approach. Since the first option would have consumed too much time, the second option was chosen in the project. Several methods were discussed for the second-best evaluation approach and finally one of them was chosen for methodological and practical reasons. Discussions were dominated by two issues: first, how to evaluate water availability if neither company-specific nor pond-specific information can be used; and second, how to link the economic evaluation to WBalMo model results in terms of withdrawal per second calculated for 100 realizations of each scenario. The solution of these two issues is: first, to use average profit data on fish farming in the study area, and second, to connect the average profit evaluation function to WBalMo via a transfer algorithm based on a pond area approach. This method was finally agreed upon in the research team.

Formally, the evaluation was executed as follows. The model ponds in WBalMo were interpreted as ponds with flexible pond area (PA). If sufficient surface water is available to ensure a monthly water withdrawal between January and August ($WW_{i,\mathrm{mt}}$) that at least meets 77% of the accumulated monthly fish-farming water demand over that period ($WD_{i,\mathrm{mt}}$) – this implies a minimum pond water level of 1m until August – then the PA coefficient ($z_{i,t}$) remains constant at 100% (Eq. 1). If, however, accumulated water withdrawal falls short of 77% of demand, the PA coefficient is smaller than one, because the PA must be adjusted in order to ensure at least a water level of 1 m over time (Eqs. (2) and (3)). This way the emergency case in the real practice of multi-pond fish farming which leads to reduced PA was reproduced in the evaluation approach by using one aggregated model pond and by calculating the percentage of lake area adjustments in order to run all ponds with a water level of 1 m. Of course, these functions are not complete with regard to economic evaluation. They are only *transfer algorithms* which connect the WBalMo model results to the final economic evaluation. Since water demand WD and water withdrawal WW are physical numbers (in cubic meters) that can be derived from the model results (in cubic meters per second) the PA coefficient $z_{i,t}$ serves as the connecting variable between WBalMo and economic evaluation.

Transfer algorithms:

$$z_{i,t} = 1 \quad \text{if} \quad \frac{\sum_{\text{mt}=1}^{8} \text{WW}_{i,\text{mt}}}{\sum_{\text{mt}=1}^{8} \text{WD}_{i,\text{mt}}} \geq 0.77 \tag{1}$$

$$z_{i,t} < 1 \quad \text{if} \quad \frac{\sum_{\text{mt}=1}^{8} \text{WW}_{i,\text{mt}}}{\sum_{\text{mt}=1}^{8} \text{WD}_{i,\text{mt}}} < 0.77 \tag{2}$$

$$\text{if } z_{i,t} < 1, \text{ then} \quad z_{i,t} = \frac{\sum_{\text{mt}=1}^{8} \text{WW}_{i,\text{mt}}}{\sum_{\text{mt}=1}^{8} \text{WD}_{i,\text{mt}}} \times \frac{1}{0.77} \tag{3}$$

with $\text{WW}_{i,\text{mt}}$: actual water withdrawal in month mt for a model pond i in year t (m^3); $\text{WD}_{i,\text{mt}}$: water demand for month mt for a model pond i to achieve the optimal water level of 1.3 m in year t (m^3); mt: month of water withdrawal and water demand, with $1 = $ January and $8 = $ August; $z_{i,t}$: fish PA coefficient for pond i to adjust the model pond to a pond area which ensures a water level of 1 m at minimum.

Based upon these transfer algorithms average discounted fish-farming profits (PFF) can be quantified (see Eq. 4). They are calculated as the product of average fish output per hectare in physical units (AQ), average profit per unit fish (AP), area of the model pond considered (AR), and the PA coefficient from the transfer algorithms ($z_{i,t}$). Data on AQ and AP were taken from a study of Klemm (2001), who determined the average output and the average profit per unit fish for the regional fish-farming activities in Saxony for 1995–1999.[3]

Evaluation algorithm:

$$\mathrm{PFF}_{i,t} = (\mathrm{AQ} \cdot \mathrm{AP} \cdot \mathrm{PA}_{i,t} \cdot z_{i,t}) \cdot (1 + r)^{-t+1} \tag{4}$$

with $\mathrm{PFF}_{i,t}$: fish-farming profit of model pond i in year t in present value (€/year); AQ: average fish output in physical units per hectare (kg/ha); AP: average profit per unit fish (€/kg); $\mathrm{PA}_{i,t}$: pond area of model pond i in year t (ha); r: discount rate.

These algorithms were integrated into the water management simulation model by means of DYN-elements such that varying water availability over space and time and its economic impacts could be modeled simultaneously for all 100 realizations of each scenario for every fish-farming unit in the model. The results in form of profit impacts of varying water availabilities are listed in output tables. Some of these results are presented in Section 4.

Concerning the development of fish profits over time different assumptions were made for the two frameworks of development (FoDs), which are outlined in detail in Chapter 4 (Messner, in this volume). With regard to the socio-economic FoD B2, which has a more regional focus on economic development and a strong environmental policy, it was assumed that subsidies for fish farming would continue to exist, wages would remain stable in real terms, and the proportion of directly marketed fish would rise. As a consequence, it was calculated that the fish price in real terms would rise by 0.20€ per kilogram up to the year 2052 and profits in real terms would increase to 560€ (€ of 2003) per hectare. With respect to the socio-economic FoD A1, which displays a stronger focus on economic liberalization and globalization and a more reactive and therefore a less stringent environmental policy, it was assumed that subsidies are halved over time and wages decline in real terms due to increased world-market competitive pressures. These assumptions lead to a declining trend in fish-farming profits amounting to about 142€ (€ of 2003) per hectare in real terms for the year 2052. These quite different sets of assumptions were used to cover a plausible range of possible and uncertain future developments in the regional fish-farming sector.

3.3. Transfer and Evaluation Algorithms for Evaluating Water Quality of Pit Mining Lakes

A good quality of water of the new mining pit lakes is a fundamental prerequisite to develop tourist activities in the Lusatia region. However, after closure of the open-cast mines the rising groundwater saturates the weathered dump materials and causes the acidification of the resulting lakes. Therefore, the best strategy to prevent the development of acid mining pit lakes is to fill clean surface water into the pits of the open-cast mines. In this way, the acid groundwater is kept in the surrounding bedrock and mining tips with only minor impacts on the lakes' water quality. In general, it can be asserted as a rule of thumb that the final lakes' water quality will be the worse the slower the filling of surface water into the mining pits is carried out and the less surface water is available in absolute terms (Grünewald, 2001; Gröschke, Uhlmann, Rolland, & Grünewald, 2002). Since tourism is one of the few options for future economic development in the study region, good water quality will in any case be produced by means of technical water treatment at least for the lakes which are projected for tourist use. Considering European water law, a good quality of all water bodies is to be achieved during the next decade according to the European Water Framework Directive (see Petry/Dombrowsky, in this volume). Designating the mining pit lakes as *heavily modified and artificial water bodies* (CIS Working Group 2.2, 2003) would diminish the water-quality requirements for some years. But such a designation could have severe impacts on the tourist use of the mining pit lakes. Thus, an acceptable level of lake water quality has to be achieved or produced during the coming years. Eventually, the effectiveness of the strategy to fill the pits with surface water and the availability of surface water to realize this strategy are crucial factors. They will determine the future lakes' water quality, the resulting water treatment efforts necessary to achieve a good water quality, and the financial costs to ensure a water quality which allows tourist use and which meets the requirements of European water law.

Considering this perspective of water quality and its impacts on society in the study region the choice of the evaluation method to be used was straight forward: a water treatment cost approach was deemed to be most appropriate. Comparing the treatment costs of the mining pit lakes with the opportunity costs of surface-water use would reveal, whether surface water should be used to fill pit lakes as quick as possible in order to keep treatment costs low or to allocate the surface water to utilizations with higher benefits. After having chosen the evaluation method, the water treatment techniques for mining pit lakes were examined. The most common technique to deal with acid water in lakes was neutralization through adding lime into the water body. For this technique cost data were available. Other, more innovative techniques which, e.g., are based on biological processes to transform the acid by means of bacteria, are still in their examination phase and no cost data were available for large-scale application. Therefore, the neutralization technique by means of inserting lime was chosen.

With regard to water quality in the pit lakes and their treatment via lime neutralization two different problem areas needed to be distinguished. The first one refers to the water treatment of pit lakes projected for tourist use. For these lakes it was assumed that the appropriate amount of lime would be added after the lake has been filled. For this case, only one treatment was planned. The second problem area relates to pit lakes that are meant to become reservoirs in the future, i.e., part of their water will time and again be released into the streams. For these mining pit lakes a permanent good quality of water has to be achieved, because, by force of water law, water released from reservoirs must meet water-quality standards. The reservoirs will mainly be used to regulate the water flows of the watershed streams through taking up water in times of abundant availability and releasing it during times of drought. Due to changing water levels in these reservoir lakes and the existing acid pressure from the groundwater the achievement of a permanent acceptable lake water quality through one lime treatment is not possible. Therefore, neutralization of acid reservoir lake water must not only take place once after the filling has been completed, but the water which is released from the reservoirs into the streams must also be treated in a continuous process.

After distinguishing these two problem areas of water treatment an evaluation algorithm could be set up straightforwardly in cooperation with the engineers and water experts of the rehabilitation firm. Eq. (5) shows the evaluation algorithm which is valid for both problem areas of water treatment described above. According to this equation the yearly costs of water treatment of a lake i in present value ($C_{\text{treat},t,i}$) amount to the discounted product of the amount of lime needed to neutralize the acid water of a lake i in year t ($v_{i,t}$) and the variable costs for lime (CL_t) and wages (W_t) in the respective year t to apply this water treatment technique to a lake.

Evaluation algorithm:

$$C_{\text{treat},i,t} = \left[v_{i,t} \times (\text{CL}_t + W_t)\right] \times (1 + r)^{-t+1} \qquad (5)$$

with $C_{\text{treat},i,t}$: water treatment costs for lake i in year t in present value (€/year); $v_{i,t}$: amount of lime required to neutralize lake i in year t (tons); CL_t: cost of lime in year t (€/ton); W_t: average wages valid in year t (€/ton lime introduced to a lake); r: discount rate.

In order to integrate this economic evaluation algorithm into the water management simulation model it was necessary to derive a transfer algorithm to connect it to the WBalMo results in terms of water flow per second. The quantity of surface water to fill a pit lake and the respective duration of time could be determined by the model. The key variable in this context was the amount of lime required to neutralize acid lake water ($v_{i,t}$). In contrast to the evaluation algorithm, the derivation of the transfer algorithm necessitates to distinguish between single water treatment (SWT), which is necessary to attain an acceptable initial water quality in any pit lake, and continuous water treatment (CWT) relating to water, which is meant to be released from reservoir lakes.

For the SWT of a pit lake the amount of lime needed to neutralize the acidic lake water was calculated based upon its acidity after conclusion of its filling with surface water. The acidity value of a lake is determined by the time needed to fill the lake with surface water and the amount of acid substances eluted from the mining-tips. The following box shows the derivation of the transfer algorithm for this case

Derivation of the transfer algorithm for SWT of a lake:
Lake acidity is determined by

$$A_i = DF_i \times mAE_i \tag{6}$$

Quantity of lime to neutralize acidity is determined by

$$v_i = A_i \times Q_{\text{lime}} \tag{7}$$

hence

$$v_i = (DF_i \times mAE_i) \times Q_{\text{lime}} \tag{8}$$

with A_i: acidity of pit lake *i* after filling with surface water is concluded (kmol); DF_i: duration of filling a pit lake *i* (month); mAE_i: mean elution of acid substances of pit lake *i* per month (kmol/month); v_i: specific amount of lime to neutralize an acid pit lake *i*;

Q_{lime}: quantity of lime required to neutralize an acidity of 1 kmol (*t*/kmol).

The water of those pit lakes which are going to be used as reservoirs in the future will probably be acidic and there will be a long-term entry of acid substances into the lakes. The amount of acid substances entering the lakes depends, among others, on the management of the reservoirs, while the quantity of lime needed for neutralization correlates to the acidity of the reservoir lake and the amount of water released from it. As a result, there exists a complex combined effect of natural water yield and reservoir water management. For a simple calculation of lime quantities needed to neutralize different acidity values, the amount of lime was related to a general acidity content of 1 mmol/L. By multiplying the amount of water released with the acidity content, the amount of lime needed for neutralization was estimated. The formal derivation of the transfer algorithm is shown in the following box.

Derivation of the transfer algorithm for CWT of a reservoir lake:
Lake acidity is determined by

$$A_i = \mathrm{RL}_{i,t} \times \mathrm{mAE}_i \tag{9}$$

Quantity of lime to neutralize acidity is determined by

$$v_i = A_i \times Q_{\mathrm{lime}} \tag{10}$$

hence

$$v_i = (\mathrm{RL}_{i,t} \times \mathrm{mAE}_i) \times Q_{\mathrm{lime}} \tag{11}$$

with A_i: acidity of water released by reservoir i in year t (mol); $\mathrm{RL}_{i,t}$: water released by reservoir i in year t (m^3); mAE_i: mean acid substance entry to reservoir i in year t (mol/m^3); v_i: specific amount of lime to neutralize water released by reservoir i; Q_{lime}: quantity of lime required to neutralize an acidity of 1 kmol (t/kmol).

Implementation of these transfer and evaluation algorithms into the water management model by means of DYN-elements enables us to determine the costs of lake water treatment. Based on a comparison of the lake water treatment costs of different scenarios with various amounts of surface water available to fill the pit lakes, the opportunity cost of surface water in terms of avoiding marginal efforts of water treatment can be modeled.

Finally, we want to add some technical information necessary to apply the presented evaluation approach. According to information of the rehabilitation firm the price of lime was 110€ per ton for the year 2002. Concerning the further background assumptions used for this evaluation of future water treatment activities, it was supposed that the price of lime will rise by the rate of inflation in both FoDs. With regard to labor costs it was assumed that real wages will remain constant in FoD B2 and will decline by 20% until 2052 in the globalization FoD A1.

However, it must be stated that the application of this evaluation approach to estimate water treatment costs of pit lakes just produces very conservative results which are likely to underestimate the actual costs. The reasons for this are twofold. First, this approach exclusively considers the water-quality problem of acidity. It is true that this water-quality problem is the most pressing one with regard to the pit lakes in the study area, but other water-quality problems related to heavy metals, nutrients and the like exist as well and are not included. Second, in practice the application of the water neutralization technique with lime does not always lead to acceptable water-quality levels in pit lakes. Sometimes, additional treatment is necessary. Hence, the results of this evaluation approach only reflect a lower cost level of water treatment.

4. RESULTS OF INTEGRATED WATER RESOURCE MODELING AND EVALUATION

The model WBalMo with integrated evaluation algorithms was applied to the river basins of the Spree river and the Schwarze Elster river and their water-scarcity situation. Five alternative water management strategies were modeled and four FoDs for global change were considered such that data for 20 scenarios were calculated in total. Results are presented according to three management strategies called *Basic*, *Filling*, and *Reduced support*, all under conditions of different FoDs. The *Basic* strategy reflects the policy strategy pursued today. The *Filling* strategy considers a different priority ranking in providing users with surface water. It gives the filling of pit mines a higher priority instead of supplying these utilizations last as it is practice today. The compliance of ecological required minimum discharges in the streams retains its high priority while the other water users receive lower priorities. In the *Reduced support* strategy it was abstained from artificially supplying small streams near the mining pits with water (see Chapter 4 (Messner, this volume) for details on scenarios). The evaluation results, which are presented for fish-farming profits and water treatment costs in pit lakes, relate to the time horizon of 2003–2052. The results in Euro are real values in € of 2003. They are mean values based on the results of the 100 realizations modeled by WBalMo for each scenario.

4.1. Impacts of Global Change and Water Management on Fish Farming

The WBalMo modeling results referring to fish farming are presented in the following figures. Fig. 5 shows the development of fish-farming profits for

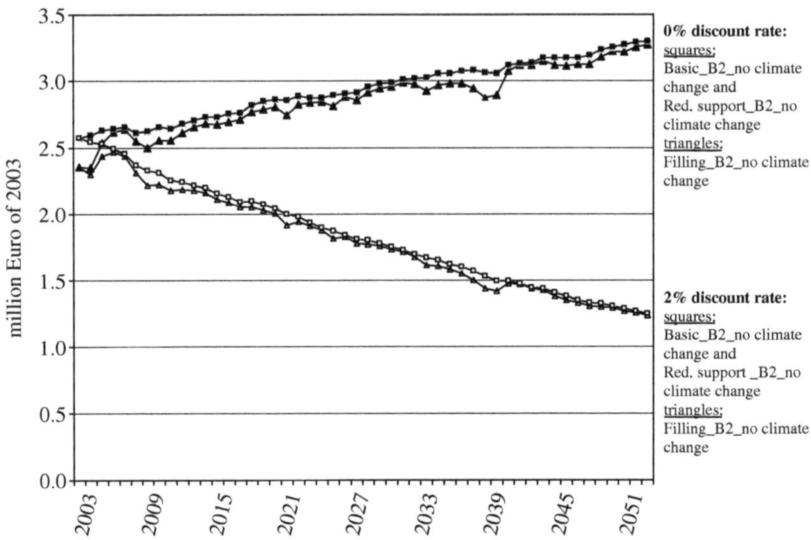

Fig. 5. Aggregated Fish-Farming Profits in Million Euro of 2003 for the Strategies *Basic, Filling*, and *Reduced Support* Under Circumstances of FoD B2 Without Climate Change and With Two Discount Rates.

the regionalization FoD B2 without climate change using a 2% and a 0% discount rate. Since the results for *Basic* and *Reduced support* are almost identical they are presented together in form of lines with squares, while the results for *Filling* are in the form of lines with triangles. It can be seen that regional fish farming profits, which are in the range of about €2.5 million in 2003, will in general increase slightly over time up to a value of about €3.3 million in 2053, if result values are not discounted (upper lines). Recalculating the results by using a 2% social discount rate in order to reflect that profits are less valuable in the future due to interest differences, the same development can be characterized by declining present value real profits with values of about €1.3 million in 2053.

Comparing the results of *Filling* with the other two water management strategies shows that a higher priority for providing surface water to the pit lakes results in yearly mean losses of about 2% to the fish-farming sector. Single deviations with larger or smaller losses in several years are due to variations in natural water yield and to the start or the end of filling activities at pit lakes, respectively. Since these results are aggregated values for the whole fish-farming sector it should be added that different fish-farming locations are affected very differently in the study area.

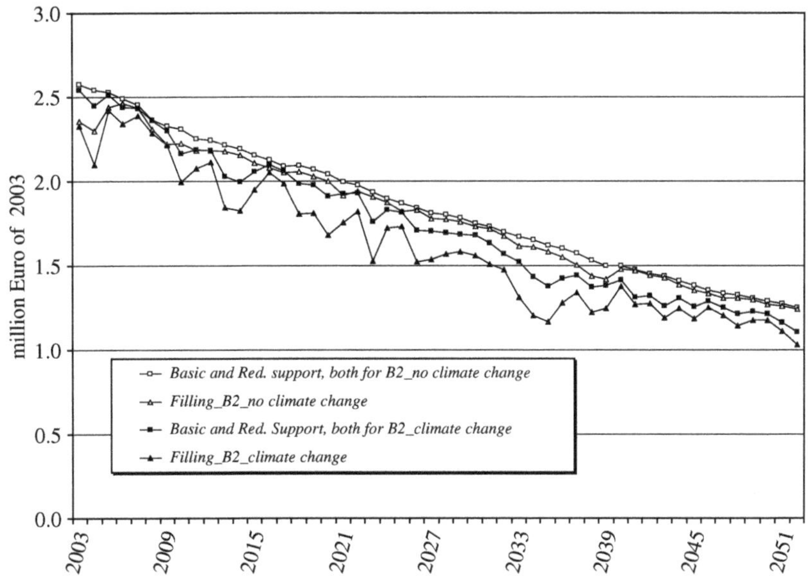

Fig. 6. Aggregated Fish-Farming Profits in Million Euro of 2003 With 2% Discount Rate for the Strategies *Basic, Filling*, and *Reduced Support* Under Circumstances of FoD B2 With and Without Climate Change.

Fig. 6 displays real and discounted fish-farming profits for the three water management strategies under the circumstances of regionalization with and without climate change (FoD B2). It is evident that reduced surface-water availability in the context of climate change leads to increased losses for fish farmers. Considering the combined effects of the *Filling* strategy and climate change amounts to average losses of about 12% compared to the *Basic* scenario without climate change. This means that climate change as assumed to take place in the GLOWA Elbe project poses a larger threat to the fish-farming sector than a change in the water management strategy.

Eventually, Fig. 7 exhibits aggregated results of 2003–2052 for the three strategies *Basic*, *Filling* and *Reduced support* under the circumstances of the four FoDs. At first glance the difference in the aggregated profit level of the B2 und A1 FoDs is outstanding. It arises from different socio-economic future assumptions for regionalization and globalization. Without question, the largest impact in this respect is the assumption relating to the future of subsidies paid to the fish-farming sector. They are kept stable in B2 due to

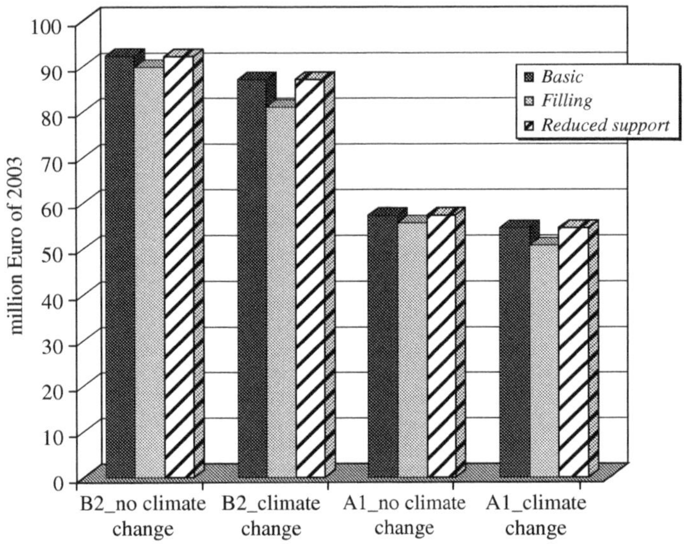

Fig. 7. Over 50 Years Aggregated Fish-Farming Profits in Million Euro of 2003 With 2% Discount Rate for the Strategies *Basic, Filling*, and *Reduced Support* for all FoDs.

realistic arguments that the subsidies may be paid in the future as compensations for environmental services for the maintenance and conservation of fish pond areas which serve as habitats for many threatened species. In contrast, it is assumed to be reasonable for A1 that the subsidies will be cut in the process of harmonizing economic fish-farming conditions in the EU. These different socio-economic assumptions lead to a distinct divergence in the aggregate profit levels being about €90 million for B2 and €55 million for A1. This difference in profits of about 40% corresponds to the amount of about 14 yearly regional fish-farming sector profits of 2003. This means for the fish-farming sector that the range of future uncertainty relating to socio-economic change is very high. Compared to this the potential profit losses due to the *Filling* strategy (0.5–2.5 yearly regional fish-farming sector profits of 2003) or to climate change impacts (1–3.5 yearly regional fish-farming sector profits of 2003) are relatively small.

In face of the fact that these results reflect mean values related to 100 realizations of each climate scenario considered over time in WBalMo, the aspect of future climate uncertainty can be taken into account in the interpretation of the evaluation results. Fig. 8 below shows the results for

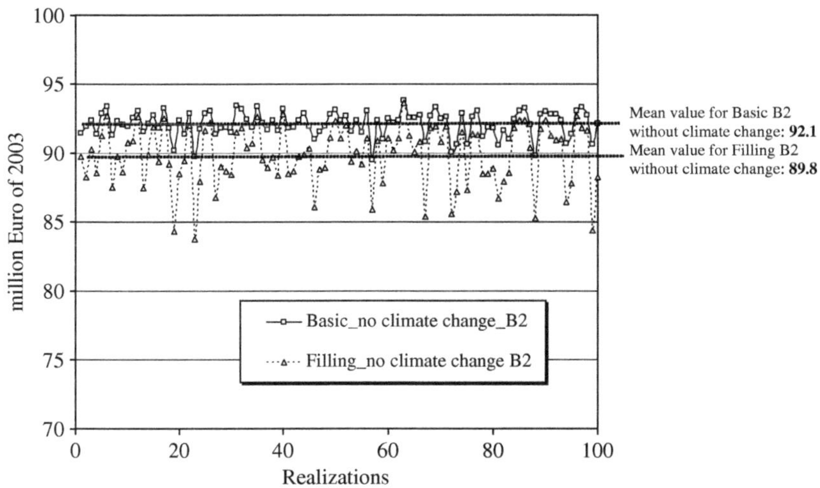

Fig. 8. Aggregated Fish-Farming Profits in Million Euro of 2003 With 2% Discount Rate for the Strategies *Basic* and *Filling*, Both in FoD B2 Without Climate Change, Relating to 100 Climate Realizations.

*Basic*_B2 without climate change and *Filling*_B2 without climate change for all 100 realizations. This figure indicates that several fish-farming evaluation results are overlapping, while there is an obvious tendency of correlated values in the realizations. In a statistical *t*-test it can be examined, whether a robust difference exists between the mean values of the scenarios in view of the varying results of the 100 climate realizations. The *t*-test produced the unambiguous result that the mean values are significantly different at a 5% confidence interval – with a *t*-value of 16.2. Therefore, it can be anticipated taken for granted that the mean values of these scenarios are different. This example reveals that the integration of evaluation algorithms into WBalMo produces the additional advantage to automatically calculate the economic results for all climate realizations. Hence, it is possible to use these results in sensitivity analyses in order to confirm the robustness of the economic evaluation results.

4.2. Impacts of Global Change and Water Management on Water Treatment Costs of Pit Lakes

The evaluation results on water treatment costs for mining pit lakes are discussed with respect to the following figures. Fig. 9 shows the development

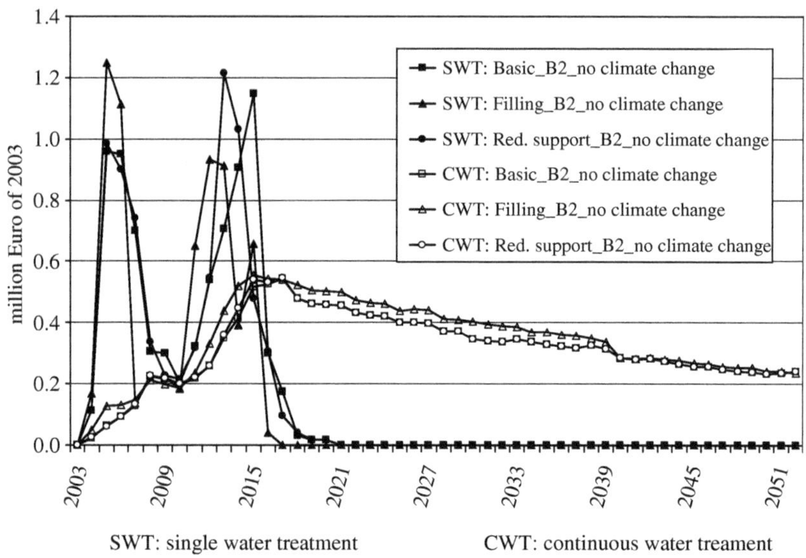

Fig. 9. Single and Continuous Water Treatment Costs for Mining Pit Lakes Projected for Tourism or Reservoir Utilization in Million Euro of 2003 With 2% Discount Rate for the FoD B2 Without Climate Change.

of costs for SWT and CWT for the strategies *Basic, Filling* and *Reduced support* for the FoD B2 without climate change.

The cost lines for SWT are characterized by an erratic course. This is due to the fact that water treatment costs emerge only once immediately after completion of the filling process. Hence, the peaks of the curves indicate the point in time when water treatment costs arise for one or several lakes. Comparing the three water management strategies shows that the strategy *Filling* is most favorable. It displays one early and high cost peak, while afterwards the cost peaks are distinctly lower than those of the strategies *Basic* and *Reduced support*. This indicates that the filling process is much faster and cost-saving compared to the other alternatives. Due to higher surface-water availability and a faster filling process in the *Filling* strategy, the inflow of acidic substances is lower and, therefore, less neutralization material is needed and water treatment costs are lower.

Considering the continuous water treatment costs for the reservoir lakes it must be stated that the strategies *Basic* and *Reduced support* are almost equal in terms of treatment costs, while *Filling* displays a higher cost level until 2040. The reason for this is that in the *Filling* strategy the lakes are

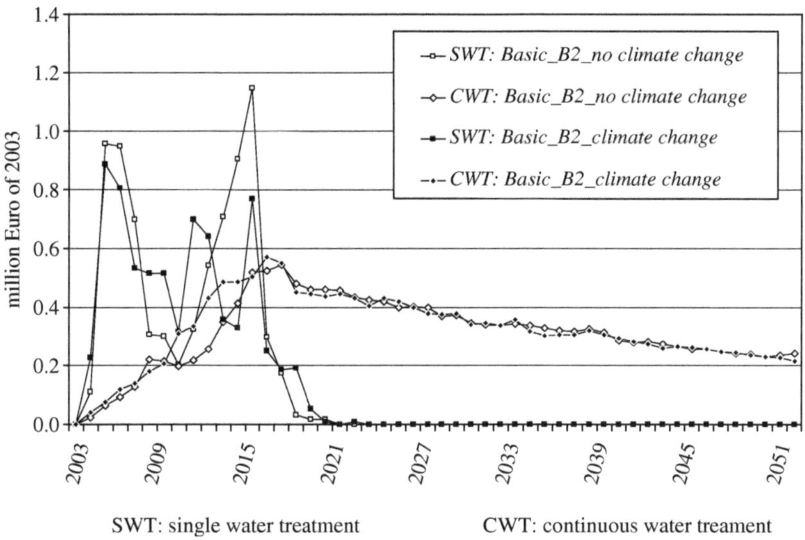

Fig. 10. Water Treatment Costs for Mining Pit Lakes With Projected Tourism and Reservoir Utilizations with Impact of Climate Change, Shown for Strategy *Basic* in FoD B2 in Million Euro of 2003 With 2% Discount Rate.

filled faster with surface water and, therefore, their operation as reservoirs can be started earlier compared to the other strategies. However, since reservoirs provide water in times of scarcity, there is always a water-utilization benefit connected to these costs which may balance this cost disadvantage.

In order to reveal the impact of climate change on single and continuous water treatment costs Fig. 10 exhibits cost lines for the strategy *Basic* in the two FoDs B2 with and without climate change.

Regarding SWT costs it must be stated that they are lower in case of climate change. Considering the findings of the climate modelers this is reasonable. The results of their calculations indicate that some areas of the region will temporarily feature higher discharges in the case of climate change. Since the Lusatian Lake District, which is the most important rehabilitation area, is affected by these higher discharges, more water is available for the lakes and, therefore, less costly SWT is necessary. It must be noticed that mining pits in other regions of the catchments receive less water under the assumed climate change conditions.

The implications for CWT are different. Due to the fact that more water is available under conditions of climate change reservoirs can start their

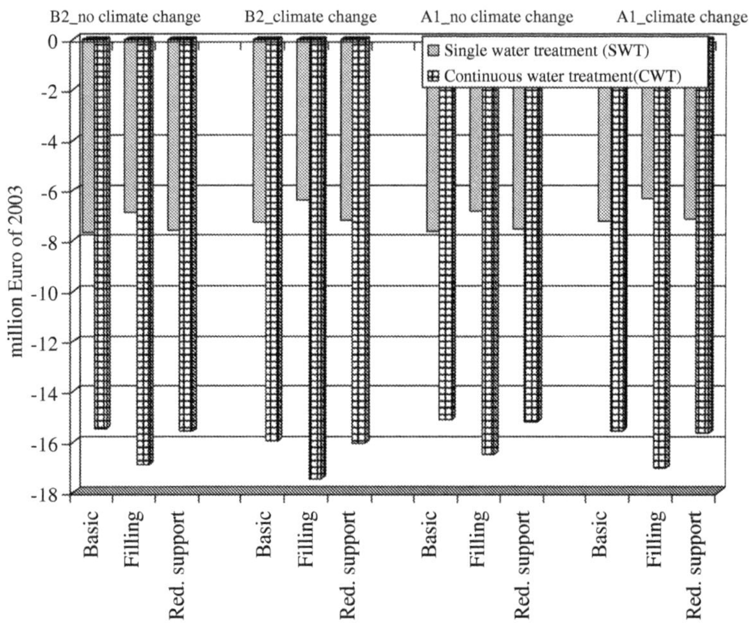

Fig. 11. Over 50 Years Aggregated Water Treatment Costs for Mining Pit Lakes Projected for Tourism and Reservoir Utilizations in Million Euro of 2003 With 2% Discount Rate for the Strategies *Basic, Filling*, and *Reduced Support* for all FoDs.

operation earlier. However, releasing reservoir water earlier also means that CWT starts earlier and produces higher costs. This time-lack effect occurs over the period 2009–2016 and is documented in Fig. 10 in terms of higher continuous water treatment costs under conditions of climate change.

Aggregated results for water treatment costs over the whole time horizon 2003–2052 are shown in Fig. 11. The SWT costs to ensure an acceptable initial water quality in all mining pit lakes amount to about €6–8 million of 2003, while continuous water treatment costs as a result of reservoir lake activities lie in the range of €15–18 million of 2003. It must be emphasized again that these results underestimate the actual costs of water treatment and can only be interpreted as minimum values.

4.3. Implications of the Results

The results presented in the above sub-sections do only cover two categories of effects related to water management and global change. Based on these results alone it is not possible to execute a final comparative evaluation of the water management strategies *Basic*, *Filling*, and *Reduced support*. For an attempt to evaluate these strategies on the basis of a more comprehensive set of monetary *and* non-monetary evaluation results see Chapter 12 (Messner, in this volume).

The major objective of this chapter was to outline an approach to integrate economic evaluation algorithms into a long-term water management model and to exemplify its application for fish farming and water treatment. The results outlined above illustrate that benefits and costs related to water management and global change may alter significantly over time due to many natural and socio-economic impacts like climate change, changing water utilization pattern in society, demographic and industrial developments, and eastward enlargement of the EU. Integrating evaluation algorithms directly into a water management model like WBalMo allows calculating a huge data set of benefits and costs simultaneously with water availability data and, thus, to automatize and standardize the computation of benefits and costs. This facilitates and accelerates economic evaluations that must be executed time and again to support decision making in water management. It also makes it possible to take information about uncertainty into account, e.g., to include the results of all 100 climate realizations of each scenario into a sensitivity analysis to test the robustness of the mean value results. In addition to this and depending on the structure of water management models, the integration of economic evaluation algorithms does also allow the computation of disaggregate spatial or temporal data which sometimes might be necessary to evaluate measures with a specific spatial or temporal focus. Last but not least, if WBalMo results on all economic effects of varying water availability are combined with its output data on physical water availability, the opportunity cost of water use in whole river basins can be calculated. Thus, the potential gains from this integrated evaluation approach are promising.

5. CONCLUSIONS

The major conclusion of this chapter reads: our attempt to integrate economic evaluation algorithms directly into the long-term water management simulation model WBalMo was successful. This paves the way for a powerful interdisciplinary tool to be applied in practice of river-basin management, e.g., to support the implementation process of the European Water Framework Directive. The basis for this success was on the one hand sophisticated interdisciplinary research. Water management modelers and economists worked intensively to explain and understand each others' scientific concepts, notions, and approaches towards evaluation and modeling. Many misunderstandings were to be discussed to set the record straight. On the other hand, only the interaction with important water use stakeholders, like the fish-farming association or the rehabilitation firm LMBV, enabled us to get an idea of the economic vulnerability and costs of single water users and to derive appropriate transfer and evaluation algorithms.

Nonetheless, in the process of this interdisciplinary endeavor we also experienced the limits to integrate economic evaluation algorithms of water use into a pre-existing water management model. Although it was possible to integrate evaluation algorithms for the most important cost and benefit categories,[4] problems emerged with regard to water use of fish farming and industry which had its roots in the general structure of the pre-existing model WBalMo Spree/Schwarze Elster. In Section 3.2 it has been mentioned that the marginal evaluation approach for fish farming could not be applied due to the structure of data contained in the model and its aggregation of individual surface-water uses to larger units. Fortunately, a second-best approach could be found for fish farming that can be considered acceptable. However, this was not the case for industrial uses of surface water. In the WBalMo model industrial water uses are aggregated on a very high level without distinguishing between the types of industry or even the types of industrial water uses like fresh water, cooling water, or process water. Thus, even if evaluation algorithms could have been derived for industrial water uses, no interfaces to the model existed to create transfer algorithms. This means, economic evaluation was not possible based on the WBalMo results on surface-water provision to industry. Therefore, a supplementary non-monetary evaluation criterion was derived in order to include the WBalMo industry results into a final assessment (see Messner, Chapter 12, in this volume).

As a result of these experiences the major lesson learned from this interdisciplinary project was that integration of economic evaluation algorithms into water management models would work out best, if the development of the water management model is performed by an interdisciplinary team right from the start. Only in this case it can be ensured that specific disciplinary aspects of evaluating effects of water availability are taken into account. In an adjacent project called GLOWA Elbe 2 we will extend our approach with regard to the evaluation of further water uses (e.g., energy production, inland navigation, irrigation agriculture) and to the modeling of the larger Elbe river basin. In this project an interdisciplinary team will work on the advancement of WBalMo – especially regarding the definition of aggregation modules for water users and the identification of evaluation and transfer algorithms. Furthermore, we will improve the economic input data of WBalMo such that surface-water demand of urban settlements, agriculture, and industry is computed by means of economic models as well. Only if the water management model has a solid basis in terms of economic demand input data and evaluation interfaces to all important water utilizations in the river basin, our interdisciplinary task will be completed adequately. This is still a considerable distance to cover.

NOTES

1. The term "water user" has to be defined, because of its different meanings in economics and water management. A water user is an actor or an entity, who withdraws or discharges surface water, or for whom water is reserved to remain in the river or in another water body. This definition includes economic actors like water utilities, industry, power stations, inland navigation, agriculture as well as official actors who manage water reservoirs and who realize water transfers. It also includes ecological systems, for which minimum streamflows are determined to remain in the river. With reference to this definition we define the term "water use" as the water-use activity which is executed by economic water users, while "water utilization" is used as a broader term, which comprises economic use *and* water use by official actors and ecological systems.

2. Unlike the other percentage figures, the 26% for Berlin do not mean that the water users in Berlin withdraw 26%. Rather, because Berlin is the last spatial module in WBalMo, this percentage reflects the amount of surface water which is demanded as inflow into the capital city.

3. According to Klemm (2001, p. 4) average fish output per hectare in physical units (AQ) was 649 kg per hectare in the late nineties and average profit per unit fish (AP) was about €0.68 per kilogram fish. As a result, profit per hectare sums up to €434.09. These figures were used for the modeling starting year 2003.

4. These most important categories were: fish-farming, water-treatment costs, tourism at mining pit lakes, and costs of providing surface water by means of pumping and diverting water from sources within and outside the river basin.

ACKNOWLEDGMENTS

This work was carried out as part of the German Research Program on Global Change in the Hydrological Cycle (GLOWA), namely GLOWA Elbe, funded by the German Federal Ministry of Education and Research (BMBF). It was supported by a number of regional organizations like environmental authorities (e.g., Landesumweltamt Brandenburg, Staatliches Umweltfachamt Bautzen) and water users (e.g., the rehabilitation firm LMBV, the energy firm Vattenfall Europe Mining & Generation, the fish-farming association, and the Lusatia Initiative of small water users). Moreover, we thank Oliver Zwirner for careful reading of the manuscript and Matthias Karkuschke for providing valuable input data from expert interviews. Finally, we thank Kirsten Hennrich for constructive comments on an earlier draft.

REFERENCES

Brouwer, R., Langford, I. H., Bateman, I. J., & Turner, R. K. (1999). A meta-analysis of wetland contingent valuation studies. *Regional Environmental Change*, *1*, 47–57.

Cech, T. B. (2003). *Principles of water resources – history, development, management, and policy*. New York: Wiley.

Chow, V. T., & Kareliotis, S. J. (1970). Analysis of stochastic hydrologic systems. *Water Resources Research*, *6*(6), 1569–1582.

CIS Working Group 2.2 (2003). *Guidance Document On Identification And Designation of Heavily Modified And Artificial Water Bodies*. 14/01/2003, http://www.sepa.org.uk/hmwbworkinggroup/Guidance/hmwb_guidance

Cunningham, A. B., & Amend, J. R. (1986). Interactive simulation of water resource systems. *Journal of Water Resources Planning and Management*, *112*(3), 326–338.

EU-WFD (2000). The European Parliament and the Council of the European Union. Directive 2000/60/EC of 23. October 2000 establishing a framework for Community action in the field of water policy (European Water Framework Directive – EU-WFD). *Official Journal of the European Communities L 327/1*, Brussels 22/12/2000, of 22 December 2000.

Gren, I.-M., Groth, K.-H., & Sylvén, M. (1995). Economic values of Danube floodplains. *Journal of Environmental Management*, *45*, 333–345.

Gröschke, A., Uhlmann, W., Rolland, W., & Grünewald, U. (2002). Hydrochemische Entwicklung Lausitzer Tagebauseen während der Flutung – Beispiel Gräbendorfer See. *Hydrologie und Wasserbewirtschaftung*, *46*(6), 256–267.

Grünewald, U. (2001). Water resources management in river catchments influenced by lignite mining. *Ecological Engineering (Special Issue)*, *17*, 143–152.

Hansjürgens, B., & Messner, F. (2002). Die Erhebung kostendeckender Wasserpreise in der Wasserrahmenrichtlinie. In: S. von Keitz & M. Schmalholz (Hrsgs.), *Handbuch der Wasserrahmenrichtlinie – Inhalte, Neuerungen und Anregungen für die nationale Umsetzung* (pp. 293–319). Berlin: Erich-Schmidt-Verlag.

Hirsch, R. M. (1981). Stochastic hydrologic model for drought management. *Journal of the Water Resources Planning and Management Division*, *107*(2), 303–313.

Kaden, S., Schramm, M., & Redetzky, M. (2004). ArcGRM: Interactive simulation system for water resources planning and management in river basins. In: R. Z. Xi, W. Z. Gu & K. P. Seiler (Eds), *Research basins and hydrological planning* (pp. 185–192). London: Taylor & Francis.

Klemm, R. (2001). *Bericht zur Wirtschaftlichkeit sächsischer Teichwirtschaften 1996/97 bis 1999/2000*. Hg. von der Sächsischen Landesanstalt für Landwirtschaft. Dresden.

Langner, N. (2002). *Results of expert interviews of April 18, April 23, and July 16, 2002*. Saxony State Bureau of Agriculture, Department of Fish farming. Königswartha.

Martinsen, D., Kraft, A., & Markewitz, P. (2001). *Szenarien zur CO_2-Minderung auf Basis des IKARUS-Optimierungsmodells*. Bericht des Forschungszentrums Jülich, Jülich.

Messner, F. (2005). The value of water in modern western societies. *IHP/HWRP-Berichte*, Number 3/2005, 7–17.

Rogers, P., & Fiering, M. B. (1986). Use of system analysis in water management. *Water Resources Research*, *22*(9), 146–158.

Rosegrant, M. W., Ringler, C., McKinney, D. C., Cai, X., Keller, A., & Donoso, G. (2000). Integrated economic–hydrologic water modeling at the basin scale: The Maipo river basin. *Agricultural Economics*, *24*, 33–46.

Vögele, S., Markewitz, P., & Martinsen, D. (2001). Die Analyse des deutschen Strommarktes mit Fokus auf die neuen Bundesländer sowie die Ableitung von möglichen Strommarktszenarien bis zum Jahr 2050. Interner Endbericht des Forschungszentrums Jülich an das UFZ.

Wurbs, R. A. (2005). *Comparative evaluation of generalized river/reservoir system models. TR-282*. College Station: Texas Water Resources Institute.

Yang, W., Khanna, M., Farnsworth, R., & Önal, H. (2003). Integrating economic, environmental and GIS modeling to target cost effective land retirement in multiple watersheds. *Ecological Economics*, *46*, 249–267.

Yeh, W. W. G. (1985). Reservoir management and operation models: A state-of-the-art-review. *Water Resources Research*, *21*(12), 1797–1818.

INTEGRATED ASSESSMENT OF WATER POLICY STRATEGIES IN THE CONTEXT OF GLOBAL CHANGE: THE INTEGRATIVE METHODOLOGICAL APPROACH AND ITS APPLICATION IN THE SPREE AND SCHWARZE ELSTER RIVER BASINS

Frank Messner

ABSTRACT

In this chapter, the integrative methodological approach (IMA) of the research project GLOWA Elbe is introduced, which represents a scientific methodology to support water management under uncertainty regarding future paths of global change. The approach paves the way for integration of research work of many disciplines, of different assessment methods, of various policy fields, and the involvement of relevant stakeholders and decision makers. IMA can be roughly described by four research elements (scenario derivation, indicator and criteria identification, model-based

Ecological Economics of Sustainable Watershed Management
Advances in the Economics of Environmental Resources, Volume 7, 265–303

ISSN: 1569-3740/doi:10.1016/S1569-3740(07)07012-5

impact analysis, and final scenario assessment based on combined benefit–cost and multi-criteria analysis), which lay the basis for the IMA activities of the global change research sequence. Its practical application is demonstrated by a case study on the Spree and Schwarze Elster river basins. Specific results of Chapter 4 (on scenario derivation) and Chapter 11 (on integrating economic evaluation into water management simulation) in this volume are picked up in order to focus on the illustration of the integrated assessment results for this German case study.

1. THE CHALLENGE OF ANALYZING WATER POLICY STRATEGIES IN THE CONTEXT OF GLOBAL CHANGE

In the year 2000, the German Ministry of Science and Education started a major and long-term research program to examine the impacts of global environmental and socioeconomic change on the water cycle for selected river basins in Europe and Africa, called GLOWA (see http://www.glowa.org/). The principal objectives of the GLOWA program were to analyze the impact of global change on major river basins, to communicate the results to the public and important stakeholders, and to identify appropriate strategies to mitigate and to adapt to its most harmful consequences in order to ensure a sustainable future use of water resources and services. The five GLOWA projects, which were initiated in 2000 and funded for at least 6 years, dealt with the analysis of highly complex systems involving many types of uncertainty. For two reasons in particular, this program was a considerable challenge for the scientific community.

On the one hand, it is already a large endeavor to improve the understanding of global change from the perspectives of natural *and* social sciences. Global change embraces aspects such as climate change, changing demographic pattern, dynamics of economic development, globalization and urbanization, the threatening shortage of freshwater all over the globe, the reduction in biodiversity, the changes in life styles and political paradigms and many other important topics of global environmental and developmental policy (WBGU, 2003). Many, partly correlated driving forces determine the development of global change. Its analysis is a highly complex and extremely interdisciplinary task. A large spectrum of time and spatial scales as well as many types of uncertainties and even ignorance need to be considered. In this context, it is not astonishing that the creation of a common harmonized interdisciplinary approach is impeded by difficulties that often arise in interdisciplinary research. For example, due to different speeds of change in the various systems to be analyzed, the choice of a general time horizon for the overall analysis is a thorny problem. While climate change takes place slowly and needs time scales on a centuries' basis (with usual time scales of about 50–200 years), changes in society and the economy are much quicker and no social scientist generally dares to make reliable predictions concerning future decades (usual time scales 1–20 years).

On the other hand, integrated and applied scientific research aiming at the analysis of the water cycle of a large river basin is also a challenging task (Thomas & Durham, 2003) – even without considering global change. Complex cause–effect relationships between natural and socioeconomic systems have to be considered and appropriate strategies for an integrated water resources management (IWRM) must be identified, preferably in co-operation with relevant stakeholders that are sometimes of quite different opinions. Many actors from water authorities, environmental ministries and water scientists, who are engaged in the current process of implementing the European Water Framework Directive (see Petry & Dombrowsky, in this volume, also Chapter 1 of Rumm, von Keitz, & Schmalholz, 2006) or IWRM approaches elsewhere in the world, are experiencing the complex task of integrating scientific understanding into the IWRM policy process. Integrated analyses of hydrological, biochemical, geological, ecological, toxicological, social and economic processes and relationships are required to determine adequate policy measures to achieve a good state of all water bodies in a river basin. Further complicating factors in this research endeavor are multi-faceted institutional and socioeconomic features of water policy and politics, which need to be considered for the practical implementation of policy measures. For example, many river basins, which are the reference units of IWRM approaches, often cross state or national borders such that many political authorities and decision makers from different states and countries are involved in the process of managing the resources of one river basin. Due to the given direction of the river flow the access to water is asymmetrical and, hence, power relations among riparian states and also between individual water users are asymmetrical as well. It is increasingly demanded by social scientists, actors of civil society, and also by modern water law that upstream–downstream conflicts about water use and water allocation should be dealt with in a participatory policy setting. This means that decision makers and important stakeholders of a river basin should be involved in the overall IWRM process in order to increase the legitimacy of the overall river basin water policy (Timmerman, 2005;

infoResources Focus, 2003; Commission of the European Communities, 2002). While already the cooperation between science and water authorities is not easy and often afflicted with frictions, the inclusion of stakeholders from civil society presents an additional demanding mission of IWRM and related research (Messner, Zwirner, & Karkuschke, 2006; Jonsson, 2005).

Hence, due to the complexity of both research fields, the combined analysis of global change and the water cycle in an IWRM policy setting is a huge challenge for the interdisciplinary environmental research community. It introduces highly complex research tasks with new research questions: How will global change affect current water problems and the effectiveness of current water policy strategies? Which new water conflicts are likely to arise and what kind of precautionary measures could be effective? How will people adapt to the consequences of global change? Will their adaptation create new problems? Which strategies are conceivable to shape global change and its effects on a local level in a constructive way? In face of large uncertainties concerning future development patterns, there are no unequivocal answers or even "optimal" solutions.

Such a complex research field requires an appropriate methodological research approach, which paves the way by showing the direction and language for an applied and structured interdisciplinary research endeavor. An adequate approach should at least embrace a common interdisciplinary language, a model system for future and policy impact analysis, a participation concept, and integrated assessment tools to identify and evaluate policy measures and strategies. For the GLOWA Elbe project such an approach was developed and improved during the first years of research. It is called the integrative methodological approach of GLOWA Elbe (IMA) and is presented in theory and practical application in this chapter. The next section introduces the most important features of IMA. First results gained in the GLOWA Elbe project by applying this approach are portrayed in Section 3, which will also bring together various GLOWA Elbe research results being described in Chapters 4 (Messner) and 11 (Messner et al.) in this volume. Section 4 concludes with some lessons learnt in the IMA-based process of interdisciplinary global change research.

2. THE INTEGRATIVE METHODOLOGICAL APPROACH OF GLOWA ELBE

The IMA has its roots in the debate about integrated assessment. This debate deals with different aspects of interdisciplinary research on complex research questions contributing to the assessment of processes or conditions in nature or society and also to the support of decisions in the policy domain. Thus, integrated assessment comprises, among other things, the multidisciplinary analysis and modeling of complex processes in nature and/or society, the development and refinement of mono- and multi-criteria evaluation methods for the evaluation of policy options in the context of complex circumstances, as well as research on the design of participatory decision processes (e.g., Hoekstra, Savenije, & Chapagain, 2001; Hope & Palmer, 2001; Alberti & Waddell, 2000; Behringer, Buerki, & Fuhrer, 2000; Hare, Letcher, & Jakeman, 2003).

The IMA of the GLOWA Elbe project is an integrative methodology that combines scenario analysis, assessment via benefit–cost analysis (BCA) and multi-criteria analysis (MCA), and participatory methods on the basis of scientific modeling. It provides a generic framework to structure a participatory evaluation process on public decision issues. Based upon the analysis and assessment concepts of Horsch, Ring, & Herzog (2001), Klauer, Drechsler, & Messner (2006), and Wenzel (1999) and referring to the drivers-pressures-states-impact-responses (DPSIR) approach of the European Environmental Agency (OECD, 1994), IMA was developed and refined in the GLOWA Elbe project in order to take the complexities of combined global change and water cycle research in an IWRM policy setting into account (Becker et al., 2001; Messner, Wenzel, Becker, & Wechsung, 2005). In its current form, IMA is a methodological instrument to support public decisions on complex environmental problems in the context of global change, affecting many people, large regions and long periods of time, involving considerable social, ecological and economic effects, and comprising significant uncertainty issues. The integrative power of IMA arises from five different types of integration: (1) integration of different scientific disciplines in one common research effort; (2) integration of stakeholders in the research process; (3) integration of the analysis of different water-related policy fields; (4) integration of various evaluation methods; and, finally, (5) integration of all results into one multi-criteria assessment setting.

Major goals of IMA are:

- to offer a framework for integrated assessment in the context of global change research, including, among other things, a consistent and integrative research sequence and a common language for interdisciplinary research;
- to support and improve the quality of environmental decision making in terms of enhanced competence and fairness.

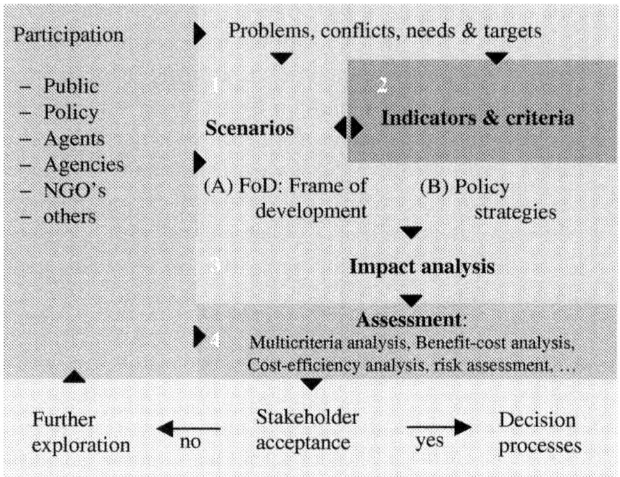

Fig. 1. A Schematic Outline of the IMA.

Regarding the first goal it must be mentioned that IMA provides a general and integrative research scheme that supports the realization of the five different types of integration in the process of assessing global change processes and policy strategies. In order to achieve this, a uniform language is used with clearly predefined definitions and abbreviations that are available in a glossary (Wenzel, 2005). In this way, the communication between scientists of different disciplines, politicians and stakeholders is improved. Thus, a major source of misunderstanding, namely the existence of different definitions for essential scientific concepts and notions, is removed or at least made explicit.

With reference to the second goal, IMA contributes to an increased competence and fairness in integrated assessment by broadening the scientific knowledge base through participation of stakeholders,[1] by including modeling results in the evaluation process, and by explicitly considering uncertainty in the different phases of integrated assessment (Messner et al., 2006).

IMA can be described by four major research elements and their interaction (see Fig. 1), being:

- first, problem analysis and scenario derivation,
- second, indicator and criteria selection,
- third, impact analysis via modeling or other effect estimation methods, and

- fourth, evaluation using benefit–cost and multi-criteria evaluation approaches.

Although this or a similar sequence of steps is indeed indispensable for any evaluation methodology (for AHP, see Forman & Gass, 2001; for BCA, see Hanley & Spash, 1993, p. 8ff.), the uniqueness of IMA arises from its specific characteristics. IMA encompasses the claim to consider uncertainties explicitly, to combine benefit–cost and MCA, and to enhance the significance of participation in all research elements in order to improve the quality of environmental decision making. In the following, the four research elements are outlined in more detail. In this context, the most important IMA specific terms are defined. When these notions are used for the first time they are printed in italics.

2.1. IMA Element 1: Problem Analysis and Scenario Derivation

The starting point in the first IMA element is a thorough problem analysis comprising the examination of the conflict and the existing institutional setting to resolve it. Literature and documents are studied to unfold the history of the problem, the parties involved, the decision-making structure to resolve the conflict, and the measures already taken. This analysis is complemented later on by a stakeholder analysis. Using semistructured qualitative interviews, actors involved in the conflict – i.e., stakeholders as well as decision makers and their executive authorities – are requested to describe their perception of the conflict and their view on how the problem could or should be resolved. This way a more comprehensive picture with a multitude of perspectives emerges: actors and information not mentioned in the literature can be revealed during interviews, informal relationships among actors and informal structures within the policy-making process can be uncovered, and local knowledge as well as internal data from authorities and enterprises becomes available to the researchers. Since stakeholder proposals to resolve the problem at hand are surveyed as well, stakeholder analysis serves as a means for bridging the initial problem analysis and the subsequent analysis for deriving scenarios.

As described in more detail in Chapter 4 (in this volume), different types of scenarios are distinguished in the context of IMA and they are briefly summarized in this paragraph. *Global change scenarios* describe the development of important driving forces and boundary conditions of natural and societal global change (like climate change, population development, water demand, etc.). These scenarios cannot be influenced by local actors. A collection of global change scenarios that contains scenarios for all important driving forces and boundary conditions is called a *Framework of Development (FoD)*. A *policy action scenario* represents the actual execution of a policy strategy over time in order to resolve specific problems or conflicts in the study region. A *policy strategy* is defined as a combination of policy options from one or several policy fields (like water policy, agricultural policy, etc.) that also includes a specific attitude regarding policy adjustment in times of societal and/or natural change events (e.g., a risk averse attitude or an affirmative attitude toward policies that are very market oriented). A policy action scenario is characterized by a set of chosen policy options with a timetable for their practical implementation. *Developmental scenarios* are eventually those scenarios that contain both, a FoD with several global change scenarios as external conditions of change and a policy action scenario reflecting the internal factors of change. These developmental scenarios are the major subjects of analysis in IMA.

The FoDs are important to reflect the uncertainty and the ignorance concerning future development, because for the success of a policy strategy it is essential to know how it will perform under different future conditions. Given the future uncertainty involved, scientific decision support should not aim at optimal policy strategies, but should rather identify robust strategies, which are effective independent of, or despite, future change processes. The assumptions underlying the different FoDs are derived by the scientists in co-operation with experts, decision makers and stakeholders. The actual modeling of global change scenarios later on is mainly a scientific task.

In addition to clarifying and understanding the network of stakeholders involved in a problem, it is also part of the stakeholder analysis to use qualitative interviews to ask stakeholders which fields of action, policy options and policy strategies they deem to be relevant for the resolution of the problem at hand. All relevant answers of the interviewees are gathered in order to get a comprehensive possibility space of policy options. This way no major predecisions – such as the exclusion of relevant options – occur in the early stage of scenario derivation. Of course, policy options and strategies can also be proposed by the scientists involved in the research.

The result of the first IMA research element is a set of scenarios that has been derived together with decision makers and stakeholders. Involving stakeholders early in the research process paves the way for resolving a regional problem in a participatory context.

2.2. IMA Element 2: Selection of Evaluation Indicators and Criteria

What is regarded as success and failure is essential for the assessment of policy options and strategies. Therefore, in an advanced phase of the interviews the actors are asked to specify the indicators they would like to use to measure and assess scenario effects. In order to prevent disputes among stakeholders and as a matter of fairness, all indicators stated to be important should be included in the assessment process – provided double counting of effects does not occur and it is feasible to estimate data for them in the third IMA element, which refers to modeling. As far as the actors accept the general policy aim of sustainable development, the inclusion of ecological, social and economic indicators should be ensured.[2] Later on, evaluation criteria must be defined based on the identified indicators, i.e., evaluation schemes must be derived for single or groups of indicators (Klauer, Messner, Herzog, & Geyler, 2001). For example, if stakeholders want to measure and assess water quality in terms of nitrate concentration in water bodies, it must be decided, among other things, which concentration levels are acceptable or unacceptable, whether concentration levels should be considered on a continuous scale or in quality classes, and which degree of spatial and time aggregation is appropriate (e.g., one could choose *one* average value over space and time for the consideration of a water body over a 10-year period or a *multitude* of specific nitrate concentrations for different locations within the water bodies and also for several time periods). Since the choice of indicators and criteria already contains value decisions, this should be cleared with the stakeholders and decision makers.

2.3. IMA Element 3: Modeling and Estimation of Scenario Effects

The third element of IMA involves the scientific modeling and estimation of scenario effects such that data for the indicators defined in IMA element 2 become available. Very different scientific modules can be used to model and estimate data for the indicators of the second IMA element (Chapter 2 of Horsch et al., 2001; Becker et al., 2001). Several requirements are crucial in this respect. First, all modeling approaches must consider the same boundary assumptions according to the FoDs. Second, the links and feedback mechanisms between the models must be defined precisely such that a flexible modular-based model system can be generated. Third, model variables that describe similar or the same impacts or driving forces must be coordinated. And, last but not least, information about model, data and future uncertainties must be considered explicitly. All modelers are requested to deliver data not only on scenario results, but also on the probability of results and the possible range of model failures due to different sources of uncertainties. Both are taken into account within the multi-criteria assessment later on (Klauer et al., 2006). For example, the water balance model in one of the GLOWA Elbe case studies was fed with 100 variants of one climate scenario in order to deliver the probability distribution connected to the climate-related uncertainty that water will be more or less available at specific locations in the future (Koch et al., 2005). Participation within the third IMA element is limited to the general discussion of models, model assumptions and model input data with experts and stakeholders. As a matter of course, local data from different local authorities is used to refine models to local conditions.

2.4. IMA Element 4: Assessment

The fourth element deals with the assessment of policy strategies in the context of global change and is divided into two parts: a preparatory mono-criteria and a final multi-criteria assessment. The mono-criteria assessment evaluates policy strategies with respect to each single criterion selected in element 2. In the context of IMA, BCA plays a major role. As a rationale of assessment, as many effects as possible are evaluated in economic welfare terms – as far as monetary evaluation is feasible, based on reliable data and accepted by decision makers and stakeholders. An advantage of using BCA in the context of MCA refers to the fact that the aggregation of monetized effects that are incommensurable in character (e.g., due to equity reasons) need not be done, i.e., the BCA approach may feed several results into the MCA. In this way, some net benefits that reflect the welfare effects of underprivileged groups may also represent social aspects of the decision. All effects that cannot be expressed in monetary terms due to methodological problems are assessed by other quantitative or qualitative criteria, using specific evaluation techniques (e.g., nitrate concentration quality classes or risk assessment for threatened species).

The results of all mono-criteria assessments enter the MCA process in the form of data for the multi-criteria matrix, which is prepared for every FoD defined in IMA element 1. The participants in this process should be selected such that all kinds of interests are represented. After having explained and discussed the results and their implications, the stakeholders and decision makers are asked to assign weights to the criteria. Using an outranking approach – for instance "extended PROMETHEE" (Klauer et al., 2006) – rankings of policy strategies are calculated for all participants and these results are subject to discussion. Most probably it will be found that some policy strategies do perform very differently using different weights or different FoDs. Therefore, it is the aim of the discussion to find a widely accepted compromise for a weighting scheme or a common risk behavior in the face of different future developments. As a result, one or a group of strategies should be identified to be the most advantageous. If it is found that none of the strategies is performing well and some additional combinations of options should be considered, an iterative process starts in IMA to take new strategies into account. Proceeding this way, MCA is not used to calculate an optimal policy strategy, but to structure the problem and the results, to reveal the uncertainties involved and to feed reliable information as an input into the participatory decision-making discourse in order to support the identification of a robust and generally agreed strategy.

2.5. The Sequence of IMA Research Tasks

The representation of the IMA by means of four research elements – scenarios, indicators, impact analysis and assessment – has the advantage that it is easily comprehensible and this facilitates the communication process with stakeholders. However, as the arrows in Fig. 1 already indicate, the four elements should not be interpreted as four successive research tasks that need to be executed step by step. Rather, many aspects of each IMA element are closely connected to aspects of other elements and depend on each other within the whole research process. Fig. 2 depicts the sequence of research tasks that need to be executed if IMA is used to analyze and assess global change effects and adaptation strategies.

The IMA elements are listed at the top of Fig. 2, and all research tasks in the ovals, which are placed underneath a specific IMA element in the same column, pertain to it. The IMA elements 2 and 4 are lumped together in order to simplify the figure and also because indicator and criteria derivation and assessment maintain a very close relationship. The 12 research tasks shown in Fig. 2 pass through three different research phases, A, B and C, which are important to distinguish in the model-based global change research context. The tasks and the three phases are briefly outlined in the following paragraphs.

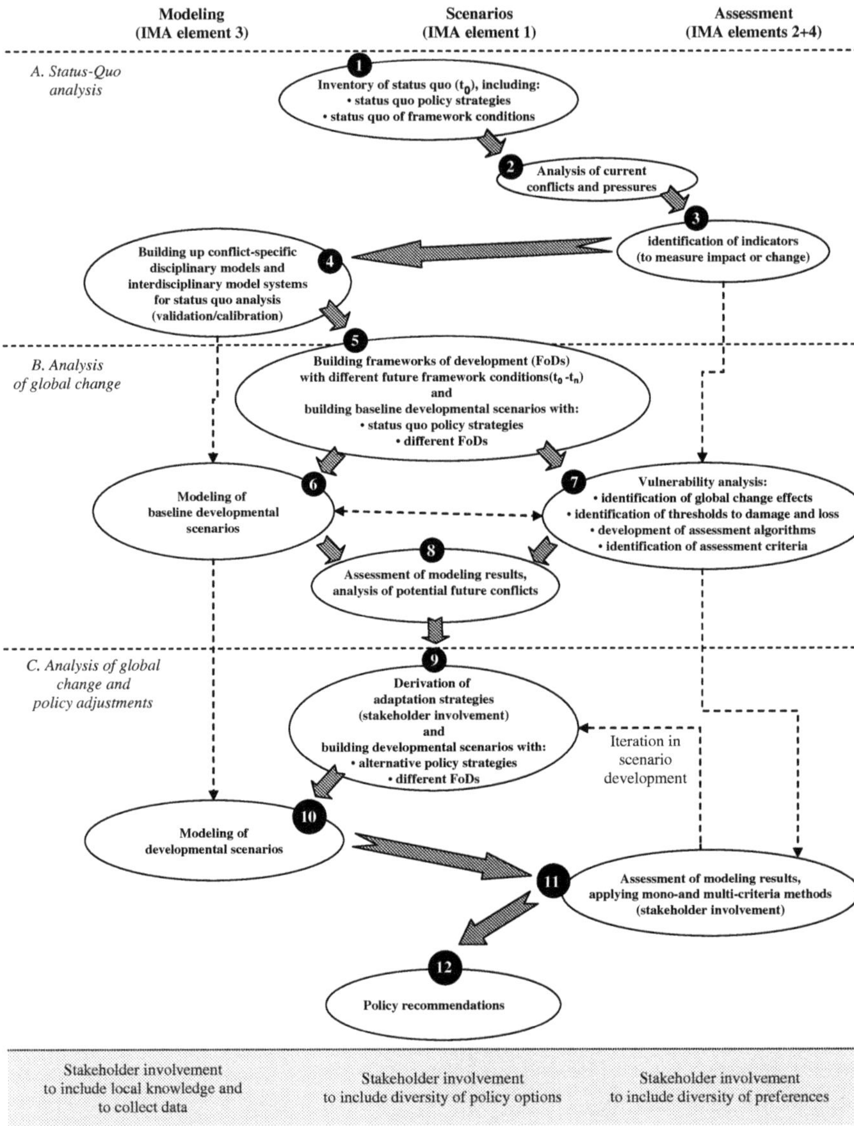

Fig. 2. Sequence of Research Tasks to Apply IMA for Global Change Analysis.

Phase A deals with status quo analysis. Current policy strategies and important institutional settings and general conditions are investigated (task 1), existing and potential conflicts are analyzed (task 2) and indicators to measure important change effects are identified (task 3). These tasks are executed by means of literature analysis, interviews and stakeholder analysis. Based upon the results of the first three tasks, models and model systems are built up or existing models are modified, validated and calibrated (task 4) such that it becomes possible to reproduce the status quo constellation through first model runs.

Phase B focuses on global change analysis. Based on scenario workshops, research discussions and interviews with stakeholders, several FoDs with different future visions are specified in the form of precise parameters and boundary conditions for the model system. Each FoD is then combined with the current policy strategy of task 1, prolonged over the time horizon considered, in order to derive *baseline developmental scenarios* (task 5). These scenarios reflect the future world in the context of global change under the assumption that no major adaptations in policy take place. These scenarios are then modeled by the model system (task 6) and, almost during the same time period, a vulnerability analysis is executed to find out which actors, institutions or systems are vulnerable to what extent and to which kind of global change effect. In this context, empirical data and interview information on historical change effects are utilized to derive vulnerability indicators and vulnerability assessment functions (task 7). The modeling results are finally analyzed and assessed by means of a set of vulnerability algorithms (task 8). This global change analysis phase is the first test bed for the model system to represent the future effects of global change. Its results offer some insight regarding the potential vulnerability of society to global change – mainly in terms of damage, loss, change and cost figures – and about potential conflict constellations that might arise in the future if a business as usual policy is pursued.

Finally, *phase C* examines both, the effects of global change and the effectiveness of policy adaptations and responses. The results of phase B are presented to decision makers and stakeholders in order to demonstrate the effects of not changing policy strategies under conditions of global change. In the following discussions, alternative policy strategies to adapt and respond effectively to the global change challenges are discussed and identified. Based on the new policy strategies the respective developmental scenarios are built up combining the new strategies with the FoDs (task 9). These scenarios are then modeled by the model system (task 10) and an assessment of the results follows, using different methods of mono-criteria assessment to evaluate single change effects and applying a multi-criteria assessment approach to aggregate the individual evaluation results (task 11). In this task, stakeholder involvement is crucial once again in order to be able to weigh the criteria, to discuss the assessment results and to produce a stakeholder view that is accepted overwhelmingly. This result might be a policy recommendation to implement a specific policy strategy (task 12), but it might also turn out that one or several new policy strategies should be analyzed by the model system to find a robust and consensual outcome.

The distribution of research time to these phases depends upon the complexity of the conflict, the amount of models applied, and the amount of models to be developed in the research process. In the first 3 years of funding of the GLOWA Elbe project, the models to be applied were already developed and only small adjustments were required, particularly regarding the adaptation to the boundary conditions of the FoDs. Therefore, less than 1 year was needed for phase A, 1 year and a few months were invested for phase B and about three quarters of a year was committed to the final phase C.

Eventually, Fig. 2 reveals that all IMA elements are closely related and all of them are needed in each research phase. Regarding stakeholder involvement, each IMA element needs some kind of participatory process to include local knowledge and data, to derive and identify policy options and strategies and to integrate the diversity of the stakeholders' preferences into the IMA process. Last but not least, it depends highly on the activity and creativity of the stakeholders to finally arrive at some form of acceptable policy strategy. The role of science in this final process is to support the discussion and decision process with sound information and analysis, result interpretation and, perhaps, moderation of the discussion process.

3. PRELIMINARY IMA RESULTS IN GLOWA ELBE

The IMA was applied for the first time in the project GLOWA Elbe for the water allocation predicament in the Spree and Schwarze Elster river basins (see Chapter 4) during the time period of 2001–2003. In the following, the results for the four IMA elements, their implications, and the special features of applying the IMA for this case study are outlined.

The derivation of *developmental scenarios* (IMA element 1) was the scientific activity that required an important portion of research time. This is not surprising, because the first 9 of the 12 research tasks of Fig. 2 are needed in order to combine FoDs and policy strategies with developmental scenarios. As described in more detail in Chapter 4, four FoDs were defined based upon the IPCC climate change story lines A1 and B2 for socioeconomic change (IPCC, 2000) and two climate scenarios, one featuring future stable climate conditions and the other featuring changing climate conditions with average increases in temperature of 1.4 K during the next five decades. The four FoDs (A1_climate change, A1_no climate change, B2_climate change and B2_no climate change) were combined with the existing policy strategy *Basic* to acquiring the baseline scenarios, which were the first scenarios to be simulated by the GLOWA Elbe models. The results of the baseline scenarios were presented to stakeholders and policy makers in a workshop and it turned out that the policy strategy at that time was not conducive to achieving the water quantity and quality goals of the water authorities. The baseline scenario simulation results revealed that neither the current nor the future water quantity and quality objectives can be achieved in the Spree and Schwarze Elster river basins, irrespective of the FoDs considered. However, the FoDs with climate change conditions resulted in a more pronounced lapse. As a consequence of these research outcomes, a highly engaged discussion arose among the stakeholders and decision makers regarding improved policy strategies and – as presented in more detail in Chapter 4 – four alternative strategies were found (*Filling*, *Reduced support*, *Transition Odra–Malxe* and *Transition Odra–Spree*). In this context, it was astonishing that strategies that were called taboo policies in some of the stakeholder interviews before – such as the transition of water from the Odra river basin to the Spree river basin, for instance – were now openly debated as possible means to achieve the water goals. After this workshop, 20 developmental scenarios could be identified for the global change and policy adjustment analysis of IMA phase C, based on five policy strategies and four FoDs (see Table 1 in Chapter 4).

The identification of *indicators* and the derivation of *criteria* (IMA element 2) started with stakeholder interviews. After talking about the problem constellation in the Spree and Schwarze Elster river basins and the problem perception of the respective interviewee, the interviewee was asked which indicators were deemed appropriate to measure and assess improvements or deteriorations of the situation. Based on the answers, a list of nine indicators was produced:

- Water availability at different gauges (in m^3/second)
- Achievement of water quality goals in the streams of the river basin
- Safety of water demand satisfaction (probability in percent that a water user gets the water he demands for a defined time period)

- Safety of water availability for aquatic ecosystems (probability in percent that the ecological minimum flow in all streams of the basin is satisfied)
- Change in economic benefits and costs to water users (in € per year)
- Number of persons who benefit from the recreational offers of the new mining lakes (persons per year)
- Time of completion of the mining lakes' filling with fresh water (date)
- Acidity of the lakes or inflow of acid material that flows into the lake (pH value of lakes or amount of acid material during time of filling)
- Impact on regional employment (number of employed persons in the region)

For most of these indicators, data could be simulated by the GLOWA Elbe models. For example, the water balance model WBalMo was able to produce estimates for water availability, water demand satisfaction probabilities and time periods for filling the mining lakes. Based on agroeconomic and energy sector models and on complementary economic vulnerability studies, the economic impacts in the form of benefits and costs could be estimated as well, while there was no regional economic model available to simulate employment effects. Therefore, it was agreed to consider employment effects only indirectly via the data on benefits and costs.

Measuring and modeling impacts by means of indicator values is one part of the story; evaluating effects by means of criteria is another one. A thorough look at the indicator list reveals that there is some form of double counting as well as interrelationships among the indicators involved. For example, water availability is closely connected to the benefits and costs of water users or, another example, the acidity of mining lakes highly depends upon the speed and time of filling the lakes with surface water. Double counting does not matter in the sphere of measuring effects, but it is highly unwanted in the sphere of effect evaluation. Therefore, criteria with unambiguous evaluation schemes needed to be identified that cover all effects and impacts mentioned during the interviews, but which do not display any kind of double counting phenomena.

Eventually, based on scientific discussions about global change and policy effects, vulnerability of water users and ecological systems, and the lack and uncertainty of information, four types of evaluation criteria were identified: first, benefits and costs to economic water users; second, satisfaction of water demand for specific activities that are difficult to monetize; third, water availability for larger regions, which cannot be subject to economic evaluation due to limitations in time and data availability; and, fourth, a veto criterion regarding the uncertainty about water quality goal violations due to water transfers. Consequently, the following criteria were chosen to be applied in the final integrated assessment:

- Net benefits of fish farming (mill. 2003 €)
- Costs of water provision (pumping and redistribution of water within a basin) (mill. 2003 €)
- Costs of acid mining lakes' neutralization (mill. 2003 €)
- Benefits of future tourism at mining lakes (mill. 2003 €)
- Percentage of satisfied water demand of industry (% per year)
- Percentage of satisfied water demand for ecological systems (% per year)
- Water inflow into the wetland region "Spreewald" (average m^3/second)
- Water inflow into Berlin (average m^3/second)
- Violation of water quality goals due to water transfers (veto criterion)

This criteria list was not the result of scientific planning at the very beginning of the project. It was the result of stakeholder interviews and scientific discussions about the possibilities, limitations and the feasibility of evaluating single effects during the global change analysis phase B. As mentioned above, evaluating as many effects as possible in monetary terms was a rationale in the GLOWA Elbe project. However, it turned out that only impacts on four major water uses could be expressed in terms of economic benefits and costs (criteria 1–4). Unfortunately, some of these four impacts could only be included partially. For example, the costs of neutralization criterion (no. 3) indeed reflects an important aspect with regard to the water quality impact on mining lakes, but it understates the total impact on the mining lakes' water quality, because other relevant types of pollution such as heavy metals were not considered (see also Chapter 11, in this volume). Furthermore, some of the water uses – especially fish farming (no. 1) – turned out to have a specific cultural meaning in the region, so to say a cultural value that is not covered directly by the net benefits of the fish farming sector. Due to these complications, it was decided for the class of benefit and cost criteria not to aggregate the figures for the four criteria, but to pass them as individual criteria into the IMA multi-criteria process in order to take the specific meaning of the figures into account. Regarding criterion number 5 – impacts on industry – it turned out during the vulnerability analysis that data on economic industry losses due to water quantity and quality problems are hard to get and – if available – they are often subject to protection of data privacy. Contrary to that it was pretty clear from the beginning that ecological impacts on water-related ecosystems (no. 6) are very difficult and complex to predict for a whole river basin and, therefore, it was decided to include this effect through the satisfaction of the ecological minimum flows, which is a clearly defined criterion applied by the German water authorities. Last but not least, it was not feasible from a financial and time perspective to calculate net benefits for all of the 20 developmental scenarios for the large regions of the wetland area "Spreewald" and the German capital city of Berlin (nos. 7 and 8). Therefore, average water inflow figures were chosen to include these aspects at least on the basis of water balance modeling. Finally, due to the uncertainty regarding the policy strategies to transfer water from the Odra basin to the Spree and Schwarze Elster river basins, a veto criterion was introduced to consider the case that the Odra water is too polluted to be used for water transfers (no. 9).

The *simulation of scenario impacts* (IMA element 3) was executed by six models or estimation techniques, respectively: the climate model STAR (Gerstengarbe & Werner, 2005), the hydrological model SWIM (Hattermann, Krysanova, Wechsung, & Wattenbach, 2004), the agroeconomic model RAUMIS (see Gömann, Kreins, & Wendland, 2004 and Chapter 6 of this volume [Kreins et al.]), the energy sector model KASIM (Martinsen, Krey, Markewitz, & Vögele, 2004), the water balance model WBalMo (see Koch, Kaltofen, Schramm, & Grünewald, 2006 and Chapter 11 of this volume [Messner et al.]), and estimation and evaluation techniques based on the results of the economic vulnerability analyses (see Messner & Kaltofen, 2004, p. 39 ff. and Chapter 11, Messner et al., in this volume). One specific feature of this noncoupled model system was the conformity of all models with regard to the external boundary factors prescribed by the four FoDs. Another highlight was the factual integration of monetary and nonmonetary evaluation algorithms into the water balance model (WBalMo), which enabled this model to generate water availability simulations and, simultaneously, to calculate impact results for the first eight criteria of IMA element 2 (see also Chapter 11, in this volume).

The results produced in the IMA elements 1–3 were passed to *IMA element 4 – assessment*. For the final assessment, it was decided to aggregate the results for the whole time horizon and put them into one multi-criteria matrix for each FoD. The outcome of this procedure is shown in Tables A1–A4 in the appendix to this chapter. These tables include criteria values as well as rank values, which indicate the rank of a policy strategy with regard to each criterion. Before presenting these results to the stakeholders and decision makers, two kinds of assessment analysis were undertaken: a qualitative analysis of the criteria results without using any MCA tool and an explorative analysis using the multi-criteria method PROMETHEE.

In the *qualitative analysis* the results for each of the nine criteria were examined without aggregating anything in order to get an initial idea about the significance of the policy strategies in the context of different FoDs. Considering the absolute criteria values for the strategies in each FoD produces the outcome that there are clear differences, indicating the significance of changing boundary conditions for the strategies' effects. On the one hand, climate change has a profound impact on water inflow for Berlin and the wetland area "Spreewald" as well as for the satisfaction of water demand for industry and ecological systems (comparison of "B2_no climate change"/ "A1_no climate change" with "B2_climate change"/"A1_climate change"). On the other hand, changing socioeconomic conditions have the largest impact on water users, who rely heavily on certain institutional circumstances such as the fish farming sector, which is highly subsidized and would lose out considerably if subsidies were to be reduced in the future (as assumed for the A1 FoDs). However, despite these absolute differences of the criteria results in the context of different FoDs, the relative changes – as displayed in the ranks of the strategies with regard to each criterion – are not exorbitantly high. On the contrary, it is striking that the ranks are similar for the FoDs "B2_climate change" and "B2_no climate change" as well as for the FoDs "A1_climate change" and "A1_no climate change". This indicates that there exists some kind of proportionality between the two A1 and the two B2 FoDs, leading to absolute changes of the criteria values in the same direction without changing the ranks. The large difference between the strategies' ranks in the A1 and B2 FoDs is due to the veto criterion on water quality (criterion no. 9). For the A1 FoDs, it is assumed that the Water Framework Directive (WFD) will not lead to a quick improvement of water quality in today's highly polluted streams – such as the Odra river – because the WFD exception rules will be used more frequently than under the B2 FoDs (see Chapter 4). As a result, the river Odra water quality will not comply with the water quality standards and will, therefore, not be usable for water transitions to other basins. Hence, in the two A1 FoDs, only three policy strategies are actually available. This of course has a large impact on the relative importance of each strategy if the A1 and B2 FoDs are compared.

The qualitative analysis of the strategies for all nine criteria does not deliver a straightforward result in the form of a clear pareto strategy. Rather, all strategies perform relatively well at some criteria and not so well at others. However, looking at the strategies' distribution of ranks among the criteria gives a first indication of their overall performance.

In the two B2 FoDs, the strategy *Transition Odra–Malxe* appears to be favorable, because it holds the first rank for most of the criteria. The strategies *Filling* and *Transition Odra–Spree* are often positioned on the second rank, while *Filling* at the same time performs worst for three criteria. The strategy *Reduced support* is found on each rank, but clearly performs worse under climate change conditions. Finally, the currently pursued policy *Basic* is mostly found on the last three ranks. This gives at least a first outcome: the alternative strategies appear to deliver an improvement under B2 conditions, because the strategy *Basic* performs relatively poorly compared to them.

In the A1 FoDs the situation is quite different, because the transition strategies are not available here. In this context, the strategy *Filling* is most often on the first rank, but for three criteria it holds the last rank. The strategies *Reduced support* and *Basic* show a relatively similar criteria rank distribution, with *Reduced support* performing slightly better under stable climate conditions and *Basic* being advantageous under climate change conditions.

In the *exploratory multi-criteria assessment analysis*, the method PROMETHEE (Vincke, 1992; Klauer et al., 2006) was used to explore the significance of strategies without using predescribed criteria weights. In a first step, PROMETHEE I is used with a uniform weighing of all criteria. All strategies are compared pairwise with regard to each criterion. Positive assessment points are distributed to a strategy if it dominates another strategy in one criterion. Negative assessment points are given to a strategy if it is dominated by the other strategy in another criterion. After the completion of this procedure with reference to every pairwise comparison of strategies with all criteria considered, every strategy achieved an amount of positive and negative points for dominating other strategies or being dominated, respectively. Based on these outcomes a ranking of strategies – a so-called partial preference order – is identified. Two strategies are considered incomparable if one strategy has a higher amount of positive points than the other strategy, but also a higher amount of negative points – irrespective of the sum of the two figures (which is called "net flow"). In a second step, the PROMETHEE I method is applied with 1,000 randomly chosen criteria weights. After completion, one can consider how often a strategy achieves which rank and then interpret the results. Finally, in a third step the weights are analyzed regarding their power to change rankings among the strategies.

The results of applying this MCA procedure to the FoDs "B2_no climate change" and "B2_climate change" are shown in Figs. 3–6. Fig. 3 shows for the FoD "B2_no climate change" that the strategy *Transition Odra–Malxe* is clearly advantageous, dominating all other strategies with regard to most

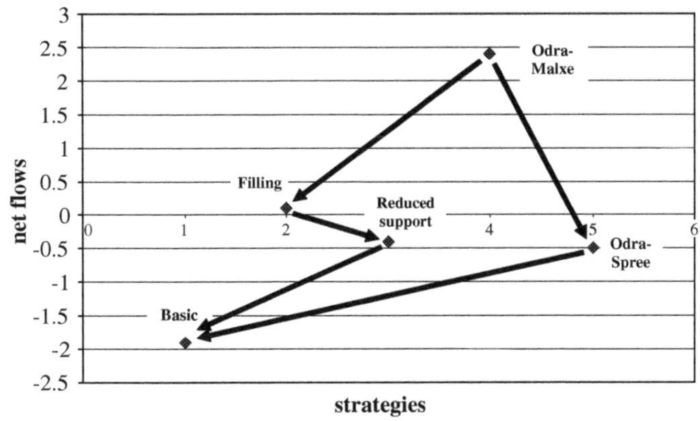

Fig. 3. Partial Preference Order for FoD "B2_no climate change".

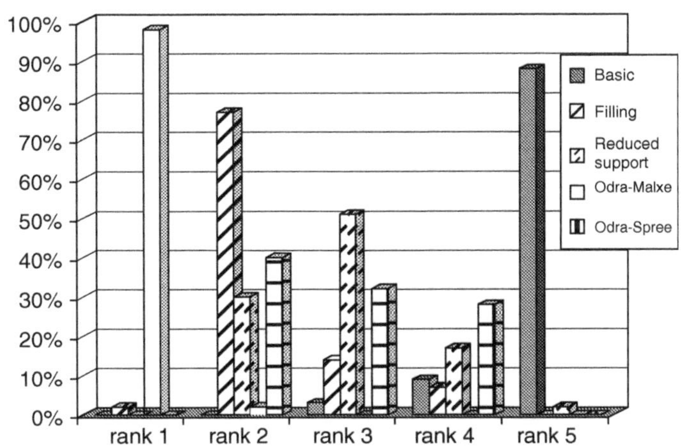

Fig. 4. Rank Distribution with 1,000 Randomly Chosen Weights for FoD "B2_no climate change".

criteria. The strategy *Filling* is dominating *Reduced support* and both are incomparable to *Transition Odra–Spree.* With reference to the currently pursued "basic" strategy it must be stated that this strategy is dominated by all other strategies in this FoD. These results are confirmed by the outcomes of the MCA procedure with 1,000 random weights (Fig. 4), which shows that *Transition Odra–Malxe* ends up at the first rank in more than 95% of all cases, while *Basic* is far behind on the last rank in more than 85% of all cases.

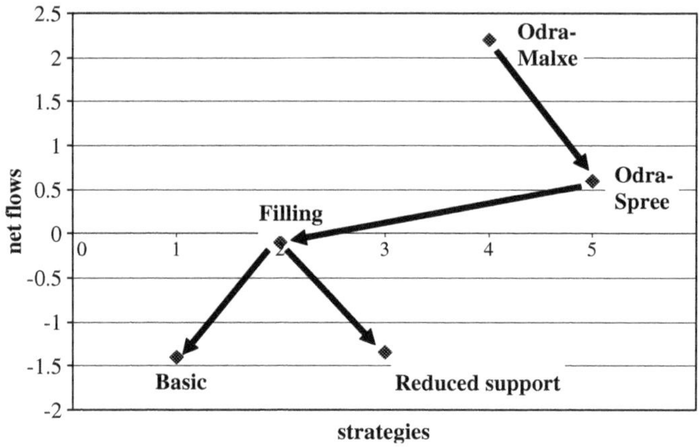

Fig. 5. Partial Preference Order for FoD "B2_climate change".

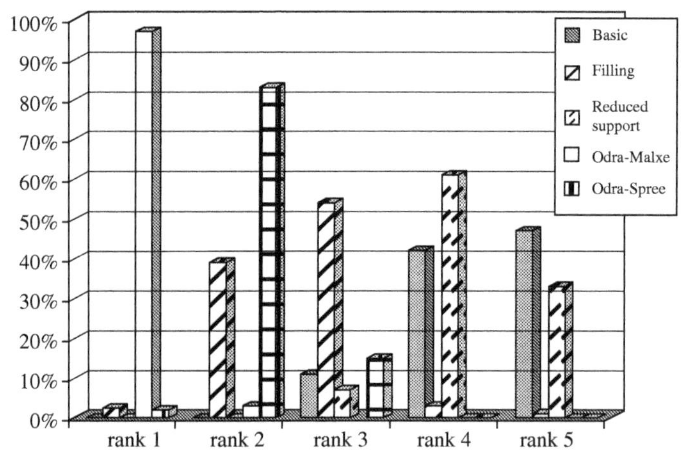

Fig. 6. Rank Distribution with 1,000 Randomly Chosen Weights for FoD "B2_climate change".

There is some change in results if the FoD "B2_climate change" is considered. While *Transition Odra–Malxe* is again clearly leading in the ranking, *Transition Odra–Spree* is now second, *Filling* third, and *Basic* and *Reduced support* are together in an incomparability situation at the last rank (Fig. 5). Again, this picture is substantiated in the procedure with 1,000 random weights (Fig. 6). Comparing the performance of the strategies of these two FoDs, it

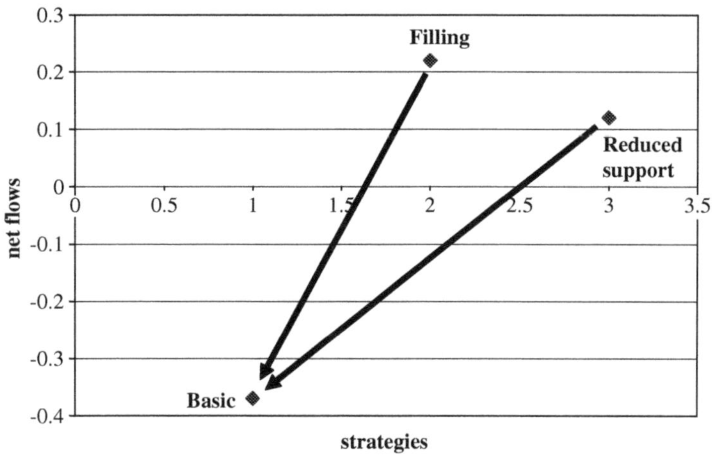

Fig. 7. Partial Preference Order for FoD "A1_no climate change".

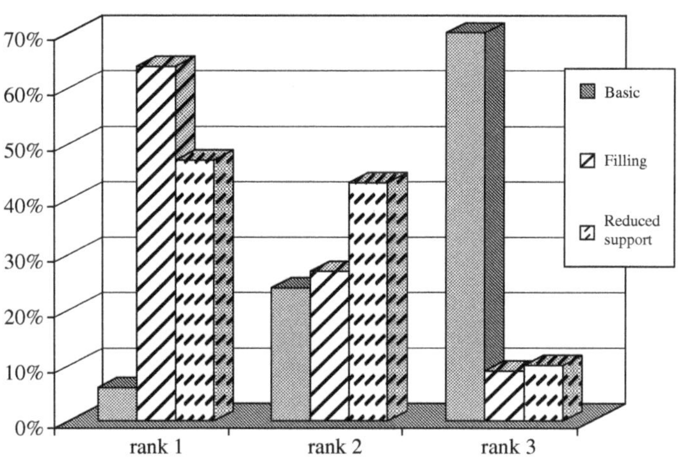

Fig. 8. Rank Distribution with 1,000 Randomly Chosen Weights for FoD "A1_no climate change".

must be stated that water transition seems to be a favorable policy under conditions of climate change, while reducing the discharges of smaller streams is rather disadvantageous in the climate change context.

The MCA results for the FoDs "A1_no climate change" and "A1_climate change" are displayed by Figs. 7–10. Since the transition strategies are not

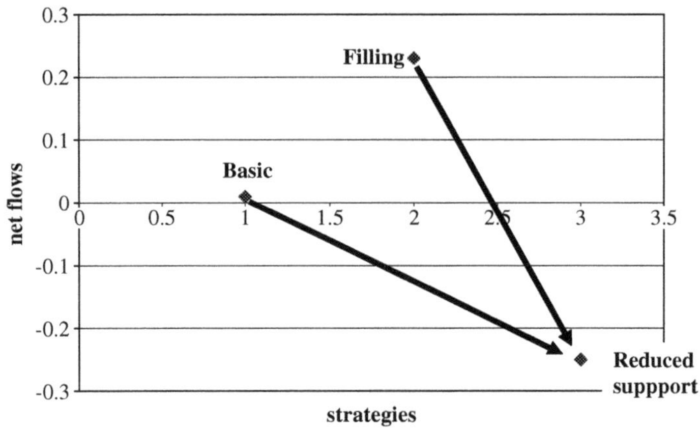

Fig. 9. Partial Preference Order for FoD "A1_climate change".

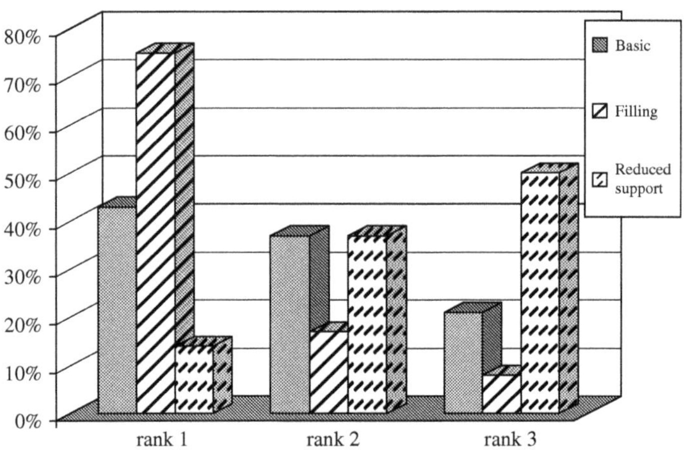

Fig. 10. Rank Distribution with 1,000 Randomly Chosen Weights for FoD "A1_climate change".

available for A1, the pictures and their implications are quite simple. Under FoD conditions with no climate change the strategies *Filling* and *Reduced support* are incomparable and both dominate the strategy *Basic* (Fig. 7). Under climate change circumstances *Filling* and *Basic* are incomparable and dominate the strategy *Reduced Support* (Fig. 9). *Filling* seems to be robust on the first rank in these two FoDs and a short look at Figs. 8 and 10 also reveals that this strategy is most often on the first rank if 1,000 randomly chosen criteria weights are considered. Concerning *Reduced support*, it again performs weakly under climate change conditions.

Eventually, the third step of the MCA analysis examined how sensitive the rank of *Basic* is to changes in criteria weights. Again, the results differ especially with regard to the socioeconomic conditions due to the impact of the veto criterion. Under B2 conditions the strategy *Basic* cannot achieve one of the two highest ranks (see also Figs. 4 and 6). At best it can achieve rank 3 by reducing the weights of those criteria for which *Basic* performs worst (criteria nos. 2 and 3, i.e., costs of water provision and costs of neutralizing acid mining lakes). This situation is different under A1 conditions. Here, *Basic* can absolutely achieve the first rank under stable climate conditions (in about 6% of 1,000 random weights, see Fig. 8) as well as under climate change conditions (in about 43% of 1,000 random weights, see Fig. 10). Again, the strategy *Basic* is preferential to the others, if criteria 2 and 3 are weighted lower than the others.

After the execution of these analyses there are already rather clear implications about the results without including any preference scheme of any stakeholder or decision maker: there are some alternative strategies that perform better than the currently pursued strategy – *if the FoD conditions are opportune*. The Odra transition strategies appear to be good strategies under the prerequisite that the water quality of the river Odra is appropriate. However, since this is uncertain for the future, these strategies might be optimal for two FoDs, but they are not robust with reference to the whole possibility space of future development. The same is true for the strategy *Reduced support*. This strategy can only bring some progress if the climate conditions remain stable. Consequently, there is only one strategy that might bring improvements for the current and future situation, irrespective of the framework development situations of the FoDs: the strategy *Filling*. It performs better than *Basic* in most of the FoDs and similarly well (but incomparable) in others.

This result – together with the simulation data and MCA calculations – was communicated to stakeholders and decision makers. It was not necessary to execute a further MCA with weights defined by different stakeholder groups. Rather, there was the demand to further specify the impacts of the strategies with regard to the criteria "water availability to industry" and "water flows" to Berlin and to the wetland area "Spreewald" (criteria nos. 5, 7 and 8). For example, it was noted that the strategy *Filling* performs rather badly regarding the water availability of industry (see criterion 5 in Tables A1–A4). If this impact should be related to high economic losses of water-related industry within the Spree and Schwarze Elster river basins, then this strategy would probably not be accepted by the basin's stakeholders. Similar statements were uttered regarding the impact to Berlin and the "Spreewald" area. Therefore, it was agreed to execute additional research in this regard to better reveal the economic benefits and costs of the criteria in question. For this reason, the results presented in this section must be considered preliminary.

Nevertheless, there is also a preliminary set of policy recommendations out of this research process:

- There are several scientific indications that the strategy *Filling* could mitigate the water allocation problems in the Spree and Schwarze Elster river basins at acceptable cost. Therefore, the water authorities should inspect the institutional prerequisites of this strategy and should start preparing its possible future implementation in order to lose as little time as possible in the event that its superiority over the strategy *Basic* can be further verified by scientific analysis.
- The Odra water transition strategies have a large potential to reduce the Spree and Schwarze Elster river basins' water problems – especially under climate change conditions. However, their political implementation might be impeded by insufficient Odra water quality. Therefore, the water quality of the Odra should be observed continuously over time and the institutional prerequisites for the transfer of Odra water should be ensured to keep the future option open to include a variant of the Odra water transition into the future water management strategy of the Spree and Schwarze Elster river basins.

4. LESSONS LEARNED

In the process of applying the IMA of GLOWA Elbe, its potential for improving the environmental decision-making process became obvious. Particularly, three of its strengths were outstanding. First of all, the coherent research structure, which was generated in the IMA process, paved the way for interdisciplinary cooperation and collaboration between researchers, decision makers and stakeholders. Hence, IMA proved to be an appropriate research approach to put applied and transdisciplinary research into practice. Second, the participatory assessment approach of IMA with the rationale to include stakeholders and decisions makers early in the research process was successful, because the stakeholders became an element of this process. They delivered valuable information and local knowledge into the research sphere, gave feedback to interim results and remained interested in the research throughout the project. Third, the IMA demonstrated the possibility and the advantages of examining the impact of policy strategies under consideration of future development uncertainties by means of considering different FoDs. This was an aspect that was especially new and innovative for the decision makers, who really appreciated this kind of analysis. In this manner, policy analysis changed in character from identifying an optimal strategy (assuming future certainty) toward identifying robust strategies, which deliver satisfactory results for a bundle of possible future development paths.

However, the GLOWA Elbe research process with IMA was not free of obstacles and shortcomings. Therefore, there are several lessons to be learned for future global change research:

- Participation of decision makers and stakeholders is a difficult process due to the multitude of different interest and power positions involved. Therefore, continuous public-relations efforts are required to keep in touch with key players on a regular basis. This was especially difficult in the case of the Spree and Schwarze Elster river study, because the institutional structures and the political responsibilities were changing rapidly and they were nontransparent for this reason. This dynamic element of participation was underestimated in the first year of research, leading to friction with stakeholders in the beginning.
- While it is an appropriate approach to derive a set of FoDs, it is sometimes not trivial to identify a compatible set of assumptions for interrelated boundary conditions. In the first years of the GLOWA Elbe research, a bundle of natural science models was applied, but only two socioeconomic models were used to simulate global change effects – the energy sector model KASIM and the agricultural sector model RAUMIS. Assumptions with regard to the general economic development and water demand of industry and households needed to be estimated by means of a simple estimation procedure. Therefore, it was decided for the second GLOWA Elbe research period to build up a coherent global change model system to better simulate the FoD development paths.
- Evaluating the water-related effects of global change and water management strategies for a whole river basin is a major research challenge, because the impacts on many water users and ecological systems need to be considered (about 400 in the Spree and Schwarze Elster river basins). We tried to overcome this problem by developing standardized (economic

and noneconomic) evaluation algorithms for all types of users and integrating them into the water management model WBalMo (see Chapter 11). However, this was only partially successful for some types of water use due to different reasons. In some cases, the economic evaluation approach did not fit with the structure of the WBalMo, because several water users were aggregated in this model such that a traditional marginal evaluation approach could not be implemented. In other cases, there was just no information available about the vulnerability of water users and no resources to survey appropriate data. As a consequence, it was decided for the coming research period to develop a specific evaluation approach for each type of water use, to involve economic *and* water management modelers into the evaluation design process and, of cause, to plan sufficient resources and time for vulnerability surveys with regard to each type of water use.

- Finally, it needs to be stated that applying IMA is neither cheap nor quick. The research described in this chapter and in Chapters 4 and 11 required about one million Euro. Therefore, such an approach is only appropriate if really large and long-term gains of scientific decision support are possible and realistic. This aspect of research costs is also an important argument for successive research, i.e., not all effects of global change and policy impacts should be quantified and evaluated in a very all-embracing – and expensive – research setting. In contrast, it is much more sensible to identify and quantify the most relevant effects first and to continue with further analysis only if the relevance of additional research for the final outcome becomes evident.

In face of the results and further potential of the GLOWA Elbe project, the German Ministry of Education and Research decided to prolong this research endeavor. As a consequence, the IMA will not only be further applied to the Spree and Schwarze river basins (about 10,000 km^2), but the whole Elbe river basin (about 150,000 km^2) will become subject to the IMA process to analyze the impacts of global change and water management. Hence, the methods developed and lessons learned in the first years of GLOWA Elbe must be considered as first research steps in the establishment of a global change research and decision support system to improve the understanding of global change and aid the finding of proper policy responses. The actual state of the GLOWA Elbe project and its further results can be observed at the project's web page (http://www.glowa-elbe.de).

NOTES

1. "Stakeholders" in the context of the IMA approach are defined as affected persons and interest groups involved in a conflict situation and/or in the process to resolve it, without formal decision power. Persons with formal decision power and their supporters from the water authorities are termed "decision makers". The term "actors" is used to indicate the whole group of stakeholders *and* decision makers.

2. In the context of the IMA approach, the so-called three-column approach to sustainable development is applied, i.e., a development is to be ensured that takes basic social, ecological and economic needs for current and future generations into account (Enquete Commission, 1998).

ACKNOWLEDGMENTS

The research was funded by the German Federal Ministry of Education and Research in the context of the GLOWA Elbe project. The author wants to express his further gratitude to the whole interdisciplinary research group of GLOWA Elbe, which struggled considerably to agree on the common IMA methodology. Furthermore, the author wants to express many thanks to all actors involved in the participatory process of the GLOWA Elbe project.

REFERENCES

Alberti, M., & Waddell, P. (2000). An Integrated Urban Development and Ecological Simulation Model. *Integrated Assessment, 1*(3), 215–227.

Becker, A., Behrendt, H., Cypris, C., Dietrich, O., Feige, H., Grünewald, U., Hansjürgens, B., Hartje, V., Ipsen, D., Kaltofen, M., Klauer, B., Leinhos, S., Messner, F., Oppermann, R., Simon, K.-H., Soete, B., Wechsung, F., Wenzel, V., & Wessolek, G. (2001). Integrierte Analyse der Auswirkungen des Globalen Wandels auf die Umwelt und die Gesellschaft im Elbegebiet (GLOWA-Elbe) – Sozio-ökonomische Konzeption. Potsdam, unpublished GLOWA Elbe document, June 2001.

Behringer, J., Buerki, R., & Fuhrer, J. (2000). Participatory integrated assessment of adaptation to climate change in Alpine tourism and mountain agriculture. *Integrated Assessment, 1*(4), 33–338.

Commission of the European Communities (2002). *Guidance on Public Participation in Relation to Water Framework Directive – Active Involvement, Consultation and Public Access to Information.* Common Implementation Strategy. Working Group 2.9, Brussels.

Enquete Commission (1998). Konzept Nachhaltigkeit – vom Leitbild zur Umsetzung. Final Report of the Enquete Commission "Protection of the human being and the environment" of the German Parlament. *Zur Sache* No. 4/98, Bonn.

Forman, E. H., & Gass, S. I. (2001). The analytic hierarchy process – An exposition. *Operations Research, 49*(4), 469–486.

Gerstengarbe, F.-W., & Werner, P. C. (2005). Simulationsergebnisse des regionalen Klimamodells STAR. In: F. Wechsung, A. Becker & P. Gräfe (Eds), *Auswirkungen des globalen Wandels auf Wasser, Umwelt und Gesellschaft im Elbegebiet* (pp. 110–118). Berlin: Weissensee Verlag.

Gömann, H., Kreins, P., & Wendland, F. (2004). Model based impact analysis of policy options aiming at reducing diffuse pollution by agriculture – A case study for the river Ems and a sub-catchment of the Rhine. *Environmental Modelling and Software*, *20*(2), 261–271.

Hanley, N., & Spash, C. L. (1993). *Cost–benefit analysis and the environment*. Brookfield: Edward Elgar.

Hare, M., Letcher, R. A., & Jakeman, A. J. (2003). Participatory modelling in natural resource management: A comparison of four case studies. *Integrated Assessment*, *4*(2), 62–72.

Hattermann, F. F., Krysanova, V., Wechsung, F., & Wattenbach, M. (2004). Integrating groundwater dynamics in regional hydrological modelling. *Environmental Modelling & Software*, *19*(11), 1039–1051.

Hoekstra, A. Y., Savenije, H. H. G., & Chapagain, A. K. (2001). An integrated approach towards assessing the value of water: A case study on the Zambezi Basin. *Integrated Assessment*, *2*(4), 199–208.

Hope, C., & Palmer, R. (2001). Assessing water quality improvement schemes: The multi-attribute technique of the UK's environment agency. *Integrated Assessment*, *2*(4), 219–224.

Horsch, H., Ring, I., & Herzog, F. (2001). *Nachhaltige Wasserbewirtschaftung und Landnutzung – Methoden und Instrumente der Entscheidungsfindung und -umsetzung*. Marburg: Metropolisverlag.

InfoResources Focus Nr. 1/03 (2003). Integrated Water Resources Management – A way to sustainability. http://www.inforesources.ch/pdf/focus1_e.pdf

IPCC (Intergovernmental Panel on Climate Change) (2000). *Emissions scenarios*. IPCC Special Report.

Jonsson, A. (2005). Public participation in water resources management: Stakeholder voices on degree, scale, potential, and methods in future water management. *Ambio*, *34*(7), 495–500.

Klauer, B., Drechsler, M., & Messner, F. (2006). Multicriteria analysis under uncertainty with IANUS – Method and empirical results. *Environment and Planning C: Government and Policy*, *24*, 236–256.

Klauer, B., Messner, F., Herzog, F., & Geyler, S. (2001). Die Ableitung von Bewertungskriterien. In: H. Horsch, I. Ring & F. Herzog (Eds), *Nachhaltige Wasserbewirtschaftung und Landnutzung – Methoden und Instrumente der Entscheidungsfindung und –umsetzung* (pp. 131–144). Marburg: Metropolisverlag.

Koch, H., Kaltofen, M., Grünewald, U., Messner, F., Karkuschke, M., Zwirner, O., & Schramm, M. (2005). Scenarios of water resources management in the lower Lusatian mining district, Germany. *Ecological Engineering*, *24*, 49–57.

Koch, H., Kaltofen, M., Schramm, M., & Grünewald, U. (2006). Adaptation strategies to global change for water resources management in the Spree River catchment, Germany. International Journal of River Basin Management, *4*(4).

Martinsen, D., Krey, V., Markewitz, P., & Vögele, S. (2004). A New Dynamical Bottom-Up Energy Model for Germany – Model Structure and Model Results. Proceedings of 6th IAEE European Conference, 2–3 September 2004, Zürich, Switzerland (CD-ROM).

Messner, F., & Kaltofen, M. (Eds). (2004). *Nachhaltige Wasserbewirtschaftung und regionale Entwicklung – Analyse und Bewertung von Szenarien zum Wassernutzungskonflikt im bergbaubeeinflussten Einzugsgebiet der Oberen Spree*. UFZ Discussion Paper 1/2004, Leipzig.

Messner, F., Wenzel, V., Becker, A., & Wechsung, F. (2005). Der Integrative Ansatz von GLOWA Elbe. In: F. Wechsung, A. Becker & P. Gräfe (Eds), *Auswirkungen des globalen Wandels auf Wasser, Umwelt und Gesellschaft im Elbegebiet* (pp. 59–70). Berlin: Weißensee Verlag.

Messner, F., Zwirner, O., & Karkuschke, M. (2006). Participation in multicriteria decision support for the resolution of a water allocation problem in the Spree River basin. *Land Use Policy*, *23*, 63–75.

OECD (1994). Environmental Indicators. OECD Core Set, Organisation for Economic Co-operation and Development, Paris.

Rumm, P., von Keitz, S., & Schmalholz, M. (Eds). (2006). *Handbuch der EU-Wasserrahmenrichtlinie – Inhalte, Neuerungen und Anregungen für die nationale Umsetzung* (2nd ed.). Berlin: Erich Schmidt Verlag.

Thomas, J.-S., & Durham, B. (2003). Integrated water resources management: Looking at the whole picture. *Desalination*, *156*, 21–28.

Timmerman, J. G. (2005). The need for participatory processes and its implication for water management information. *Regional Environmental Change*, *5*(4), 162–163.

Vincke, P. (1992). *Multicriteria Decision-Aid*. Chichester: Wiley.

WBGU (2003). *Welt im Wandel: Energiewende zur Nachhaltigkeit. Hauptgutachten 2003*. Berlin: Springer Verlag.

Wenzel, V. (1999). Ein integrativer Algorithmus zur Unterstützung regionaler Landnutzungsentscheidungen. In: H. Horsch, F. Messner, S. Kabisch & M. Rode (Eds), *Flußeinzugsgebiet und Sozioökonomie: Konfliktbewertung und Lösungsansätze*, UFZ-Bericht 30/1999 (pp. 75–86). Leipzig: Umweltforschungszentrum Leipzig-Halle GmbH.

Wenzel, V. (2005). Der Integrative Methodische Ansatz im stringenten Sprachkalkül. In: F. Wechsung, A. Becker & P. Gräfe (Eds), *Auswirkungen des globalen Wandels auf Wasser, Umwelt und Gesellschaft im Elbegebiet* (pp. 71–81). Berlin: Weißensee Verlag.

APPENDIX: IMPACT MATRICES FOR THE DEVELOPMENTAL SCENARIOS FOR FOUR FODS

Table A1. Impact Matrix for FoD "B2_no Climate Change".

Strategies	Criteria							
	1		2		3		4	
	Net benefit fish farming (mill. 2003 €, 2% discount rate)[a]	Rank	Net benefit water provision (mill. 2003 €, 2% discount rate)[a]	Rank	Net benefit of cleaning acid mining lakes (mill. 2003 €, 2% discount rate)[a]	Rank	Net benefits of future tourism at mining lakes (mill. 2003 €, 2% discount rate)[a]	Rank
Basic	0	3	0	5	0	4	0	4
Filling	−2.33	5	10.57	2	−0.61	5	2.75	1
Reduced support	−0.01	4	41.35	1	0.01	3	0.75	2
Transition Odra–Malxe	0.04	1	$<0.67^b$	3	0.71	1	0.56	3
Transition Odra–Spree	0.03	2	$<0.53^b$	4	0.16	2	0.00	4

[a]Net benefit refers to the difference to the "basic" strategy. Therefore, the net benefit for "basic" is always zero.
[b]These figures do not include investment and maintenance costs of the Odra transitions.

Criteria								
5		6		7		8		9
Percentage of satisfied (maximum) water demand of industry	Rank	Percentage of satisfied water demand of ecological systems	Rank	Average water inflow into the wetland region "Spreewald" (m^3/second)	Rank	Average water inflow into Berlin (m^3/second)	Rank	Violation of water quality goals due to water transfers (Veto criterion) (yes/no)
89.9	2	99.1	4	14.1	3	25.7	4	No
88.0	5	99.5	1	14.2	2	25.8	2	No
89.9	2	99.2	2	14.0	5	25.6	5	No
90.0	1	99.2	2	14.3	1	25.9	1	No
89.9	2	99.1	4	14.1	3	25.8	2	No

Table A2. Impact Matrix for FoD "B2_climate Change".

Strategies	Criteria							
	1		2		3		4	
	Net benefit fish farming (mill. 2003 €, 2% discount rate)[a]	Rank	Net benefit water provision (mill. 2003 €, 2% discount rate)[a]	Rank	Net benefit of cleaning acid mining lakes (mill. 2003 €, 2% discount rate)[a]	Rank	Net benefits of future tourism at mining lakes (mill. 2003 €, 2% discount rate)[a]	Rank
Basic	0	3	0	4	0	3	0	5
Filling	−5.99	5	13.62	2	−0.63	5	3.17	1
Reduced support	−0.04	4	41.53	1	−0.02	4	0.67	3
Transition Odra–Malxe	0.31	1	<-1.07[b]	5	0.94	1	0.74	2
Transition Odra–Spree	0.18	2	<0.74[b]	3	0.30	2	0.01	4

[a]Net benefit refers to the difference to the "basic" strategy. Therefore, the net benefit for "basic" is always zero.
[b]These figures do not include investment and maintenance costs of the Odra transitions.

Criteria								
5		6		7		8		9
Percentage of satisfied (maximum) water demand of industry	Rank	Percentage of satisfied water demand of ecological systems	Rank	Average water inflow into the wetland region "Spreewald" (m^3/second)	Rank	Average water inflow into Berlin (m^3/second)	Rank	Violation of water quality goals due to water transfers (Veto criterion) (yes/no)
86.4	3	96.9	3	12.1	3	18.3	4	No
81.3	5	98.0	1	12.4	2	18.6	3	No
86.4	3	96.9	3	12.0	5	18.2	5	No
87.0	1	97.1	2	12.5	1	18.7	2	No
86.6	2	96.9	3	12.1	3	18.9	1	No

Table A3. Impact Matrix for FoD "A1_no Climate Change".

Strategies	Criteria							
	1		2		3		4	
	Net benefit fish farming (mill. 2003 €, 2% discount rate)[a]	Rank	Net benefit water provision (mill. 2003 €, 2% discount rate)[a]	Rank	Net benefit of cleaning acid mining lakes (mill. 2003 €, 2% discount rate)[a]	Rank	Net benefits of future tourism at mining lakes (mill. 2003 €, 2% discount rate)[a]	Rank
Basic	0	3	0	5	0	4	0	5
Filling	−1.60	5	10.57	2	−0.60	5	11.51	1
Reduced support	−0.01	4	41.35	1	0.01	3	3.13	2
Transition Odra–Malxe	~~0.02~~	~~1~~	~~<0.67[b]~~	~~3~~	~~0.70~~	~~1~~	~~2.41~~	~~3~~
Transition Odra–Spree	~~0.01~~	~~2~~	~~<0.53[b]~~	~~4~~	~~0.15~~	~~2~~	~~0.01~~	~~4~~

Shaded area: Due to the "yes" of the veto criterion (no. 9) the two strategies with transition of Odra water are not considered in the final assessment (quality of Odra water is not sufficient). Therefore, the criteria values of the other criteria are crossed out.

[a]Net benefit refers to the difference to the "basic" strategy. Therefore, the net benefit for "basic" is always zero.

[b]These figures do not include investment and maintenance costs of the Odra transitions.

Criteria								
5		6		7		8		9
Percentage of satisfied (maximum) water demand of industry	Rank	Percentage of satisfied water demand of ecological systems	Rank	Average water inflow into the wetland region "Spreewald" (m^3/second)	Rank	Average water inflow into Berlin (m^3/second)	Rank	Violation of water quality goals due to water transfers (Veto criterion) (yes/no)
89.9	2	99.1	4	14.1	3	25.7	4	No
88.0	5	99.5	1	14.2	2	25.8	2	No
89.9	2	99.2	2	14.0	5	25.6	5	No
~~90.0~~	~~1~~	~~99.2~~	~~2~~	~~14.3~~	~~1~~	~~25.9~~	~~1~~	Yes
~~89.9~~	~~2~~	~~99.1~~	~~4~~	~~14.1~~	~~3~~	~~25.8~~	~~2~~	Yes

Table A4. Impact Matrix for FoD A1_climate Change".

Strategies	Criteria							
	1		2		3		4	
	Net benefit fish farming (mill. 2003 €, 2% discount rate)[a]	Rank	Net benefit water provision (mill. 2003 €, 2% discount rate)[a]	Rank	Net benefit of cleaning acid mining lakes (mill. 2003 €, 2% discount rate)[a]	Rank	Net benefits of future tourism at mining lakes (mill. 2003 €, 2% discount rate)[a]	Rank
Basic	0	3	0	4	0	3	0	5
Filling	−3.73	5	13.62	2	−0.61	5	13.35	1
Reduced support	−0.04	4	41.53	1	−0.02	4	2.78	3
Transition Odra–Malxe	0.17	1	<−1.07[b]	5	0.92	1	3.21	2
Transition Odra–Spree	0.09	2	<0.74[b]	3	0.30	2	0.06	4

Shaded area: Due to the "yes" of the veto criterion (no. 9) the two strategies with transition of Odra water are not considered in the final assessment (quality of Odra water is not sufficient). Therefore, the criteria values of the other criteria are crossed out.

[a]Net benefit refers to the difference to the "basic" strategy. Therefore, the net benefit for "basic" is always zero.

[b]These figures do not include investment and maintenance costs of the Odra transitions.

Criteria								
5		6		7		8		9
Percentage of satisfied (maximum) water demand of industry	Rank	Percentage of satisfied water demand of ecological systems	Rank	Average water inflow into the wetland region "Spreewald" (m^3/second)	Rank	Average water inflow into Berlin (m^3/second)	Rank	Violation of water quality goals due to water transfers (Veto criterion) (yes/no)
86.4	3	96.9	3	12.1	3	18.3	4	No
81.3	5	98.0	1	12.4	2	18.6	3	No
86.4	3	96.9	3	12.0	5	18.2	5	No
~~87.0~~	~~1~~	~~97.1~~	~~2~~	~~12.5~~	~~1~~	~~18.7~~	~~2~~	Yes
~~86.6~~	~~2~~	~~96.9~~	~~3~~	~~12.1~~	~~3~~	~~18.9~~	~~1~~	Yes

PART V:
WATERSHED POLICY INNOVATION AND DESIGN

URBAN METABOLISM AND PAYMENT FOR ECOSYSTEM SERVICES: HISTORY AND POLICY ANALYSIS OF THE NEW YORK CITY WATER SUPPLY

Melinda Kane and Jon D. Erickson

ABSTRACT

The interaction of urban cores and their rural hinterlands is considered from an ecological–economic perspective. The concept of 'urban metabolism' motivates discussion of urban dependence on geographic regions outside their borders for both sources of inputs and as waste sinks. The U.S. Environmental Protection Agency's 1989 Surface-Water Treatment Rule forces cities to consider the ecosystem services preserved by appropriate land-use management inside suburban and rural watersheds used for urban water supplies. A case study of New York City and its water supply from the Catskill–Delaware watershed system is used to explore these themes. Compensation from the city to watershed communities may be an effective way to motivate protection of those ecosystem functions. Both direct payments and investment in economic development projects consistent with water quality goals are reviewed as policy instruments.

Ecological Economics of Sustainable Watershed Management
Advances in the Economics of Environmental Resources, Volume 7, 307–328

ISSN: 1569-3740/doi:10.1016/S1569-3740(07)07013-7

1. INTRODUCTION

> Great cities are planned and grow without any regard for the fact that they are parasites on the countryside which must somehow supply food, water, air, and degrade huge quantities of wastes.
>
> – Eugene Odum (1971)

The challenge of providing potable water to cities has motivated major engineering achievements in both the ancient and modern worlds. While water supply failure is most acute in the developing world, home to most of the 1.1 billion people who lack access to an improved water supply, the developed world is not without its own share of water concerns (WHO/UNICEF, 2000). Major urban areas, in particular, encounter a unique set of problems in acquiring adequate supplies for their concentrated populations. Many very large cities, both in the United States and throughout the world (for example, New York, Los Angeles, Seattle, Caracas, Mexico City, and Cape Town, South Africa) acquire their supplies from distant sources, adding significant political challenges and often legal battles to water supply management.

Several cities in the United States have claimed distant water sources, choosing them for their purity when clean, local sources were not available. These cities have historically managed to avoid some of the costs of water treatment, in part because the water quality laws and institutions in place allowed them to do so. However, the 1986 amendments to the Safe Drinking Water Act (SDWA) changed the rules regarding these supplies and have required water departments in the affected cities to consider more explicitly the ecosystem services that have allowed them to avoid these costs. The Surface-Water Treatment Rule (SWTR), promulgated in 1989 pursuant to the 1986 SDWA amendments, requires that all surface-water supplies be filtered unless the water obtained from those sources meets or exceeds all existing water quality standards. Currently about 100 communities fulfill the criteria required to receive a filtration avoidance determination (FAD) from the U.S. Environmental Protection Agency (EPA). Source-water protection, rather than filtration, is the primary water quality strategy for these locales.

The recent changes in the law are a response to the general threat of degradation of water supplies. While some improvements have been made in controlling point sources of pollution, relatively little progress can be claimed in the mitigation of non-point sources. Increasing urban populations, sprawling suburban communities, and the loss of water purification functions of disappearing wetlands continue to threaten water quality across the country. As sprawling cities occupy more acreage around their former cores, a new era of problems has emerged in the form of leaking septic systems, contaminated runoff from pesticide-laden lawns, and parking lots and driveways coated in hazardous materials. Following industrial discharges, urban and suburban runoff is the second most prevalent source of water quality impairment in the nation's estuaries (U.S. EPA, 1998). Outside urban areas, agricultural runoff is also of major concern as fertilizers and other agricultural chemicals run off into rivers and streams leading to eutrophication and, in the worst cases, hypoxic "dead zones" at the mouth of major rivers like the Mississippi (Vitousek et al., 1997; Carey et al., 1999). Most recently, the presence of residuals from personal-care products and pharmaceuticals has added cause for concern because of the uncertain impacts on human and animal populations at concentrations of only a few parts per trillion (Raloff, 1998; Daughton, 2002).

The SWTR does not impact only urban water supplies, however statistics on community water systems (CWS) from the U.S. EPA (2001) confirm that urban supplies are disproportionately affected. CWS are defined as those which supply water to the same population of at least 25 people year-round. This includes all municipal water utilities, as well as some private water sources. Public CWS served almost 264 million people in 2000, the vast majority of the U.S. population. Only 7 percent of the over 54,000 CWSs serve 81 percent of that population, while the very largest of them, just 0.7 percent (353), serve 44 percent. These include those that serve the largest urban centers. Of all the CWSs in the country, the vast majority (78.9 percent) obtains water from groundwater sources, but these provide water for only 32.5 percent of the population served by CWSs. In contrast, although only 21.1 percent of CWSs obtain their water from surface supplies, these systems provide water for 67.5 percent of the population served by CWSs.

The amendments to the SDWA place the issue of urban water supplies squarely in a regional framework by emphasizing the city's relationship with and control over the source-water region. Cities are forced to look for the most cost-effective ways to protect the quality of the water they consume, and are thus attempting to place a value on these ecosystem services in order to compare the costs and benefits of source-water protection versus end-of-pipe treatment. Even municipalities with filtration infrastructure are examining the benefits of source-water protection measures in order to relieve some of the pressure on their systems (U.S. EPA, 1999).

This chapter examines some of the economic issues surrounding the urban appropriation of distant water sources. We begin by discussing an urban metabolism framework from which to consider interdependencies between city and country. This is followed by an examination of the experience of New York City, perhaps the most well-known case on these issues, in attempting to protect the ecosystem functions of the watersheds of its principal supply sources. New York's main challenge has been to motivate the watershed's largely rural resident population of just over 50,000 to adopt a stricter set of land-use rules and regulations in order to benefit an urban population of over 8 million.[1] The main concern of the watershed residents, however, has been to maintain their economic opportunities, often viewing the new regulations as an infringement on those prospects. A settlement was reached after a protracted negotiation process, although challenges persist in the execution of that agreement. An understanding of the interdependence between New York and the watershed communities contributes to the ongoing study and potential application of this model for other urban places.

2. URBAN METABOLISM AND PAYMENT FOR ECOSYSTEM SERVICES

Wolman (1965) first coined the term "urban metabolism," the conception of a city as a living organism that takes in energy and materials, transforms them through metabolic processes into usable goods and services, and excretes waste. The increasingly obvious problems of pollution and waste disposal of this era, coupled with the public's new awareness of the planet as a closed material system, prompted him to contemplate the management of cities as open systems and the waste-disposal problems thus created. He estimated the water and fuel requirements for a growing U.S. urban population and discussed the need for better management of pollutant emissions into hydrologic and atmospheric media. More recent studies have attempted to quantify the flows of carbon and other nutrients in addition to water, waste, and material flows. For example, Baccini (1996) studied the Swiss Lowlands, comparing the per capita metabolism of the human system in 1800 and 1995. In their survey of energy and material flows for the world's 25 largest cities, Decker, Elliott, Smith, Blake, and Rowland (2000) show that water is by far the largest flow in major cities' metabolisms, comprising more than 90 percent of material flows.

Fig. 1 shows one schematic of the urban metabolic model, introduced by Newman (1999).[2] The notion of urban metabolism raises the question of the source of inputs and location of waste sinks for areas that are not

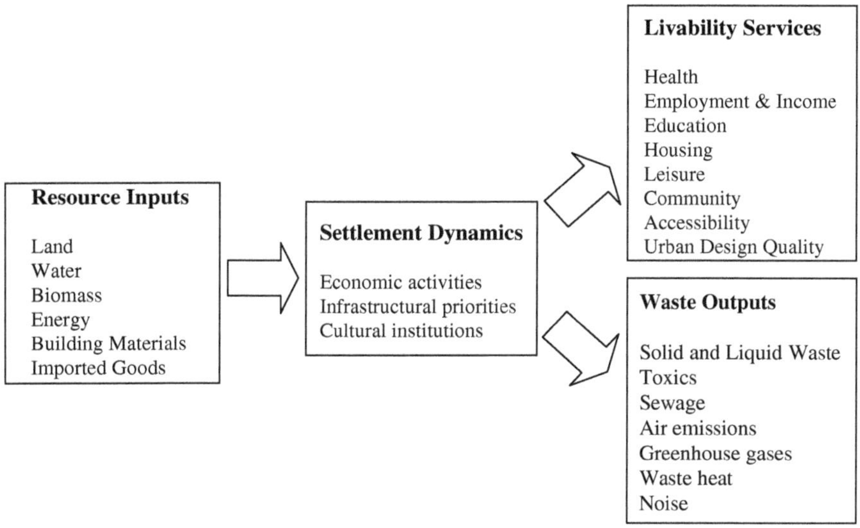

Fig. 1. Metabolism Model of the City (Adapted From Newman, 1999).

self-sufficient, and is at the heart of questions of sustainability (Baccini, 1997). Central place theory and other spatial theories in economics (see Chapter 9 in this volume) propose that rural hinterlands are dependent on urban cores for the provision of certain economic commodities. The production of these goods and services require large potential markets that otherwise would not be available in rural communities. However little work has been done in the other direction, showing the dependence of urban cores on the rural hinterland (Rees, 1992). Cronon's *Nature's Metropolis* (1991) is one important exception. Using data from bankruptcies and estates in Chicago banks, Cronon followed the economic linkages between creditors and debtors to show the resource flows in grain, livestock, and lumber in the latter half of the 19th century. He argues that the increasing commodification of these biotic resources, coupled with the infrastructural development of the railroads, helped to raise the metabolism of the city of Chicago. These developments made the city dependent on a very large and distant hinterland for its own development and its firms' profitability.

Such urban dependence on the rural natural-resource base is not limited to marketable goods. The case of the watershed of New York City's water supply illustrates urban dependence on non-marketed ecosystem services. When the

U.S. EPA initially granted a temporary FAD for New York City under the SWTR, renewal depended on the ability of the city to prove that it could control the further degradation of water quality in its source watersheds. Negotiations between the city and the communities within the watershed led to the adoption of a historic agreement to compensate watershed communities for the provision of ecosystem services. The city agreed to invest over $1 billion in water protection programs and other means of compensation for the watershed communities, and in turn avoided an estimated $6 billion in capital costs to build water treatment infrastructure, plus annual operating costs for the same. In some cases, individuals were directly compensated for changing their activities. For example, farmers were paid between $100–150 per acre to take riparian land out of crop production and grazing use, and individual property holders have decreased their tax burden by selling conservation easements to the city. Other benefits of the agreement are more dispersed, for example when the city funds the maintenance and repair of local wastewater-treatment plants, benefiting all residents of the sewer district without regard to their marginal burden on the system.

Several authors have cited the case as an example of how economic incentives can generate solutions to environmental problems (Budrock, 1997; Heal, 2000). The next section reviews the history that led to this historic case, discusses its relevance in the expanding area of urban metabolism, and evaluates the policy significance of this leading example of establishing payments for ecosystem services.

3. EVOLUTION OF THE NEW YORK CITY WATER SUPPLY

New York City's quest for clean, abundant water has, at various times, advanced the practice of civil engineering, contributed to legends of political corruption, and in general brought out both the best and worst in individuals struggling between the temptations of personal greed and advancement and the desire for public well-being and long-lived legacies. Today's New York City water supply draws from two principal watersheds, as shown in Fig. 2. The older of the two, known as the Croton system, provides approximately 10 percent of the city's daily water demand. It lies east of the Hudson River and consists of a collection of small reservoirs and controlled lakes; The Catskill–Delaware system, comprised of 5 larger reservoirs lying west of the Hudson River, provides 90 percent of the city's daily supply and

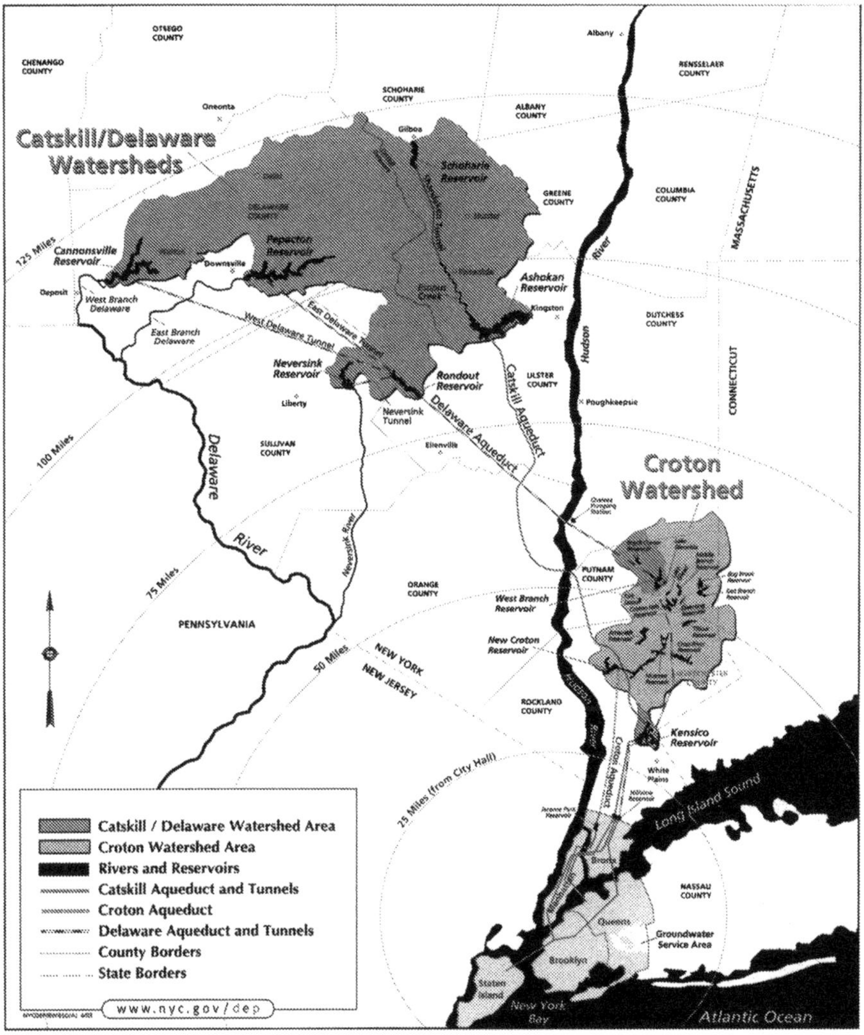

Fig. 2. Catskill and Delaware Watershed Systems. *Source*: Reproduced with permission of the New York City Department of Environmental Protection.

is the focus of the water quality efforts discussed in this chapter. The system's evolution involves some of the most famous and most obscure characters of U.S. history, as well as countless immigrants who worked along the aqueduct lines. A brief re-telling of the main themes of this ongoing saga merits treatment here, if only to appreciate how far the city has gone in its quest for water.

Three major histories of New York City's drive for water exist in the literature. Koeppel's *Water for Gotham* (2000) covers the period from the settlement of the island of Manhattan to the opening of the Old Croton Aqueduct in 1842; Weidner's *Water for a City* (1974) covers the period from Manhattan's settlement to the late 1940s; and Galusha's *Liquid Assets* (2002) covers the period from British takeover of the original Dutch settlement to the signing of the Memorandum of Agreement (MOA) between the upstate communities and the city in 1997. The quest for ever more distant water supplies was urged on by a few key phenomena, including disease, population growth, and increasing per capita use. Early on, both quantity and quality concerns spurred the search for water beyond the boundaries of the city, but later, the search for ever more distant water supplies could be boiled down to a race with the city's ever increasing population.

Early European settlers on the island today known as Manhattan relied on shallow wells and collected rainwater for their water supplies. Both groundwater and surface water were quickly polluted as the population grew. A large pond known as "The Collect" was initially tapped to provide an alternative, cleaner source of water, but public wells continued to serve most of the population. Early entrepreneurs sold water from The Collect by the cask or the pail in town. As the population grew, however, this source was surrounded by a growing assortment of slaughterhouses, tanneries, and other polluting industries on its banks. The wealthy could afford spring water shipped by the barrel from the northern part of island, but for the masses dirty water was a fact of life.

Disease was a main consequence of the lack of pure water until the mid 19th century when the first upstate water was delivered to the city. A yellow fever outbreak in the summer of 1702 claimed 12 percent of the population, while smallpox claimed 6 percent in 1731. Another yellow fever outbreak in the 1790s prompted the first proposal for an off-island water supply (choked off by private interests) and yet another in 1805, though less deadly, caused over one third of city residents to flee to the country. A massive outbreak of Asiatic cholera in 1832, which claimed nearly one in 50 lives in the city of nearly a quarter million, finally spurred the civic leaders to action on the search for a pure, public supply after decades of stalling. Sporadic efforts at stricter enforcement of sanitation laws improved conditions at times, but never for more than a few decades.

As populations grew, another negative consequence of the city's lack of water infrastructure was the devastation of major fires. The city built several rain-collecting cisterns to provide water for fire-fighting, but because of frequent drought they were still often under-supplied. River water was used early on when the settlements were confined to the lower parts of Manhattan where the island is relatively narrow. But as settlements spread northward where the island widens significantly, this strategy grew less viable. The beginnings of New York's public water system were laid in the process of addressing this lack of fire-fighting resources. In 1829, the Common Council proposed and began funding a public system with wells, a tank reservoir, and an iron-pipe distribution system which later served as the core of the distribution system for the upstate water supplies. In the midst of the planning for the first upstate supplies, the city experienced one of the worst fires in its history, the Great Fire of 1835, destroying a twenty-block area covering 52 acres and estimated at 10 percent of the assessed value of all the property in the city.

Distrust in the ability of evolving governmental institutions to build and manage a massive water project led first to hesitation, and later to the abdication of this responsibility to private interests. In April of 1799, a private company was chartered with the purpose of providing New York with a clean and abundant water supply. Through skillful (and perhaps underhanded) politics, state Assemblyman Aaron Burr laid out the design of the Manhattan Company and secured its approval in the legislature.[3] Similar to the state's two private canal companies, the Manhattan Company was given substantial rights of eminent domain over lands, rivers, and streams, but without the usual obligations of repairing streets torn up in pipe-laying and providing free water to the city to fight fires. Another significant departure from past corporate charters was that the company's charter was perpetual, provided that within 10 years it fulfilled its main purpose of supplying pure water to the city. But the most significant difference, which was to change corporate law forever, was its permission to use surplus capital for its own benefit.

The charter had in effect created a bank, and the new company's banking efforts were clearly a higher priority than its water supply obligations. Instead of developing off-island sources, as the city assumed it would do, it opted to further develop The Collect, a far cheaper plan. The company was slow to fulfill its water obligations, but much more aggressive in its banking business. Though it opened its first banking office that September, in its first two and half years in business, it could claim only about 1,700 household water customers in a city of over 60,000. By the early 1820s, though the city's population and settled area had tripled over the previous two decades, the company had added only a few hundred customers and 40 miles of pipe.

As late as 1830, Manhattan Company water was still only available to about one third of the city's population.

Fire, disease, and the city's experience with the Manhattan Company all pointed to the need for an improved public supply, though there were few possibilities for adequate supplies on the island. For years the city had been commissioning studies of available on-island and off-island sources. Finally in 1832, a report by Col. DeWitt Clinton Jr. convinced the city's Common Council of the need to pursue off-island sources from the Croton River in Westchester County. Clinton estimated the costs of construction at just $2.5 million for a quantity of water that he claimed would last the city for many centuries. Construction began in 1837 and in the end cost the city over $11 million and was adequate for just 40 years, but his work set in motion the city's appropriation of ever more distant water sources. The first musings of upstate–downstate conflict over water began with demands for compensation and general opposition to the project on the part of many Westchester residents.

Croton River water was eventually delivered in the city on July 4, 1842. Though celebrated with great fanfare, the city did not take to its new water source quickly. By 1844, Croton water was still largely a public amenity, providing mainly fire protection, free street hydrants, and fountains, but most homeowners and landlords did not install private service pipe because of the cost. In 1848, when the High Bridge over the Harlem River was finally completed and the water pressure in the city was consequently boosted, business for city plumbers finally boomed.

The city's population increased by about 20,000 per year during the 1840s and by nearly 30,000 per year in the 1850s. Not long before, the water closet had been patented in the U.S. and during this time was being widely adopted.[4] As a result of increasing per capita use and strong population growth, New York City's daily demand for water increased from 12 million gallons per day (mgd) in 1842 to 35 mgd in 1849, and further to 60 mgd by the mid 1870s. By the 1880s, per capita daily use was over 100 gallons, higher than anywhere else in the world at that time. Just 40 years after its completion, the capacity of a system intended to last for centuries had been reached.

The fast-growing consumption motivated an increase in water rates in 1850, though this did little to curb the city's demand. In 1852 the Croton Aqueduct Department began installing commercial meters, but the measure was not widely implemented until 1878. No such incentive for personal water conservation existed as residences were charged by a complicated street frontage formula that remained in effect until the city adopted a universal metering system in 1986. The city seemed to be under continual threat of water shortage between 1850 and 1890, but concentrated on increasing supply rather than on limiting consumption.

Persistent droughts and demand-induced shortages led the city on a search to develop additional sources in the Croton watershed that could be linked to the original Croton Reservoir and thus transported via the aqueduct. During the remainder of the 19th century, the city constructed an additional eight reservoirs in Westchester and Putnam Counties and acquired three controlled lakes in Putnam County. The capacity of the original aqueduct was quickly exceeded, so in 1881, with a severe drought in progress, chief engineer Isaac Newton presented a report recommending a new, larger Croton Aqueduct and reservoir to deliver 250 mgd. The new aqueduct had been in service less than a year when city officials reported that water consumption had jumped by 50 percent from 110 to 165 mgd; 4 years later, in 1895, consumption was at 183 mgd. This prompted the city to move forward with plans to enlarge the original Croton Reservoir. Three additional reservoirs were constructed before the entire Croton system was completed.

Reports by the chief engineer and president of the Aqueduct Commission in the early 1890s stated that the Croton supply would be sufficient for years to come, but clearly these estimates were made with the assumption that "New York City" would continue to be confined to Manhattan and the Bronx. In 1898, the boundaries were expanded to include Brooklyn, part of Queens, and Staten Island. The city's population instantly increased from 2 million to 3.5 million, and annual increases in population numbered around 135,000. When the five boroughs were consolidated, Manhattan and the Bronx were getting their water from the two Croton aqueducts, while Brooklyn was pumping its supply from groundwater sources on Long Island. Queens and Staten Island were supplied from local wells, mostly owned and operated by private water companies.

By 1905, New York was the second largest city in the world, second only to London. The search for additional water supplies continued unabated. Reports commissioned by the city recommended a variety of sources, including several rivers and streams in Dutchess County, the Upper Housatonic River in western Massachusetts and Connecticut, the Hudson River, and the Esopus and Catskill Creeks across the Hudson. The authors of these reports also noted the massive waste of water in the city, and recommended universal metering and more stringent plumbing regulations. But the culture of the city's water agencies was biased heavily towards developing additional supplies, and so they turned next to the Catskill Mountains, having dismissed the other recommended supplies because of unresolved legal and jurisdictional issues.

In 1905, the newly formed Board of Water Supply submitted initial plans for the development of the Catskill system. When the first Catskill water arrived in the city from the Ashokan Reservoir in the mid-1910s, many argued the next phase of the plan, the Schoharie project, was unnecessary, because the city was lacking money and its population was temporarily decreasing as many recent immigrants returned to Europe to fight in World War I for their native countries. Despite this, the city decided to proceed with its plans and in 1928, when at full operation, the Catskill system provided an average of 614 mgd of New York City's daily consumption of 879 mgd. Average daily per capita usage rate had reached 135 gallons.

By 1930, there were nearly 7 million people calling New York City home. Impending shortages, even after the construction of the Catskill system, sent the city looking for additional upstate sources. In March 1923, the city had set in motion a venture to tap Delaware River sources. That month, the state legislature passed an act establishing a commission to work with representatives of New Jersey, Pennsylvania, and the federal government to formulate a treaty outlining the conservation, use, and development of the Delaware River drainage basin. The East and West Branches of the Delaware River originate in Delaware County, NY, while the main branch forms the boundaries between New York and Pennsylvania and between Pennsylvania and New Jersey. The commission developed a 24-article pact that included plans to develop dams to supply not only New York City, but also Philadelphia and northern New Jersey. The agreement was never ratified, having been rejected by the legislatures in Pennsylvania and New Jersey not once, but twice.

New York continued to pursue rights to the water, arguing that 70 percent of the water flowing past Tri-State Rock at Port Jervis originated in, and so belong to, New York State. The city was prepared to allow communities along the new aqueduct to take water from it, and anticipated building sewage treatment plants for a number of villages in the region. Following the stock market crash of 1929 and the beginning of the Great Depression, communities welcomed any economic stimulus the construction might drive, but were wary of similar community destruction and relocation as had occurred at other reservoir sites. In a suit that went all the way to the U.S. Supreme Court, New Jersey sought a permanent injunction against the taking of water from the Delaware, arguing that New York was already wasteful, should install meters to curb wastage, could further develop its existing upstate supplies; and that drawing from the Delaware would damage the state's coastal estuaries and thus its oyster industry. The city countered that New Jersey would actually benefit from improved regulation of stream flow, enhanced electricity generating potential, improved sanitary conditions, and recreational opportunities. The Supreme Court sided with New York, but permitted only 440 mgd (rather than the 600 mgd planned) and required releases to maintain downstream flows. This issue would return to the Supreme Court in 1954, but in the meantime, both the Great Depression and World War II significantly slowed progress.

Population pressure and increasing contamination of groundwater sources on Long Island led to the accelerated development of what was supposed to be phase 2 of the Delaware River plans, the Pepacton Reservoir. At 140 billion gallons, the Pepacton's capacity was 40 percent bigger than the combined volume of all the Croton System reservoirs. Meanwhile, New York's thirst for water was growing rapidly. Though the city had used an average of over 1 billion gallons per day in 1930, consumption decreased temporarily during the Depression, but went over the 1 billion gallon mark for good in the early 1940s. The state commission approved the further expansion of the Delaware system in 1950, prompting the city to request an amendment from the 1931 Supreme Court decision, approved in 1954. Reporting on the last service of a local Methodist church, the *West Branch Courier* quipped on July 5, 1962, that "The Lord may have dominion over the heavens and the earth, but he isn't doing to well against the U.S. Supreme Court and the New York City Board of Water Supply" (quoted in Galusha, 2002, p. 222).

With these projects completed, the city closed its chapter of reservoir building for the time being, though city water infrastructure is still under construction in the form of City Tunnel #3, a large new distribution pipeline that will take an estimated 50 years to complete. In recent decades instead of developing new water supplies the city has focused on conservation measures. In 1986, the city finally adopted a universal metering system and began implementation in 1988. By the end of 1997, 456,000 meters had been installed, finally replacing the flat-rate, street-frontage system of billing that dated from the middle of the 18th century. The city also replaced 1.3 million toilets with low-flow models. Overall savings amounted to 273 mgd over 9 years, cutting daily consumption from nearly 1.5 bgd to just over 1.2 bgd. The city's Department of Environmental Protection announced in 1997 that:

> ... based on the results of metering, toilet replacement, leak detection, public information and other conservation programs achieved to date and expected in the future, it is projected that no additional water sources will be necessary for the next 50 years. (Galusha, 2002, p. 229)

4. WATER CONFLICT AND COMPENSATION

As the city battled against its own growth and nature's occasional droughts, it also fought over water rights and legal claims with upstate residents. Galusha (2002, p. 41) notes:

> The battle between stubborn nature and shovel-wielding workmen mirrored the one between the city and the people who lived along those [lakes and streams]. It was a fight that continued for decades.

Frustration did not end with the completion of construction projects, but was ongoing as the city worked to protect its investments in water purity. The legislature's passage of the Webster Act of 1893 gave the city authority to condemn watershed property in the name of water-quality protection. A directive from the Common Council established a 300-foot buffer around reservoirs and feeder streams through ordering the compensated evacuation of homes, removal of barns and pigsties, and the burning of privies. In preparation for the development of the Catskill system, guarantees were put in place to protect property owners seeking redress for damages, but at the same time the city was allowed to take possession of land and/or buildings upon payment of one half of the assessed value. Though later amendments would further clarify the city's responsibilities, residents were generally displeased with the level, timing, and bureaucracy of compensation or actions taken by specially formed Appraisal Commissions. For example, litigation from owners of 954 parcels relating to the construction of the Ashokan reservoir took 36 years to complete and resulted in 366 volumes of commission hearing transcripts. Since their inception in 1935 the Delaware Commissions of Appraisal settled 6,700 claims totaling $26.8 million. Some of these cases still were not resolved as late as the 1980s.[5]

Galusha's (2002) accounting of the tangible sacrifices incurred by the upstate communities over the duration of construction projects includes 36 communities flooded, 10,307 residents displaced, and 11,580 graves reinterred. This is surely an underestimate, since many of the reservoirs, particularly in the Croton system list no such impacts under their headings. The list of uncompensated, intangibles include loss of connections to historic family land holdings and disruption of community life both directly by forced relocations, and indirectly, via the introduction of thousands of migrant laborers to be absorbed into local schools and churches.

Given the sacrifices endured over the generations, lingering resentment towards the city was to be expected when, in September of 1990, the city published a new set of draft watershed rules and regulations pursuant to the EPA's 1989 SWTR without much consultation with watershed communities.[6] Previously water quality was loosely managed under the purview of a set of 1953 rules for land use and activity, based on the New York State Sanitary Code. In January 1993, the city received its first FAD effective for one year, with renewal contingent upon its ability to prove progress towards preventing further degradation of the water supply. The general response in watershed communities was typified by the statement of a Delaware County farmer at one of the hearings: "For generations, Jersey cows have paid our bills. Is it fair that now they can't drink out of our brook that we pay taxes on?" (quoted in Galusha, 2002, p. 257).

The following March, representatives from the towns and villages in the five affected counties met in a local school cafeteria and agreed to form the Coalition of Watershed Towns (CWT). The election of Rudolph Giuliani as New York City major that autumn and the subsequent appointment of new Department of Environmental Protection commissioner Marilyn Gelber helped to ease tensions. Gelber won respect and even some affection from upstaters by making several visits to the region to talk with local representatives. Meanwhile, the Ad Hoc Task Force on Agriculture and NYC Watershed Regulations formed the Watershed Agricultural Council (WAC), and garnered $4 million for pilot programs in "Whole Farm Planning." Success with those pilot programs led to the allocation of an additional $35.2 million from the city. WAC convinced fellow farmers of the value of the program, and by 1997 had signed on more than 85 percent of eligible farms. The city's original FAD was renewed in December 1993, effective through December 1996.

Despite these successes, tensions still ran high. In 1994, similar efforts at engaging in "Whole Community Planning" failed when DEP released its final version of the proposed new watershed rules and regulations, including the federally mandated acquisition of 80,000 acres of land. The city did not rule out the possibility of acquiring land through condemnation, and the CWT was not willing to negotiate without assurances that such methods would not be exercised. These circumstances and the looming deadline of the renewed FAD brought the intervention of the state governor, whose office sponsored several more months of intense negotiations. An agreement in principle was announced on November 2, 1995, but it took another 10 months for the details of the pact to be assembled. The final New York City Watershed MOA was released the following September, signed in January, and went into effect in April of 1997.

The agreement comprised three major pieces, for which the city committed to spend $1.2 billion over the next 7–15 years: a Land Acquisition Program, Watershed Rules and Regulations, and Partnership & Protection Programs. Of that $1.2 billion, nearly half was slated for infrastructure and physical water quality improvements, while approximately $270 million was for partnership programs in the Catskill–Delaware watershed. As an added incentive, the city issued nearly $10 million in "Good Neighbor" payments to towns, villages, and counties based on their respective acreages in the watershed, which could be used for any purpose other than tax reduction. Among the various Watershed Partnership and Protection Programs is the Catskill Fund for the Future (CFF), a $60 million fund set aside to provide loans and grants for economic development projects consistent with the MOA's water quality goals.

In 1997, the EPA issued another 5-year FAD, and in November of 2002, that determination was again renewed. As of August 2002, the city had solicited the owners of more than 281,000 acres in various priority areas (out of 350,050 total acres to be solicited over 10 years) under the land-acquisition program. Spending on other elements of the Watershed Partnership and Protection Programs amounted to $288 million as of December 2001 (NYC Department of Environmental Protection, 2001). Table 1 shows the type of projects funded with loans and grants from the CFF. Additional programs administered by the WAC also seek to create some economic benefits for watershed farmers and forest-based firms. For example, one such effort links farmers with food outlets (restaurants, bakeries, grocery stores, and the like) and directly with New York City residents in programs like a Taste of the Catskills (WAC, 2003).

Table 1. Loans and Grants Made by the Catskill Fund for the Future as of December 2001.

Economic Sector	Loans $	Loans #	Grants $	Total $
Natural resource-based industry	525,000	3	34,108	559,108
Retail/service businesses	1,989,120	13	22,764	2,011,884
Manufacturing	2,111,968	6	115,251	2,227,219
Tourism (including lodging and restaurants)	1,672,400	13	1,173,108	2,845,508
Village/hamlet/main street revitalization			185,057	185,057
Totals	6,298,488	35	1,530,288	7,828,776

Source: NYC Department of Environmental Protection, 2001.

5. TOWARDS A NEW REGIONAL ECONOMY

The negotiations surrounding the 1997 MOA brought the watershed counties a long-awaited seat at the table to discuss their own vision of the future. That future will likely include a much more symbiotic relationship with the city that drinks its water, and the degree to which that symbiosis develops will depend on strengthening existing economic relationships and forging new economic ties. The Catskill economy can benefit from the tremendous consumer demand of New York City, while New York can continue to benefit from continued supply of water supply from the Catskill/Delaware watershed.

The west-of-Hudson watershed region includes counties that are generally less well off than the rest of the state. Table 2 shows the per capita personal income (PCPI) in each county in absolute terms and relative to the state and national averages, along with its ranking (out of 62 counties in New York State). The economy of the Catskills has mirrored that of the national economy in terms of its changing structure and composition. Manufacturing, once a strong presence, has given way to a more service-oriented economy (Kane & Gowdy, 1998). Even within the counties, the watershed areas show some marked differences compared with the non-watershed areas. An analysis of 1997 ES-202 data from the NYS Department of Labor shows that roughly three-quarters of employment in the five-county region is located outside the watershed boundaries. Approximately 11 percent of the employment was in firms located inside the watershed. The remainder could not be pinpointed geographically (Kane & Gowdy, 1998). There is also a notable difference in the industrial composition inside the watershed compared to the five-county region as a whole. For example, in 1997 the

Table 2. Per capita personal income in the watershed counties.

County	PCPI in 2001 $	State Ranking	PCPI as Percent of State Average ($35,878) %	PCPI as Percent of National Average ($30,413) %
Delaware	21,692	51	60	71
Greene	24,315	34	68	80
Schoharie	22,813	44	64	75
Sullivan	25,544	27	71	84
Ulster	26,023	23	73	86

Source: U.S. Department of Commerce, Bureau of Economic Analysis, 2003.

government sector represented 19 percent of employment inside the watershed, while it comprised only 10 percent of employment in the five counties.

The New York City economy, with total industrial output of nearly $548 billion and $259.2 billion in total personal income in 1998, obviously dwarfs the economy of the watershed counties whose total output in 1998 was just $12 billion and whose total personal income was $8.2 billion. In terms of income, however, it includes some of the most and least wealthy areas of the state. New York County (Manhattan) had a 2001 PCPI of $92,984 (ranking 4th in the state), while Bronx County was $19,896 (ranking 58th).

In the end, the watershed region needs income opportunities and New York City needs water. The SWTR provides a powerful incentive for municipalities and water utilities to adopt protective or restorative measures within their source watersheds and to explicitly consider the economic value of the services those ecosystems provide. By attempting to compensate them for constraints on their economic opportunities, New York City has tried to motivate the watershed residents to be better stewards of the ecosystems that purify its drinking water. To the extent that some mutual dependence exists, acknowledging it might help facilitate negotiations over urban water appropriation and the economic consequences of that appropriation for the watershed region. For this vision to be applicable, the economic benefits gained (whether measured in terms of income opportunities, jobs, or tax revenues) must accrue to the residents of the watershed region. This would be especially true for the agricultural sector, since it forms such a significant part of the economic dependence between the city and the watershed, and is such a dominant feature of the human-occupied landscape.

The watershed region does not function as a true economic region, and thus as currently arranged it is likely that the economic benefits from investments made in the region will also not remain there. Some benefits of the city's investment will accrue, but the multiplicative effects (the spending and re-spending of each round of income) will likely be drawn out of the watershed toward other urban cores (Kane, 2003). The places where benefits are most likely to accrue are those closest to the city, and within its sphere of economic influence. The watershed area in the two counties that fall closest to the city's borders, Ulster and Sullivan Counties, is largely protected by virtue of its location inside the state-protected Catskill Park. In contrast, in Delaware County, where the bulk of farms inside the watershed are located, stronger incentives are likely needed to encourage watershed stewardship, and yet the indirect economic benefits of New York City's spending may be more diffuse.

The lessons for the nation from this case could come in the form of mitigated spending requirements on water and sewer infrastructure, for which the accumulated gap between actual and needed spending is estimated by the EPA to exceed $650 billion in the next 15–20 years, with some other estimates reaching $1 trillion (Revkin, 2002). While regional economic policies have been in decline since their heyday in the 1960s and 1970s (Miernyk, 1982; Leven, 1985), they could prove a cost-effective way to compensate the residents of key watershed basins to motivate the necessary ecosystem service preservation. Before this is feasible, more research must be done to determine the types of economic activities that are consistent with preserving the ecosystem water purification functions, and other technological means (e.g., Best Management Practices) of preventing water pollution in the first place. Still, this notion is one that deserves future consideration given the fact that national clean water policy initiatives such as the SWTR are simultaneously driving the need for greater water infrastructure spending (much of which is expected to be borne by the federal government) and the increased interest in source-water protection.

The simultaneous goals of maintaining a pure water supply, avoiding the need for filtration, and revitalizing the regional economy have brought the city and its watershed region through the difficult process of assessing the problems that face a large variety of stakeholders. Negotiating the MOA was the first step in this fragile experiment. But it remains to be seen whether the efforts will bear fruit. In particular, ongoing debates over the future vision of economic development in the watershed region and the role that the city will play in influencing the scale of development will need to be resolved. Despite the lack of evidence that the negotiated settlement will provide long-term relief from the SWTR's filtration requirement, the New York City case has been cited as a model for urban water management conflicts in other parts of the country and the world. Before it is applied, water managers and civic leaders must be careful to examine their own situations carefully, paying attention to the spatial distribution of hydrologic phenomena and economic activity, and analyzing the water quality impacts of the economic development strategies they choose to employ.

NOTES

1. A population for the watershed of 52,500 or 32.2 persons per square mile is approximated by zip codes using U.S. Census data (2003). Other reports have estimated the watershed population at 65,000, though it is unclear whether these

estimates include seasonal populations. Though the New York City Metropolitan Statistical Area has a population of about 9.4 million, this study focuses on the city proper, and thus reports only on figures as they correspond to the five boroughs of the city.

2. This schematic is rather unidirectional, although presumably some of the waste products of the city's metabolism would impact in some way the resource inputs available to the city, especially those that are not imported.

3. The Manhattan Company would later come to be known as Chase Manhattan Bank. Its waterworks at Chambers Street were emptied soon after Croton water became available and were demolished in the early 1900s, though a bank employee ceremoniously pumped water at the site daily until 1923 for fear of losing its state charter. In 1965, a national charter was finally written.

4. Weidner (1974) gives this description of the water supply problems caused by the widespread introduction of the water closet, based on a Sears, Roebuck & Co. advertisement: "To test its efficiency and durability, one of the company's display items, a fern-green toilet, was flushed 52,365 times, *the estimated number of times it would be used in five years*. At this rate one toilet would consume 115 to 120 gallons of water daily, or about 42,000 gallons per year. Multiply this by the number of units in a city the size of New York and the result is no longer simply an exercise in mathematics but a problem for city planners and engineers" (p. 55, emphasis is the author's).

5. By the middle of the 1980s more than 800 Delaware system business damage and indirect real estate claims had been filed but not prosecuted, and the city called for anyone with pending claims to come forward. The bulk were dismissed because the plaintiffs had died in the preceding three decades, or had lost the records to prove their cases.

6. It is interesting to note that in the initial planning stages of the Catskill reservoir system construction, the city planned on building a water filtration plant. But with water quantity issues becoming a more pressing problem, and the adoption of chlorine and other chemicals for disinfection purposes, these expensive filtration plans were shelved.

REFERENCES

Baccini, P. (1996). Understanding regional metabolism for a sustainable development of urban systems. *Environmental Science and Pollution Research*, *3*(2), 108–111.

Baccini, P. (1997). A city's metabolism: Toward the sustainable development of urban system. *Journal of Urban Technology*, *4*(2), 27–39.

Budrock, H. (1997). *Summary guide to the terms of the watershed agreement*. Arkville, NY: Catskill Center for Conservation and Development.

Carey, A. E., Pennock, J. R., Lehrter, J. C., Lyons, W. B., Schroeder, W. W., & Bonzongo, J. C. (1999). *The role of the Mississippi river in gulf of Mexico hypoxia*. Tuscaloosa, AL: Environmental Institute, University of Alabama.

Cronon, W. (1991). *Nature's metropolis*. New York, NY: W.W. Norton & Company.

Daughton, C. G. (2002). Environmental stewardship and drugs as pollutants. *Lancet, 360*, 1035–1036.

Decker, E. H., Elliott, S., Smith, F. A., Blake, D. R., & Rowland, F. S. (2000). Energy and material flow through the urban ecosystem. *Annual Review of Energy and the Environment, 25*, 685–740.

Galusha, D. (2002). *Liquid assets: A history of New York City's water system*. Fleischmanns, NY: Purple Mountain Press.

Heal, G. M. (2000). *Nature and the marketplace: Capturing the value of ecosystem services*. Washington, DC: Island Press.

Kane, M. (2003). *Shared futures of city and country: A total flows analysis of urban–rural ecological economic dependence in New York City's Catskill Delaware watershed*. Ph.D. Dissertation, Department of Economics, Rensselaer Polytechnic Institute, Troy, NY.

Kane, M., & Gowdy, J. (1998). *Compatible development in the Catskills: An economic profile of the Catskill region (report prepared for the Catskill Center for Conservation and Development, Arkville, NY)*. Troy, NY: Rensselaer Polytechnic Institute.

Koeppel, G. T. (2000). *Water for Gotham*. Princeton, NJ: Princeton University Press.

Leven, C. L. (1985). Regional development analysis and policy. *Journal of Regional Science, 25*(4), 569–592.

Miernyk, W. (1982). *Regional analysis and regional policy*. Cambridge, MA: Oelgeschlager, Gunn & Hain.

New York City Department of Environmental Protection. (2001). *New York City's 2001 watershed protection program summary, assessment and long-term plan*. New York, NY.

Newman, P. W. G. (1999). Sustainability and cities: Extending the metabolism model. *Landscape and Urban Planning, 44*, 219–226.

Odum, E. (1971). *Fundamentals of ecology*. Philadelphia, PA: Saunders.

Raloff, J. (1998). Drugged waters. *Science News, 153*, 187–189.

Rees, W. (1992). Ecological footprints and appropriated carrying capacity: What urban economics leaves out. *Environment and Urbanization, 4*(2), 121–130.

Revkin, A. C. (2002). Federal study calls spending on water systems perilously inadequate. *New York Times*, 10 April 2002, p. A22.

United States Environmental Protection Agency. (1998). National Water Quality Inventory: 1996 Report to Congress (p. 521). Washington, DC.

United States Environmental Protection Agency. (1999). *Protecting sources of drinking water: Selected case studies in watershed management*. Report No. EPA 816-R-98-019. Washington, DC, Office of Water.

United States Environmental Protection Agency. (2001). *Factoids: Drinking water and groundwater statistics for 2000*. Report No. EPA 816-K-01-004. Washington, DC, Office of Water.

United States Census Bureau. (2003). State and County Quick Facts. http://quickfacts.census.gov/qfd/

United States Department of Commerce Bureau of Economic Analysis. (2003). Regional Accounts Data; Local Area Personal Income, Bureau of Economic Analysis. http://www.bea.gov/bea/regional/bearfacts/

Vitousek, P. M., Aber, J. D., Howarth, R. W., Likens, G. E., Matson, P. A., Schindler, D. W., Schlesinger, W. H., & Tilman, D. G. (1997). Human alteration of the global nitrogen cycle: Sources and consequences. *Ecological Applications, 7*(3), 737–750.

Watershed Agricultural Council. (2003). Organizational web site, www.nycwatershed.org

Weidner, C. H. (1974). *Water for a city: A history of New York city's problem from the beginning to the Delaware river system*. New Brunswick, NJ: Rutgers University Press.

Wolman, A. (1965). The metabolism of cities. *Scientific American*, *213*, 179–190.

World Health Organization and UNICEF Joint Monitoring Programme for Water Supply and Sanitation. (2000). Global Water Supply and Sanitation Assessment 2000 report. Geneva, Switzerland, World Health Organization/UNICEF. (www.who.int/water_sanitation_health/Globassessment/GlobalTOC.htm)

FISCAL TRANSFERS FOR COMPENSATING LOCAL ECOLOGICAL SERVICES IN GERMANY

Irene Ring

ABSTRACT

Provision of ecological goods and services at the local level is often related to benefits at higher governmental levels. On the one hand, sustainable watershed management and biodiversity conservation are strongly connected to local land-use decisions. On the other hand, related conservation activities and protected areas are frequently associated with regional, national or even global public goods. Therefore, spatial externalities or spillover effects exist that – if not adequately compensated – lead to an under-provision of the public goods and services concerned. This chapter investigates fiscal transfers as an innovative instrument for compensating local jurisdictions for the ecological goods and services they provide across local boundaries. From a public finance perspective, fiscal transfers are a suitable instrument for internalising spatial externalities. A case study is presented that investigates the present and potential use of fiscal transfers for ecological public functions in the German federal systems. Analysis of the German system of fiscal equalisation at the local level shows that, so far, mostly end-of-the-pipe activities are currently considered with resource protection and nature conservation being widely underrepresented.

Ecological Economics of Sustainable Watershed Management
Advances in the Economics of Environmental Resources, Volume 7, 329–346

ISSN: 1569-3740/doi:10.1016/S1569-3740(07)07014-9

1. INTRODUCTION

Sustainable watershed management requires a variety of conservation efforts and services, not least at the local level. Dealing with environmental pollution is as necessary as focusing on precautionary tasks such as soil, water and biodiversity conservation. The environmental quality of a watershed is closely linked to its land-use pattern and the type of management performed by public jurisdictions and private land users. However, there are few incentives for local actors to encourage in conservation activities, when ecological benefits cross local boundaries. This is the case for a number of ecological services, for example water protection or nature reserves. Decisions on the designation of respective protected areas are often taken by institutions above the local level, whereas the concrete consequences in terms of restrictions in land use are to be born by local actors, often without any or sufficient compensation.

The aim of this contribution is to present innovative instruments that are able to address this basic problem. We will analyse the role intergovernmental fiscal transfers can play in federal systems for compensating local ecological services. From a public-finances perspective, it is the "value-added" of local ecological services, i.e., the benefits crossing the boundaries of local jurisdictions that are of special interest. Without internalising these positive spatial externalities, adequate provision of ecological goods and services at decentralised levels of government cannot be secured. In this context, fiscal transfers as part of the system of fiscal equalisation at the local level have been increasingly discussed for Germany (SRU, 1996, 2000; Rose, 1999; Ewringmann & Bergmann, 2000; Ring, 2001a, 2001b, 2002; Perner & Thöne, 2002). Therefore, the first part of this chapter gives a short introduction into public finances principles as relevant to ecological goods and services. The second part presents a case study from Germany, focusing on the status quo of and perspectives for integrating ecological public functions into intergovernmental fiscal transfers.

2. FISCAL FEDERALISM AND LOCAL ECOLOGICAL SERVICES

2.1. Basic Principles of Fiscal Federalism

The basic task of fiscal federalism is one of effectively and efficiently assigning public functions, expenditures and revenues to the central, state and local governmental levels in federal systems, or, in other words, determining the optimal size of jurisdiction for the various public functions concerned. As Oates (1999, p. 1120) puts it, "... we need to understand which functions and instruments are best centralised and which are best placed in the sphere of decentralised levels of government". Concerning the allocation function of public sectors, the basic principle of fiscal decentralisation has been put forward (Musgrave, 1959; Oates, 1972). Provision of most public goods and services is more efficiently guaranteed when production and consumption are limited to the lowest governmental level possible. In this way, the regionally differing preferences of the population can be more adequately reflected (Tiebout, 1956).

The general decentralisation rule for allocation of public goods and services only applies in the absence of economies of scale. In the presence of economies of scale, the provision of public goods and services concerned should be moved to the cost-efficient centralised level (Postlep & Döring, 1996). In addition, due to the characteristics of non-rivalry and non-excludability of many public goods, some of them are associated with spatial externalities or spillovers between jurisdictions. Here, the principle of fiscal equivalence applies, which advocates achieving a "match between those who receive the benefits of a collective good and those who pay for it" (Olson, 1969, p. 463). Social welfare is increased through the differentiation of public services in accordance with local costs and preferences. The implementation of fiscal equivalence may require the shifting of competence to a more centralised level of government. However, regional co-operation, e.g., in the form of negotiations between the parties concerned, can also bring along the potential for an efficient Coasian type of resolution of jurisdictional spillovers (Bergmann, 1999; Oates, 2001). Furthermore, the formation of administrative institutions mapping the spatial range of costs and benefits are also discussed to internalise spatial externalities (e.g., Breton, 1965; Frey, 1996). Olson (1969) has suggested still another solution to this kind of problem. Provided diseconomies of large-scale operation call for local provision, spillovers can be internalised through government grants from more centralised levels. In this way, fiscal transfers in the form of grants compensate the local government for the external benefits of its expenditures.

Intergovernmental fiscal grants play a substantial role in fiscal federalism that can serve different functions. The literature emphasises the role of the internalisation of spillover benefits to other jurisdictions. Based on normative considerations of equity, they can also serve the purpose of fiscal equalisation among different jurisdictions. These equalising grants play an important role in the fiscal system of Germany, as well as in other federal systems such as Canada and Australia (Oates, 1999). Before moving on to the German case study, we will first introduce the field of environmental federalism and discuss how the principles of fiscal federalism generally apply to ecological goods and services.

2.2. *Fiscal Federalism and Ecological Public Functions*

Environmental federalism links environmental issues with the basic theory of fiscal federalism. Both the general principle of sustainable development as adopted by international conventions – or regarding Germany also by European legislation – and the numerous ecological public functions[1] as already assigned to the various governmental levels within nation states call for consideration of ecological goods and services in intergovernmental fiscal relations. On the one hand, ecological public functions consist in the conservation and sustainable use of resources and landscapes. These precautionary type functions comprise fields such as soil, water and biodiversity conservation. However, they also include activities aiming at the conservation of nature as a sound living basis for humans, including recreational aims. On the other hand, ecological public functions include discharging activities such as sewage and waste disposal or the rehabilitation of contaminated sites and landscapes, in short, dealing with all aspects of environmental pollution. Implementation of the concept of sustainable development requires consideration and adequate financing of these ecological public functions at appropriate governmental levels (Ring, 2002).

What are the consequences of the decentralisation rule and the principle of fiscal equivalence for environmental issues? Following the general decentralisation rule for the allocation function of public goods and services, provision of ecological public functions should be assigned to lower levels of government where appropriate. However, due to the characteristics of natural resources and environmental quality, the implementation of this rule requires a differentiated approach. This is reflected in the on-going discussion on the competencies of the national or even supranational governmental level versus the state or local level in setting environmental standards, or carrying out other ecological public functions (Döring, 1997; Scheberle, 1997; Oates, 1999, 2001).

In the European Union, fiscal decentralisation is connected to the term "subsidiarity". According to the subsidiarity principle as consolidated and adopted by the Treaty of Maastricht on European Union of 1992, public policy and its implementation should be allocated to the member states whenever the latter are competent to achieve the objectives. Since the Maastricht Treaty, environmental federalism has been rediscovered and widely debated (e.g., Huckestein, 1993; Hansjürgens, 1996; Döring, 1997; Oates, 1998). Despite the fundamental strengthening of the subsidiarity principle in the new Article 3b of the Treaty, a fair amount of leeway is left for interpretation. Any concrete implementation of environmental policy has to consider the specific details of the subject matter.

For example, the basic research function, the dissemination of information on environmental damages and pollution control techniques or the effectiveness of various environmental policy instruments need to be assigned to a more centralised level of government, for this kind of public good tends to be under-provided at decentralised levels (Oates, 2001). Further issues pointing to a fundamental role of centralised governments relate to global change problems such as climate change. Highly mobile environmental compartments and associated pollutants that easily cross national boundaries create far-reaching spatial externalities. The depletion of the ozone layer, the emissions of carbon dioxide and other air pollutants associated with climate change require more centralised if not global emission policies.

In contrast, ecological public functions associated with less mobile environmental compartments are better suited for assignment to decentralised levels of government (Ring, 2002). This is due to the lower probability of causing spatial externalities. Problems of land use and soil conservation, as well as public functions associated with inland waters, can usually be solved within national boundaries. In the federal system of Germany, the provision of public goods and services related to land-use planning, water resources and nature conservation are only subject to framework regulation at the national level. Practical implementation is delegated to the various German states (*Länder*). State laws regulate the respective functions of state and local levels of government.

2.3. Considering Spatial Externalities

Despite the general qualification of land-use questions to be assigned to lower governmental levels, spatial externalities may require appropriate solutions. This is especially the case for priority areas, e.g., for the protection of natural resources, that may cause costs within the concerned jurisdiction, but externally also benefit others. In contrast to certain local costs, be it in terms of land-use restrictions or measures for keeping up and improving the quality of the respective reserves, benefits from some of these activities cross local boundaries.

For example, water protection zones are often located in rural areas, mostly providing drinking water far beyond local demand. Especially urban agglomerations and capital regions with high population densities and industrial activities heavily rely on water resources lying outside own municipal borders. In the case of water resources, an important task consists in properly valuing these resources and their functions which then, as far as possible, should be reflected in water prices (Hansjürgens & Messner, 2002; Unnerstall & Messner, in this volume). However, for various reasons this option is not yet fully implemented, and, concerning certain tasks of long-term resource protection might even not be a feasible solution.

The conservation and sustainable use of biodiversity is another example for the widespread existence of spatial externalities (Ring, 2004a). On the one hand, the loss of biodiversity belongs to the very serious global change problems, demanding centralised standard setting and policies. This is reflected in the Convention on Biological Diversity and related activities. On the other hand, decentralised activities related to local land use have – if accumulated – a tremendous influence on the state of biodiversity world-wide. Reflecting the value of ecological services in market prices is even more difficult if not impossible for many fields of biodiversity conservation. This is especially true for benefits related to non-use values such as existence and option values that may accrue for people everywhere. The practical consequences of spatial externalities related to species protection are illustrated by an empirical study of List, Bulte, and Shogren (2002). They found in their study of federal and state spending under the Endangered Species Act in the U.S. a free-riding behaviour on the part of the states. States tend to spend less (relative to the federal government) on those species that demand a large habitat area and those whose preservation causes conflicts with economic development. Perrings and Gadgil (2003) address a number of reforms necessary to reconcile both local and global public benefits of biodiversity conservation. One of them is adjusting incentives to allow local communities to be rewarded and paid for their conservation efforts.

In the following case study, the focus for solving such discrepancies will be on fiscal transfers. Provided diseconomies of large-scale operation call for local provision, which is usually the case for public goods and services associated with land use, spillovers can be internalised through government grants from more centralised levels. These grants compensate the local government for external benefits of its expenditures or restrictions to be born. This is especially necessary for social benefits accruing in the long term where public and private actors are emerging for today's costs. In this way, the "value-added" of local ecological goods and services is socially acknowledged, which at the same time can provide an incentive for local actors to engage in more conservation activities.

3. FISCAL TRANSFERS FOR LOCAL ECOLOGICAL SERVICES IN GERMANY

3.1. The German System of Fiscal Equalisation at the Local Level

In the federal system of Germany, basic regulations concerning intergovernmental fiscal relations are part of the German Constitution. Public functions of a general, usually nation-wide character are carried out at the federal level. State and local level authorities are responsible for regional and local development issues. Federal and state governments decide upon their own budgets and bear individual responsibility for their implementation. Intergovernmental fiscal relations between the state and the local level are regulated in the 13 different fiscal equalisation laws of the various German Länder (*Kommunaler Finanzausgleich*). Apart from own revenues such as local taxes and charges, fiscal transfers from fiscal equalisation represent a considerable source of income for local jurisdictions in Germany. In eastern Germany (former GDR), more than half of the local average income is obtained from these fiscal grants (relevant to the German states of Brandenburg, Mecklenburg-Western Pomerania, Saxony, Saxony-Anhalt and Thuringia). Own revenues are higher in West Germany (Baden-Württemberg, Bavaria, Hesse, Lower Saxony, North Rhine-Westphalia, the Rhineland-Palatinate, Saarland and Schleswig-Holstein), but grants from fiscal equalisation still make up for 30% of average local income (Karrenberg & Münstermann, 2000, p. 14). Therefore, fiscal grants play a crucial role in local development by way of securing financial resources to local jurisdictions to carry out their various public functions.

In most German states, the larger share of these vertical transfers to the local level of government is allocated in the form of "unconditional grants". These lump-sum transfers can be used in any way the recipient wishes. They are predominantly assigned on the basis of the fiscal need of a local jurisdiction in relation to its fiscal capacity (e.g., own revenues based on local taxes). The main indicator for calculating the fiscal need is given by the number of inhabitants of a jurisdiction, often even multiplied by a weighting factor that increases with the population. Correspondingly, the more inhabitants a local jurisdiction has, the higher the lump-sum transfers. Some of the states also have additional approaches for allocating lump-sum transfers. They take into account more specific indicators based on certain central functions, the number of schoolchildren, or the social burdens of the local jurisdictions. Mostly, the indicators for additional approaches are related to the socio-economic public functions of the jurisdictions.

The remaining share of vertical transfers is represented by "conditional grants" for specific purposes. They are often given in the form of matching grants where the grantor finances a specified share of the recipient's expenditure. Regarding the various purposes for conditional grants, again great consideration of socio-economic functions is to be noticed in the various fiscal equalisation laws of the German states. For example, transport and road construction, social burdens, health services, education and cultural investments are clearly addressed as a possible motivation for application.

To sum up, the dominance of an inhabitant-based indicator and related spillovers generally favour urban areas as opposed to rural and remote areas due to the lower population densities and socio-economic public functions of the latter. Densely populated areas in Germany presently take advantage from the system of intergovernmental fiscal relations due to their socio-economic and cultural functions and the importance of inhabitants as the main indicator for allocating lump-sum transfers. Conversely, remote and rural areas receive a much smaller share of the overall financial flow due to their low population densities. However, the latter areas usually provide a variety of ecological goods and services for society as a whole such as drinking-water protection, biodiversity conservation, resource provision and recreational purposes. Consequently, any sustainable development strategy, such as implementing sustainable watershed management, has to address both these areas. Actually, they have to consider the imbalance between urban and rural areas concerning the specific land uses in place (SRU, 1996; Ewers, Rehbinder, & Wiggering, 1997; Ring, 2001a).

3.2. Investigating Fiscal Grants for Ecological Public Functions

Ecological public functions are – up to a certain extent – already part of fiscal equalisation at the local level. In the following, empirical results based on an analysis of all fiscal equalisation laws at the local level as in force for the year 2002 will be presented for the states of the Federal Republic of Germany.[2] Some preliminary remarks must be made with respect to the scope of analysis. Apart from fiscal grants as part of fiscal equalisation at the local level, all German states are familiar with a large number of additional earmarked grants, mostly in the form of incentive programmes for environmental or conservation-oriented purposes. These programmes can be exclusively implemented at the state level, or they are jointly managed by the federal and the state level, others are combined with agri-environmental programmes at the European level (Frank & Ring, 1999; Hartmann, Thomas, Luick, Bierer, & Poppinga, 2003; Unnerstall, 2004). Eligible applicants vary for the different programmes; they can include local jurisdictions, private land users, farmers or various types of associations. The main purpose of this investigation, however, is to present the status of ecological public functions within fiscal equalisation at the local level in Germany. Here, local jurisdictions are in the centre of interest and grants from fiscal equalisation are among the most important income sources for local jurisdictions in Germany. There are numerous – albeit not sufficiently financed (Hampicke, 2005) – support programmes for private land users to compensate foregone economic benefits due to environmentally sound land uses. To the contrary, there is no compensation mechanism yet for public jurisdictions at the local level that, for example, focuses on long-term land-use restrictions due to protected areas and reward-related conservation activities. Furthermore, fiscal equalisation serves additional purposes from the angle of public finance (Ring, 2002). Within the system of fiscal equalisation at the local level, there is the option to have both unconditional and conditional (i.e., earmarked) grants. Fiscal grants can be based on ecological indicators referring to the fiscal need for local ecological services where respective revenues do not necessarily have to be used for environmental purposes. Furthermore, "fiscal grants can effectively address spillover effects related to ecological public functions just as spillover effects are currently addressed with respect to socio-economic and cultural functions of urban agglomerations" (Ring, 2002, p. 422). In this context, the following investigation aims firstly at identifying the kinds of ecological functions considered (or neglected) within fiscal equalisation at the local level, and secondly at analysing the types of grants and indicators currently used for ecological functions as a basis for further recommendations.

3.3. Fiscal Grants Based on Area-Related Indicators

Area-related indicators for allocating intergovernmental fiscal grants can constitute a first step towards acknowledging ecological functions

(Ring, 2002). This is due to the significance of area and its related land uses for many ecological functions. The consideration of area is especially important for large jurisdictions (e.g., district councils) as opposed to smaller communities that are part of the wider district (local councils). The relevance of area-related indicators also increases with decreasing population densities, respectively the distance from urban agglomerations. In terms of land use, these remote areas are often characterised by a higher proportion of agricultural land and forestry, as well as designated areas for water or nature conservation purposes. From an economic point of view, the cost relevance of area for financing specific local public functions is of special interest. The larger the community or district area in association with a lower population density, the more the provision of certain public goods and services may cost. Bergmann (1999) and Henneke (2001) refer to extra costs related to both ecological and socio-economic public functions such as nature conservation, waste disposal, water supply, public transport, education and health services.

In Germany, the majority of state fiscal equalisation laws already consider area-related indicators. Area is used as a main or additional indicator for the distribution of lump-sum transfers in Brandenburg, Mecklenburg-Western Pomerania, the Rhineland-Palatinate and Saxony-Anhalt. Some states use area as one indicator among others to distribute lump-sum transfers for municipal investments. However, most German states do not use area as an indicator for lump-sum transfers. They choose conditional grants and directly concentrate on selected public functions that become more expensive with a larger area of the jurisdiction concerned. These conditional grants cover tasks such as road construction and maintenance, transport for schoolchildren, public transport or sewage disposal. Even though theoretical explanations clearly cover ecological functions, existing regulations concerning area as an indicator in the various fiscal equalisation laws at the local level mostly concentrate on socio-economic functions (Ring, 2002). For the acknowledgement of local ecological goods and services with positive spillovers, it would be necessary to directly connect area-related indicators with specified ecological public functions. For example, the quantity of protected areas (e.g., in hectares) in relation to the overall area of a local jurisdiction could serve as a straightforward indicator suitable for integration into fiscal transfers, as far as these protected areas provide benefits beyond local boundaries. This is usually the case for water protection zones and large or highly protected nature conservation areas such as national parks. This also holds for protected areas complying with international standards, e.g., UNESCO's Biosphere Reserves or the Natura 2000 Network according to the European Union's Fauna-Flora-Habitat Directive.

3.4. Direct Consideration of Ecological Public Functions

Most of the German states already consider ecological public functions within their fiscal equalisation laws at the local level in one way or another. Due to the sovereignty of the states to set up their own system of fiscal equalisation at state level, different realisations can be noticed.

Bavaria, Baden-Württemberg, Saxony and Mecklenburg-Western Pomerania earmark a certain amount of funds in advance for selected ecological purposes (Ring, 2002). In this way, the total amount of finance available for distribution to local government is divided such that ecological functions are taken into account before any indicators come into play for further distribution of grants. Bavaria, for example, allows certain funds for supporting the construction of waste disposal plants. In Saxony, there is an option to earmark such money with respect to special communal requirements related to water supply and sewage disposal. Usually, the monies set aside are rather small in relation to the total amount of finance available. Nevertheless, many states use this option for various purposes, and there are authors recommending to consider this option also for ecological purposes (Ewers et al., 1997; Rose, 1999).

Another possibility for considering ecological public functions consists in using them as a basis for calculating the fiscal need in the course of determining lump-sum transfers. One has to say, though, that the principal approach for determining lump-sum transfers is related to inhabitant-based indicators. In comparison with this principal approach, the additional approaches based on ecologically related indicators clearly are of subordinate importance. In this context, Hesse, the Rhineland-Palatinate and Saarland use them for local public functions related to recreation (spas), Saarland has an additional approach for communes suffering from mining damage (Ring, 2002).

So far, the most common way to consider ecological public functions in Germany is to directly address them by means of conditional grants. Table 1 provides a picture of supported fields and measures in the states concerned (Ring, 2001a). Many fiscal equalisation laws include measures related to water management, namely sewage disposal and water supply. Conditional grants for waste disposal or the prospecting and remediation of

Table 1. Ecological Functions in Fiscal Equalisation at the Local Level in Germany.

Field	Supported Measures	German States
Soil	Prospecting and remediation of contaminated sites, recultivation	BAV, BW, HE, NRW, THUR
	Water protection	HE
Water	Water supply	BAV, BW, HE, MWP, RP, SAAR, SAX, THUR
	Sewage disposal	BAV, BRG, BW, HE, MWP, NRW, RP, SAAR, SAX, THUR
Nature conservation	Nature protection and landscape conservation	BRG, HE, MWP, NRW
Recreation	Spas	BW, NRW, RP
	Recreation and tourism	BRG, MWP, RP, SH
Waste disposal	Waste disposal plants	BAV, HE, MWP, RP, SAAR, THUR
Energy	Energy saving measures	HE

BAV, Bavaria; BRG, Brandenburg; BW, Baden-Württemberg; HE, Hesse; MWP, Mecklenburg-Western Pomerania; NRW, North Rhine-Westphalia; RP, Rhineland-Palatinate; SAAR, Saarland; SAX, Saxony; SH, Schleswig-Holstein; THUR, Thuringia.
Source: Ring (2001b).

contaminated sites are also quite widespread. Concerning conservation and precautionary type measures, the field of recreation is most commonly addressed. Local public functions related to nature and resource protection or landscape conservation can only be occasionally found in a few fiscal equalisation laws. Brandenburg mentions functions related to agriculture and tourism, while Mecklenburg-Western Pomerania considers landscape maintenance. North Rhine-Westphalia uses grants for the preservation of cultural landscapes and natural monuments, including the ecological rehabilitation and landscaping of the Emscher-Lippe area. In the matter of conditional grants, Hesse used to support projects in the areas of biotope protection and biotope networks. However, in recent years Hesse ceased to actually assign monies in their state budget to this type of grant. The case of Hesse shows that a continued and more thorough analysis must not only look at whether the legal text itself mentions ecological public functions, but should consider state budgets and the monies actually spent on the various ecological tasks.

Most ecological public functions as currently implemented in German fiscal equalisation laws are only represented by conditional grants or the provision of loans for local government. There is also a general tendency to support end-of-the-pipe infrastructure such as sewage and waste disposal. Apart from functions related to (drinking) water and recreational purposes, resource protection and nature conservation activities are hardly supported. With the exception of area as an indirect indicator for some ecological functions, there is no indicator which generally takes into account ecological functions comparable to the consideration of inhabitants for socio-economic functions. Acknowledgement of protected areas with significance across local boundaries is still completely absent. As a result, many of the ecological functions provided by rural and remote areas are still underrepresented in fiscal equalisation at the local level, and therefore respective jurisdictions are not compensated for external benefits of neither local restrictions to be born nor related activities and expenditures. Insufficient spatial coincidence of costs and benefits is likely to lead to an under-provision of the public goods and services concerned (Bergmann, 1999). In recent years, the German Federal Agency for Nature Conservation became interested in the potential of fiscal equalisation at the local level for nature conservation and issued a corresponding report (Perner & Thöne, 2002). However, the ideas developed in this respect are still quite far from being implemented in Germany.

4. PERSPECTIVES FOR COMPENSATING LOCAL ECOLOGICAL SERVICES

A suitable way of counteracting potential under-provision of local ecological goods and services would be to integrate ecological functions into intergovernmental fiscal transfers to the local level. Concerning suitable types of ecological functions, precautionary ecological functions must be stressed whose benefits cross local boundaries, such as nature conservation and water protection. These ecological services are mostly provided by designating more or less large protected areas that play a significant role in sustainable watershed management and biodiversity conservation in the long run. Usually, local governments have little scope for influencing decisions made on the designation and maintenance of a large proportion of the area set aside for protection. Due to the regional, national or even international importance of these areas, municipalities are obliged to accept decisions made at higher levels of government. In this way, local sovereignty in land-use planning and management is restricted in the long term. These decisions also affect the ability to develop productive activities and to generate revenue in a variety of ways, both for private land-users and local governments.

In this chapter, the focus was on the role of public institutions, i.e., the local government in its need to be compensated for the ecological goods and services it provides. The "forced" provision of ecological goods and services in terms of protected areas without compensating for positive spillovers is neither effective nor efficient. Provided relevant framework regulation exists, concrete decisions on the immobile factor "land" are best to be taken at most decentralised levels. This is reflected in the decentralisation of land-use planning in many countries where concrete implementation is mostly assigned to the local level. From an economic view, it is rational for local governments not to be interested in or even be against water and nature protection areas if associated costs are to be born locally whereas a number of benefits cross local boundaries. Apart from intrinsic motivation in local conservation activities that shall not be overlooked here, the majority of municipalities will not support the existence of protected areas within their territory. Even though protected areas might exist, lack of enforcement, control or just information can easily lead to the deterioration of the quality of these areas. Therefore, a prerequisite for long-term sustainable watershed management consists of the integration of protected areas with positive spillovers into intergovernmental fiscal transfers to the local level. This would keep concrete decisions on land use at the most suitable local level. The financial acknowledgement of the provision of ecological services across local boundaries would raise local awareness for the transboundary significance of these protected areas. By way of internalisation of positive spatial externalities, it brings local interests in line with supra-local interests, thereby making incentives for local behaviour consistent and contributing to economic efficiency.

This German case study has shown that various mechanisms already exist for acknowledging ecological goods and services in the German system of intergovernmental fiscal transfers. However, only very few states have implemented area-related indicators in their fiscal equalisation laws. Especially the relevance of protected areas has not yet been recognised. Therefore, the majority of German municipalities still perceive protected areas as an obstacle to development (Bauer, Abresch, & Steuernagel, 1996, p. 334; Stoll-Kleemann, 2001).

Although this chapter presented a national case study, the general message can be transferred to other federal systems. However, the kinds of recommendations to be made for considering protected areas of supra-local significance strongly depend on the type of federal system investigated, the general role and functions of different jurisdictions within these systems, and the specific environmental legislation in force.

Köllner, Schelske, and Seidl (2002), for example, present a case study for integrating biodiversity into intergovernmental fiscal transfers for Switzerland. In Brazil, a few states started compensating municipalities for the existence of protected areas within their territory already during the 1990s. Ecological indicators were introduced into the system of redistributing the value-added tax from the state to the local level. In the meantime, the effects are significant, both in terms of increased protected areas and changing revenues to the local level (Grieg-Gran, 2000; May, Veiga Neto, Denardin, & Loureiro, 2002; Ring, 2004b). Especially municipalities with a high share of protected areas considerably benefit from the new ecological fiscal transfers, and therefore, mostly appreciate the ecological services they provide across local boundaries. So far, Brazil seems to be the only federal system where at the state level, ecological fiscal grants have been implemented to a significant extent. It would be worthwhile to study this positive example more in detail in order to investigate its transfer potential to other federal systems.

Research at the interface of implementing sustainable watershed management and the economic theory of federalism is still more or less in its infancy. There are relatively few studies that investigate intergovernmental fiscal relations for their potential to adequately consider ecological aspects in terms of public functions and appropriate financing (Ring, 2002). Whereas economic and social public functions have a rather long tradition in intergovernmental fiscal relations of federal systems, ecological functions have only been considered comparatively recently. Both the theoretical analysis of principles of the economic theory of federalism related to spillovers of protected areas and the respective empirical investigation of the German system have shown that there still is a great need for adequately rewarding ecological services provided by the local level.

NOTES

1. See Ring (2002) for a more detailed definition of "ecological public functions".
2. The city states Berlin, Bremen and Hamburg are excluded from the analysis. Here, local and state public functions can hardly be separated and so no fiscal equalisation laws exist.

REFERENCES

Bauer, S., Abresch, J.-P., & Steuernagel, M. (1996). *Gesamtinstrumentarium zur Erreichung einer umweltverträglichen Raumnutzung*. Stuttgart: Metzler-Poeschel.

Bergmann, E. (1999). Lenkung der Flächenausweisung über Zuweisungen? *Informationen zur Raumentwicklung*, *9*, 555–564.

Breton, A. (1965). A theory of government grants. *The Canadian Journal of Economics and Political Science*, *31*(2), 175–187.

Döring, T. (1997). *Subsidiarität und Umweltpolitik in der Europäischen Union*. Marburg: Metropolis.

Ewers, H.-J., Rehbinder, E., & Wiggering, H. (1997). Umweltbezogene Erweiterung des Finanzausgleichs. *Zeitschrift für Gesetzgebung*, *2*, 135–147.

Ewringmann, D., & Bergmann, E. (2000). *Möglichkeiten und Grenzen einer Funktionalisierung des Finanzausgleichs für eine dezentrale Agrarumweltpolitik*. Berlin: Analytica.

Frank, K., & Ring, I. (1999). Model-based criteria for the effectiveness of conservation strategies – An evaluation of incentive programmes in Saxony, Germany. In: I. Ring, B. Klauer, F. Wätzold & B. Å. Månsson (Eds), *Regional Sustainability. Applied Ecological Economics Bridging the Gap Between Natural and Social Sciences* (pp. 91–106). Heidelberg: Physica.

Frey, B. (1996). FOCJ: Competitive governments for Europe. *International Review for Law and Economics*, *16*, 315–327.

Grieg-Gran, M. (2000). *Fiscal Incentives for Biodiversity Conservation: The ICMS Ecológico in Brazil*. Discussion Paper 00–01, International Institute for Environment and Development, London.

Hampicke, U. (2005). Naturschutzpolitik. In: B. Hansjürgens & F. Wätzold (Hrsg.), *Umweltpolitik und umweltökonomische Politikberatung in Deutschland. Sonderheft der Zeitschrift für angewandte Umweltforschung*, (Vol. 15, pp. 162–177). Berlin: Analytica.

Hansjürgens, B. (1996). Föderalismustheorie und europäische Umweltpolitik. Ökonomische Kriterien für die Verteilung umweltpolitischer Kompetenzen in der EU. In: R.-D. Postlep (Hrsg.), *Aktuelle Fragen zum Föderalismus: Ausgewählte Probleme aus Theorie und politischer Praxis des Föderalismus* (pp. 73–98). Marburg: Metropolis.

Hansjürgens, B., & Messner, F. (2002). Die Erhebung kostendeckender Wasserpreise in der Wasserrahmenrichtlinie. In: S. von Keitz & M. Schmalholz (Hrsg.), *Handbuch der Wasserrahmenrichtlinie – Inhalte, Neuerungen und Anregungen für die nationale Umsetzung* (pp. 293–319). Berlin: Erich-Schmidt-Verlag.

Hartmann, E., Thomas, F., Luick, R., Bierer, J., & Poppinga, O. (2003). *Kurzfassungen der nach der Verordnung EG 1257/1999 kofinanzierten Agrarumweltprogramme der Bundesländer*. BfN-Skripten 87, Bonn – Bad Godesberg.

Henneke, H.-G. (2001). Der kommunale Finanzausgleich 2001 unter besonderer Berücksichtigung der Landkreise. *Der Landkreis*, *71*(2/3), 210–220.

Huckestein, B. (1993). Umweltpolitik und Föderalismus – ökonomische Kriterien für umweltpolitische Kompetenzen der Europäischen Gemeinschaft. *Zeitschrift für angewandte Umweltforschung*, *6*(3), 330–339.

Karrenberg, H., & Münstermann, E. (2000). Gemeindefinanzbericht 2000. *Der Städtetag*, *4*, 4–16.

Köllner, T., Schelske, O., & Seidl, I. (2002). Integrating biodiversity into intergovernmental fiscal transfers based on cantonal benchmarking: A Swiss case study. *Basic and Applied Ecology*, *3*, 381–391.

List, J. A., Bulte, E. H., & Shogren, J. F. (2002). "Beggar thy neighbor": Testing for free riding in state-level endangered species expenditures. *Public Choice*, *111*(3), 303–315.

May, P. H., Veiga Neto, F., Denardin, V., & Loureiro, W. (2002). Using fiscal instruments to encourage conservation: Municipal responses to the 'ecological' value-added tax in

Paraná and Minas Gerais, Brazil. In: S. Pagiola, J. Bishop & N. Landell-Mills (Eds), *Selling Forest Environmental Services: Market-based Mechanisms for Conservation and Development* (pp. 173–199). London: Earthscan.

Musgrave, R. M. (1959). *The theory of public finance*. New York: McGraw-Hill.

Oates, W. E. (1972). *Fiscal federalism*. New York: Harcourt Brace Jovanovich.

Oates, W. E. (1998). Environmental policy in the European community: Harmonization or national standards? *Empirica, 25*, 1–13

Oates, W. E. (1999). An essay on fiscal federalism. *Journal of Economic Literature, 37*, 1120–1149.

Oates, W. E. (2001). *A reconsideration of environmental federalism*. Discussion Paper 01–54, Resources for the Future, Washington, DC.

Olson, M., Jr. (1969). The principle of "fiscal equivalence": The division of responsibilities among different levels of government. *American Economic Review, 59*, 479–487.

Perner, A., & Thöne, M. (2002). *Naturschutz im Finanzausgleich. Erweiterung des naturschutzpolitischen Instrumentariums um finanzielle Anreize für Gebietskörperschaften*. Studie im Auftrag des Bundesamtes für Naturschutz, Bonn. Finanzwissenschaftliches Forschungsinstitut an der Universität zu Köln, Köln.

Perrings, C., & Gadgil, M. (2003). Conserving biodiversity: Reconciling local and global public benefits. In: I. Kaul, P. Conceição, K. Le Goulven & R. U. Mendoza (Eds), *Providing global public goods: Managing globalization* (pp. 532–555). Oxford: Oxford University Press.

Postlep, R.-D., & Döring, T. (1996). Entwicklungen in der ökonomischen Föderalismusdiskussion und im föderativen System der Bundesrepublik Deutschland. In: R.-D. Postlep (Hrsg.), *Aktuelle Fragen zum Föderalismus: Ausgewählte Probleme aus Theorie und politischer Praxis des Föderalismus* (pp. 7–44). Marburg: Metropolis.

Ring, I. (2001a). Nachhaltige Entwicklung und kommunaler Finanzausgleich: Status quo und Entwicklungsoptionen. In: H. Horsch, I. Ring & F. Herzog (Hrsg.), *Nachhaltige Wasserbewirtschaftung und Landnutzung – Methoden und Instrumente der Entscheidungsfindung und -umsetzung* (pp. 387–408). Marburg: Metropolis.

Ring, I. (2001b). Ökologische Aufgaben und ihre Berücksichtigung im kommunalen Finanzausgleich. *Zeitschrift für angewandte Umweltforschung. Sonderheft, 13*, 236–249.

Ring, I. (2002). Ecological public functions and fiscal equalisation at the local level in Germany. *Ecological Economics, 42*, 415–427.

Ring, I. (2004a). Naturschutz in der föderalen Aufgabenteilung: Zur Notwendigkeit einer Bundeskompetenz aus ökonomischer Perspektive. *Natur und Landschaft, 79*(11), 494–500.

Ring, I. (2004b). *Integrating local ecological services into intergovernmental fiscal transfers: The case of the ICMS-E in Brazil*. UFZ Discussion Papers No. 12/2004, UFZ Centre for Environmental Research Leipzig-Halle GmbH, Leipzig.

Rose, M. T. (1999). Überlegungen zur Berücksichtigung der ökologischen Ausgleichsfunktion ländlicher Räume im kommunalen Finanzausgleich – dargestellt am Beispiel des Landes Nordrhein-Westfalen. *Zeitschrift für angewandte Umweltforschung, 12*(2), 267–279.

Scheberle, D. (1997). *Federalism and environmental policy: Trust and the politics of implementation*. Washington, DC: Georgetown University Press.

SRU (Der Rat von Sachverständigen für Umweltfragen). (1996). *Konzepte einer dauerhaft-umweltgerechten Nutzung ländlicher Räume. Sondergutachten*. Stuttgart: Metzler-Poeschel.

SRU (Der Rat von Sachverständigen für Umweltfragen) (2000). *Umweltgutachten 2000*. Stuttgart: Metzler-Poeschel.

Stoll-Kleemann, S. (2001). Barriers to nature conservation in Germany: A model explaining opposition to protected areas. *Journal of Environmental Psychology*, *21*(4), 369–385.

Tiebout, C. M. (1956). A pure theory of local expenditures. *Journal of Political Economic*, *64*, 416–424.

Unnerstall, H. (2004). Rechtliche Grundlagen gegenwärtiger und zukünftiger ökonomischer Instrumente im Naturschutz und in der Flächenhaushaltspolitik. In: R. Döring & M. Rühs (Hrsg.), *Ökonomische Rationalität und praktische Vernunft. Gerechtigkeit, Ökologische Ökonomie und Naturschutz* (pp. 249–274). Würzburg: Königshausen & Neumann.

COST RECOVERY FOR WATER SERVICES ACCORDING TO THE EU WATER FRAMEWORK DIRECTIVE

Herwig Unnerstall and Frank Messner

ABSTRACT

The requirement of full cost recovery for water services including environmental and resource costs in accordance with the polluter pays principle in Art. 9 EU-Water Framework Directive is a unique provision in the history of the European environmental law. The wording of the provision is a compromise between the Council's and the Parliament's versions that mirrors different conceptual ideas on how to internalize environmental and resource costs. Art. 9 now contains a two-step concept for the achievement of the aim. The uniform implementation of the full cost-recovery calls for common accounting standards for the calculation of financial cost and a common methodology for the estimation of environmental and resource costs on the European level. In Germany, the requirements of the first step are partly fulfilled, but necessities of the second step are not being met at the moment.

Ecological Economics of Sustainable Watershed Management
Advances in the Economics of Environmental Resources, Volume 7, 347–383

ISSN: 1569-3740/doi:10.1016/S1569-3740(07)07015-0

1. INTRODUCTION

The European Water Framework Directive (European Community, 2000) requires in its Art. 9 acknowledgement of the principle of cost recovery[1] for water services[2] in accordance with the "polluter pays" principle. The cost recovery is achieved through the prices the consumers of the water service have to pay to the provider directly *and* through any tax, charge or levy that is imposed on said service, and is borne by the consumer directly or indirectly. Full cost-recovery pricing of water services according to the "polluter pays principle" fulfills three basic economic functions: an information, a steering and a financing function.

- The *information* function consists in the fact that consumers of water services are made aware of *all* the costs caused by their consumption of water services, especially the use of water. This ought to make users appreciate how precious water is and encourage them to treat it more carefully, e.g., by reducing the amount of water consumption and waste-water discharge.
- The information function is closely linked to the *incentive or steering* function. Users are to be charged for all the direct and indirect costs of their specific use of water, ranging from the costs of abstraction, distribution and disposal to environmental and resource costs, so that they can decide on that basis whether payment of these costs is justified for the use concerned, or whether their money would be better spent on something else (*opportunity-cost principle*). This process of informed consideration should lead to water only being used when it makes economic sense for the purpose in mind. Charging the full costs to the users results in *efficient water service use.*
- The *financing* function is based on the fact that charging water users generates additional funding for sustaining long-term provision of this water service and for financing accordant measures for protecting water resources.

Without application of the polluter pays principle, the cost-recovery pricing would only fulfill the financing function. Therefore, the polluter pays principle serves as an efficiency rule (Hansjürgens, 2001).

"Costs" may be classified as financial, economic, social, external, environmental, resource, indirect, induced or opportunity costs. There is no coherent system of classification and "full cost recovery" can imply different meanings. However, neglecting one of the relevant cost components hampers the above described economic functions of (full cost recovery) prices, and this leads to sub-optimal and inefficient water use and to distortions of competition in water-related markets. If, for example, irrigation agriculture is subsidized by means of low water costs as is the case in many arid or semi-arid countries, water is overused and the prices of agricultural goods do not reflect the economic, social and environmental costs of the underlying agricultural practices (European Commission, 2000a, p. 22). As a consequence, an agricultural practice is supported which cannot be sustained over the long term. What is more important is the fact that the resulting unsustainable agricultural products may put sustainable rival products (without subsidies) out of the market. This is obviously an inefficient and undesirable outcome. A similar effect may occur if the water service provider considers all costs, but the "polluter pays" principle is not applied. Then, some water users pay too much for the water service and others do not pay enough – the former cross-subsidizing the latter.

Certainly the concept of full cost recovery for water services according to the "polluter pays" principle is an ideal one. There is no country in the world which has implemented this all-embracing water-pricing approach. Still, different water regulation policies have been practiced in several countries to get the prices for water services right. Some policy instruments like the reduction of subsidies, command-and-control regulations for ensuring environmentally benign provision of water services and the introduction of environmental charges or taxes do directly aim at the cost structure of the water services.

In this chapter we want to highlight some of the economic and legal perspectives and challenges for implementing full cost-recovery prices of water services in Germany and the EU. The chapter is organized as follows: In the following Section 2 we give an overview over the history of water service pricing in European and German (Water) Law and the use of economic instruments for correcting water prices. Section 3 will outline the requirements of Art. 9 WFD regarding full cost recovery, starting with the legislation process. In Section 4 the economic and legal challenges and obstacles to implementation of the WFD in Germany will be discussed. Section 5 closes with a summary and an outlook.

2. ECONOMIC INSTRUMENTS AND COST RECOVERY IN EUROPEAN AND GERMAN WATER LAW BEFORE THE WFD CAME INTO FORCE

The *European Community* realized the importance of economic instruments from the beginning of its environmental policy. In its first program of action on the environment (European Community, 1973) the Council stated that common rules for the imputation of the cost of environmental protection are to be established. Member states and Council emphasized the role of the "polluter pays" principle serving as a leading principle for the establishment of economic instruments promoting environmental objectives. Standards and charges, or a possible combination of the two, were regarded as the major instruments of action available to public authorities for the avoidance of pollution (European Community, 1975a). Later the Commission suggested using charges not only for the financing of environmental protection activities, but for the decreasing, as well, of other taxes that are perceived as distorting the economy, such as labor taxes (European Commission, 1997a, No. 4, and similar European Commission, 1993).

However, the Community has only contested original competencies for community taxes or charges according to Art. 269 EC (Treaty Establishing the European Community),[3] and has used it only in marginal cases (income tax for the staff), as it requires the unanimous adoption of the Member States in accordance with their respective constitutional requirements. More important are the competencies as regards the "harmonisation of legislation concerning turnover taxes, excise duties and other forms of indirect taxation" (Art. 93 EC) and in general as regards "fiscal provisions" (Art. 94 with Art. 95 (2) EC).[4] This competence is not extended by Art. 175 (2) EC.[5] As a result, the environmental taxes and charges have not become reality in European environmental law.

Even more surprisingly there are only few cases in environmental legislation, whereby the Member States were obliged or allowed to introduce environmental taxes or charges. In Art. 14 of the Waste Oil Directive (European Community, 1975b) the Member States were allowed to impose a charge on products, which after use are transformed into waste oils or on waste oils – "in accordance with the 'polluter pays' principle". The most important activity with respect to environment-related taxes and charges has occurred in the field of harmonization of those taxes and charges already levied in the Member States. The EC has established several Directives to harmonize exercise duties or charges:

- Council Directive 92/12/EEC of 25 February 1992 on the general arrangements for products subject to excise duty and on the holding, movement and monitoring of such products.[6]
- Council Directive 92/82/EEC of 19 October 1992 on the approximation of the rates of excise duties on mineral oils.[7]

- Directive 1999/62/EC of the European Parliament and of the Council of 17 June 1999 on the charging of heavy goods vehicles for the use of certain infrastructures.[8]

Finally, the European Commission has issued two different proposals for a CO_2/energy-tax (1992, 1995) based on Art. 93 EC (and 175 (2) EC) (Jans, 2000, p. 61), but the Council could not enter into an agreement, as unanimity was required. With the newly established greenhouse gas emission allowance trading scheme[9] the CO_2/energy-tax is likely to be obsolete by now with regard to industrial emissions.

Especially in the field of water law, the different Directives[10] previous to the WFD did not contain anything like cost recovery, the imposition of charges or taxes for water services or water uses. The notions "costs", "recovery of costs", "polluter pay principle" cannot be found. Art. 9 WFD is therefore almost unique in the history of the environmental and water-related law of the EC. However, the Community on the basis of Art. 93 and 94 EC may undertake a harmonization of taxes and charges established in the Member States due to the implementation of Art. 9 WFD in the further future, but this requires unanimity within the Council.

The legislative and administrative competencies in water-use regulation in *Germany* are traditionally split up between the Federation and the federal states ("Länder") with changing weights (cf. Unnerstall & Köck, 2004). In that history charges and fees have a long tradition. For a long time the federal states and their predecessors imposed charges for water extraction and the use of river for transportation purposes (cf. Kloess, 1908, 2ff. and Wüsthoff, 1962, 14ff.). In the German Constitution of 1871 (Art. 54), fees for the use of waterways were prohibited apart from fees for special installations like locks. At the beginning of the 20th century some states also abandoned water extraction charges,[11] while they remained possible in other states (Bavaria and Baden).[12] In the course of the newly established framework legislation competence on the federal level in the Basic Law of the Federal Republic of Germany, water extraction charges were again discussed in the 50s of the last century to be launched in the Federal Water Act. The German Parliament (Bundestag) refused to introduce them, but did not prohibit the states to impose them (BVerfG, 1995, p. 341). Until 1987 no state was charging water extraction fees, when Baden-Württemberg (re-) introduced them. Several other states followed this example, but some abandoned them again (e.g., Hesse). In 2004, there are water extraction charges in 10 of 16 states,[13] usually having different rates for groundwater and surface water extraction and for various water use purposes (e.g., drinking water supply, irrigation, cooling).[14] There is no regional differentiation of rates according to geographical differences of water availability or water quality.

As regards waste-water discharge a fee was introduced in 1978 by the Wastewater Charge Act (WWCA).[15] The fee is imposed according to the load of dirt of the discharge, determined by the chemical and biological oxygen demand (§ 3 WWCA). For technical reasons and for reduction of levying costs, the fees are not calculated on the basis of the real load determined by continuous monitoring of the discharge, but on the basis of the value fixed in the discharge permission which is only monitored selectively (§ 4 WWCA). There is no possibility to differentiate the rate according to the quality of the body of water where the waste water is discharged or according to geographical differences in the assimilation capacity of water bodies. Since 1986 the WWCA also allows for a deduction of investment costs for improving the treatment facilities (§ 10 WWCA). A reduced rate (after 1999: 50%) is applied if the load of dirt is reduced according to the best available technology (§ 9 (5) and (6) WWCA). The revenues of the charge have to be used for measures to maintain or improve the quality of water resources, but can also be used to cover the charging costs (§ 13 WWCA).[16] Against the background of the design of the waste-water charge, economists complain that the economic incentive of the charge has been lost (Gawel, 1993).

Water extraction charges, waste-water charges and fees for the use of waterways are not the only means of setting up economic incentives and achieving cost recovery. Water supply, especially drinking water supply and the treatment of urban waste water have been regarded as public services to be provided by the municipality[17] and to be financed by the citizens/users by fees that cover the costs of the services. The accounting standards for these services have developed and changed in the last centuries and all Municipality Charges Acts refer to business accounting standards, but many details and basic principles are still highly contested (cf. Gawel, 1995, 1999), e.g., whether there are generally accepted accounting standards, regardless of the aims of accounting (different costs for different purposes) and whether there are specific aims for public enterprises providing services of general interest. As municipal enterprises are not usually allowed to make profits, the current law and its interpretation do not allow the integration of environmental and resource costs into the calculation of local rates (cf. Gawel, 1995, p. 100 and Wolfers, 2004, p. 120) and restricts them to the recovery of financial costs. However, a rate of return on the necessary operating capital between 6% and 8% is often included (Wolfers, 2004, p. 118). In addition there is also a long tradition in subsidizing investments in waste-water treatment facilities or water supply facilities by the federal states for the municipalities (cf. Ewringmann, 2002, p. 285). Although the percentage of the overall investments is regarded as low (Kahlenborn, Buck, & Kraemer, 1999, p. 35), this is a violation of the full cost recovery and the polluter pays principle. There are also frequently subsidies for ongoing purposes (cf., e.g., RP Gießen, 2002, p. 44). Cross-subsidizing among the different branches of municipality-based public services is also well-known, e.g., from water supply to public local transport (cf. Ewringmann, 2002, p. 285 and Ewers, Botzenhart, Jekel, Salzwedel, & Kraemer, 2000, p. 13), now to be disclosed according to Art. 3a Transparency Directive (European Community, 2000a).[18] This recent development in EU law opens the possibility for a review of the above-mentioned financial transfers with respect to the European State aid provisions (Geiger & Freund, 2003, p. 491). Thus, the financing and subsidizing of services of general (economic) interest are regarded more and more critically in this respect,[19] and a stronger focus is laid on the application of adequate economic incentives.

As drinking water supply and urban waste-water treatment are technically and legally organized on the local or regional level, the different charges partly reflect the different natural conditions (Kahlenborn et al., 1999, pp. 36, 46).

There is another problem with regard to local fees: The coverage of financial costs on the municipal or service area level is not required of an individual but merely on an overall level, i.e., the overall revenue shall cover the overall costs (Kahlenborn et al., 1999, p. 18; critical Gawel, 1995, p. 174). This allows especially for tariffs where the *variable* fraction of the *charges* is higher than the *variable* fraction of the *costs* (Kahlenborn et al., 1999, pp. 35, 46) and tariffs that discriminate between different user groups, granting mass consumers a discount per used quantity.

Before we discuss what challenges Art. 9 WFD pose on this situation we will review what the exact content of Art. 9 WFD is.

3. THE EU-WFD DEMAND FOR COST RECOVERY FOR WATER SERVICES

The integration of economic aspects in the river basin management and especially the idea of the introduction of a cost-recovery principle with regard to water services have been a feature of the WFD from the first drafts onwards. The final adopted provision is the following:

Article 9: Recovery of costs for water services

1. Member States shall take account of the principle of recovery of the costs of water services, including environmental and resource costs, having regard to the economic analysis conducted according to Annex III, and in accordance in particular with the polluter pays principle.

Member States shall ensure by 2010:

- that water-pricing policies provide adequate incentives for users to use water resources efficiently, and thereby contribute to the environmental objectives of this Directive.
- An adequate contribution of the different water uses, disaggregated into at least industry, households and agriculture, to the recovery of the costs of water services, based on the economic analysis conducted according to Annex III and taking account of the polluter pays principle.

Member States may in so doing have regard to the social, environmental and economic effects of the recovery as well as the geographic and climatic conditions of the region or regions affected.

...

3. Nothing in this article shall prevent the funding of particular preventive or remedial measures in order to achieve the objectives of this directive.
4. Member States shall not be in breach of this Directive if they decide in accordance with established practices not to apply the provisions of paragraph 1, second sentence, and..., for a given water-use activity, where this does not compromise the purposes and the achievement of the objectives of this Directive. Member States shall report the reasons for not fully applying paragraph 1, second sentence, in the river basin management plans.

The definitions for "water services" and "water use" are found in Art. 2 WFD:

38. 'Water services' means all services which provide, for households, public institutions or any economic activity:
    (a) abstraction, impoundment, storage, treatment and distribution of surface water or groundwater,
    (b) waste-water collection and treatment facilities which subsequently discharge into surface water.

39. 'Water use' means water services together with any other activity identified under Article 5 and Annex II having a significant impact on the status of water.

This concept applies for the purposes of Article 1 and of the economic analysis carried out according to Article 5 and Annex III, point (b).

3.1. The Development of the Cost-Recovery Principle in the Legislation Process

In the legislation process the adequate wording for this idea has been highly contested, especially the question of the binding force and the extent to which the polluter pays principle was to be applied (Kaika & Page, 2003). The European Commission in its initial draft (1997b) suggested to

> "ensure full cost recovery for all costs for services provided for water uses overall and by economic sectors, broken down at least into households, industry and agriculture".

The "services provided" remained undefined, whilst "water use" was fully defined as

> (a) abstraction, distribution and consumption of surface water or groundwater;
> (b) emission of pollutants into surface water and waste-water collection and treatment facilities which subsequently discharge into surface water;
> (c) any other application of surface water or groundwater having the potential of a significant impact on the status of water.

The European Parliament (1999) in its first reading only amended Art. 12 (1) by adding "including abstraction" to "water uses overall" as an illustration, but actually already part of the definition of "water use". More important are the amendments of Art. 12 (1a) and (1b). Art. 12 (1a) introduced an instrumental view on water charges:

> "Where it is not possible, or impractical, to calculate the full environmental costs of water use, charges shall be set at a level which encourages the attainment of the environmental objectives of this Directive."

Art. 12 (1b) claims

> "that water users faced with a need to treat their water as a result of another's polluting activities can fully recover their additional costs from the polluter."

This provision implies that the polluting activities mentioned were not covered by "water use" – a proposition that was true for diffuse pollution, as it was neither an emission nor an application of surface water or groundwater. It was also relevant for point sources of pollution as the "cost recovery" defined in Art. 2 (33) (European Commission, 1997b) did not include environmental and resource costs. These costs were only mentioned in Art. 12 (2) and should be included in the prices at a later date. In addition, the Parliament's amendment Art. 12 (1b) could be read as setting the standard for the distribution of cost-recovery charges among different uses according to their causal contributions, i.e., the establishment of the polluter pays principle that was neither mentioned in Art. 12 (1), nor in Art. 12 (2), nor in the definition of "full cost recovery" in Art. 2 (33).

However, the Council did not adopt this differentiated concept of financial costs and environmental and resource costs in the Common Position (European Council, 1999), but merged it in a general provision on cost recovery including all costs and in accordance with the polluter pays principle. In addition the Council weakened the degree of obligation significantly from "shall ensure" to "shall take account of". It also did not accept the use of charges as purely an economic instrument to control the water use. The Council introduced the differentiation between "water services" and "water use". It restricted "water services" to the abstraction, emissions and the waste-water issues like Art. 2 (32) lit. (a) and (b) of the initial draft. At the same time the Council expanded "water use" to any activity having significant impact on the status of water (including the "water services"), unlike in Art. 2 (32) lit. (c) of the initial draft.

In its second reading the European Parliament (2000b) reinforced its instrumentalist approach by abandoning the restriction to the impossibility or impracticability of calculating environmental costs. The Parliament also stuck to the binding character of the cost-recovery principle ("ensure" instead of "take into account") but set a generous time limit (2010). It also weakened the strict division of costs between households, industry and agriculture to an "adequate contribution", but it did not alter the Council's limitation in the definition of "water services". Finally, the Parliament did not pursue further the former amendment Art. 12 (1b), although it was proposed by the Environmental Committee (European Parliament, 2000a). However, its basic idea can be derived from the new version of Art. 9 (1) 2nd indent (European Parliament, 2000b) that requires one to "take into account the polluter pays principle" (like in the Common Position), as application of the polluter pays principle is not restricted to those activities directly producing the costs of the water services. Instead, it covers all kinds of activities, including those that can have significant effect on the status of water bodies, e.g., pollution from diffuse sources. Here, there is no difference between the Common Position and the Parliament's amendments and any reference to the notion "water use" seems unnecessary. This argument is not in contradiction with the fact that the broader term "water use" is only used in the economic analysis, i.e., that its application is restricted to Art. 1, Art. 5 and Annex III 3 (b) (acc. to Art. 2 (35) 2nd sentence Common Position). The non-reference to "water use" cannot be understood as restricting the application of the polluter pays principle. Both the

Parliament's and the Council's concept were sound – regarded separately, covering a wide range of activities as "water services", whose costs have to be recovered, embedded in the wider range of "water use" used as the basis for economic analysis.

3.2. The Finally Adopted Version of Art. 9 WFD and its Interpretation

However, the Council and Parliament could not enter into a common text during the conciliation process. Therefore, they simply merged their particular textual proposals in Art. 9 WFD: Art. 9 (1) 1st subpara. (being identical with the Common position, European Council, 1999) stems from the European Council, the 2nd subpara. stems from the Parliaments amendments to the Common position (European Parliament, 2000b), with one alteration introducing the reference to "water use" in the 2nd indent of Art. 9 (1) 2nd subpara. WFD. As there are no documents about the discussion of the Conciliation Committee, one can only guess what the reasons were for the different actors to favor one wording over the other. As seen from the legislation process it is likely that an analysis will reveal overlapping normative contents between the different parts of Art. 9 (1) WFD especially between 1st subpara. and the 2nd subpara. 2nd indent WFD.

An initial problem arising from the newly inserted reference to "water use" is whether the definition in Art. 2 (39) is applicable within Art. 9 WFD, although it is not mentioned there. However, there are no alternative definitions and Art. 9 refers to Art. 5 which is the basis for Art. 9 WFD. There is no reasonable alternative to reading "water use" in Art. 9 in accordance with the definition in Art. 2 (39) WFD. The restriction must be seen as an editorial mistake by the Conciliation Committee. According to the interpretation of the Parliament's version, the introduction of "water use" does not change the normative content of the provision. However, it seems that "services" and "use" are excluding each other in Art. 9 WFD, but the definitions in Art. 2 (38) and (39) show that the "services" are part of the "use". How can this contradiction be solved? The most plausible answer is that it is only an attempt to emphasize the fact that not only those who receive the water services (cf. Art. 12 of the first draft) have to bear the costs for the services, but also those whose activities ("water use") can have significant impact on water services' costs, a proposition already derivable from the polluter pays principle as seen above. Although key ideas of Art. 9 (1) 1st subpara. and 2nd subpara. 2nd indent WFD are equivalent (cost recovery for water services according to the polluter pays principle), there are some differences at the first sight. The scope of application has proved to be the same: as shown above it covers water services and water use. However, there is a crucial difference in the degree of binding force and the relevant time of application. From 2003 on, cost-recovery has only been "taken into account" as a "principle". This weakening allows for a cost-recovery rate significantly below 100%. The threshold value can only be determined arbitrarily but anything that is – let's say – below 70% cost recovery is not in accordance with the principle any more and can only be justified with arguments based on Art. 9 (1) 3rd subpara. or Art. 9 (4) WFD. From 2010 on, the object "cost recovery" is not weakened. The qualification by "adequate contribution" relates not to the overall cost recovery but only to distribution among the causers, as it is (only) specified by the reference to the polluter pays principle. Therefore any rate that is not close to 100% cost recovery is not in accordance with the 2nd subpara. 2nd indent of Art. 9 (1) WFD and again can only be justified in accordance with Art. 9 (1) 3rd subpara. or Art. 9 (4) WFD. While a normative request regarding the rate of cost recovery rises, the rigor of the application of the polluter pays principle is to be declining by 2010, since it has to be only "taken into account" and cost recovery does not have to be "in accordance" with it any more. This limitation conserves an area for Member States' discretion regarding the distribution of costs that is similar in effect to one reached in the 1st subpara. by granting discretion as regards the overall rate of cost recovery. Another difference concerns the scope of application of the polluter pays principle. In Art. 9 (1) 1st para. no level is mentioned and in Art. 9 (1) 2nd subpara. 2nd indent it is the level of consumer groups (households, agriculture and industry). The former has to be read that the polluter pays principle is to be applied on the lowest level, i.e., the level of individuals are at least on the level of single households or enterprises. The latter distinction is a minimum level, which has to be read in light of the former one. Therefore, the level of single households must not be ignored completely and the polluter pays principle has also to be applied *within* the explicitly mentioned groups in Art. 9 (1) 2nd subpara. 2nd indent WFD after 2010.

Another difference could be seen in the fact that only Art. 9 (1) 1st subpara. refers to the environmental and resource costs. In the Parliament's draft (European Parliament, 2000b) these costs are not mentioned and in the proposal of the Environmental Committee of the Parliament (European Parliament, 2000a) they were only mentioned indirectly, as it amended Art. 9 (1a): "where it is not possible, or impractical, to calculate the full environmental costs of water use,...". Although this amendment was not accepted by the Parliament, it clearly proves that the Committee was proceeding from the assumption that environmental and resource costs are included in Art. 9 (1) 2nd subpara. 2nd indent WFD. In addition, the reference to the polluter pays principle confirms this interpretation, since the notion "polluter pays principle" in economic theory usually only refers to economic costs and not solely to financial costs.

This interpretation remains true unless the alterations in the definitions for "water services" and "water use" in the conciliation process lead to a different result. Art. 2 (38) and (39) WFD have been changed from Common Position (Art. 2 (34) and (35)) although they were undisputed by the Parliament and its Environmental Committee. No relevant information can be derived from the report on the joint text of Parliament's delegation to the Conciliation Committee (European Parliament, 2000c), but the Commission had revised the definitions in its report on the amendments of the Parliament to the Common Position (European Commission, 2000). It defined "water services" as

> (a) all services providing abstraction, impoundment, distribution and treatment of surface water or groundwater; (b) waste water collection, waste water treatment and waste water disposal into surface water. (European Commission, 2000, pp. 7f.)

"Water use" was broadly defined as including

> the main economic sectors such as domestic, agriculture and industry, amenities or other legitimate uses of the environment together with any other activity identified under Article 5 and Annex III having a significant impact on the status of water. (European Commission, 2000, pp. 7f.)

The Commission justified the revision only as a necessary adaptation to its amendment to Art. 9 WFD, which essentially adopted the Parliaments' ideas, but not the wording.

What are the differences between the Common Position and the final version? In the definition of "water services" in the Common Position the technical infrastructure, the facilities were excluded from the main focus as regards water abstraction and distribution (Art. 2 (34) lit. (a)) but not as regards "waste water" (Art. 2 (35) lit. (b)). The costs of these facilities were therefore not subjected to the cost-recovery principle. In the WFD now the focus lies on the facilities as it is exemplified in Art. 2 (38) lit (b) which no longer includes "emission of pollutants into surface water". Similarly in the WFD, water extraction itself is no water service, but only in connection with technical means, which changes the water in key characteristics or if it is used for an economic activity. But not only the facilities constitute the services – otherwise environmental and resource costs would only be those of the material and energy used to produce the water "services". It must also include the abstraction of water used for the production of services and, therefore, the environmental and resource costs of the abstraction. Similarly, the costs of waste-water collection and treatment do comprise the environmental and resource costs of discharge of waste water, if there are any. Abstraction for irrigation purposes is a "water service" and environmental and resource costs are those due to the irrigation driven degradation of the soil (e.g., salinization) or the loss of wetlands due to drainage (WATECO, 2002, Annex IV.I.40 and similar Kahlenborn, 1999, pp. 19f.), both aspects are not covered by the definition of "water services" in the Common Position, which focused on the supply side. The EU Working Group 2.6 on Water and Economics of the Common Implementation Strategy (WATECO) explains "water services" as "an intermediary between the natural environment and the water use itself", where "key characteristics of natural waters are modified (i.e. the service offered is this modification)" or "key characteristics of water 'discharged' by users are modified" (WATECO, 2002, Annex II.III.1). "Water services" contain modifications of the spatial or temporal distribution, of the height of waters, of the chemical composition or temperature of water. "Water services" include also hydromorphological changes serving water supply and flood protection purposes (reservoirs) (Brackemann, Ewens, Interwies, Kraemer, & Quadflieg, 2002, p. 39), navigation or energy production purposes (WATECO, 2002, Annex IV.I.40 and LAWA, 2002, now dissenting LAWA, 2003, Part 3, p. 80) or drainage for agricultural purposes (WATECO, 2002, Annex IV.I.40) and consequently as well for mining activities. There is no difference between publicly and privately run services (WATECO, 2002, Annex II.III.2) and private services such as industrial–commercial water supply (own production), agricultural water supply (irrigation) and industrial–commercial waste-water disposal are not only relevant if they have a significant (considerable) influence on the water balance as the LAWA claims (2003, Part 3, 80f.). As a result, it can be stated that the content of "services" and "use" changed significantly in the conciliation process, but not in a way that makes necessary a revision of the above interpretation of Art. 9 WFD.

Another difficult question concerning the content of "water pricing polices" in Art. 9 (1) 2nd subpara. 1st indent WFD is, whether they include only water supply services or also waste-water related services.[20] Looking from the aim of contributing to the environmental objectives, only both aspects of the "water services" can be meant, as the chemical quality of water is a major aspect of said objectives. This holds true even if "use water resources" does not mean "water use" in the technical sense. Generally, using water pricing as an incentive is possible beyond the recovery of costs for water services and beyond the application of the polluter-pays-principle. From an economic point of view this is tricky, as "full cost recovery" in itself serves as an incentive to use water resources efficiently, especially in case the consumers did not have to pay for the water services before. But if this 1st indent was restricted to this degree of incentive, it would not have any normative content of its own. As the legislator cannot be imputed to adopt a provision lacking substance, it has to be understood to allow for the use of water pricing policies beyond the cost-recovery principle and the polluter pays principle. Economically speaking, this understanding of "incentives" leads to an inefficient use of water, making Art. 9 (1) 2nd subpara. 1st indent self-contradictory. The legislator probably did not stick to pure economic theory, but this concept is at least compatible with the standard-price approach (cf., e.g., Baumol & Oates, 1971). However, as far as cost recovery and incentive approaches contravene each other, it opens room for Member States' discretion for implementation, but does not allow deviation from the aim of full cost recovery.

"Water pricing" covers all aspects relevant to the final price that costumers have to pay including levies and taxes that are imposed on the consumption of water services and water uses. "Water services pricing" is not restricted to the provision of water services by public enterprises. The polluter pays principle does not have to be applied on the individual level. It may be applied only on the level of consumer groups (households, industry and agriculture), but it is prohibited to disregard completely its application within these user groups according to Art. 9 (1) 1st subpara. WFD.

The WFD does not contain any definition of costs, especially not of environmental and resource costs. In the first draft the WFD (European Commission, 1997b) enclosed a definition of costs in Art. 2 No. 33 containing operation and maintenance costs, capital maintenance costs, capital costs (principal and interest payments) and reserves for the future and extensions. In its communication on water pricing policies the Commission repeated almost all these components as a definition of financial costs, adding only "return on equity where appropriate" and withdrawing "reserves" (European Commission, 2000a, p. 10). The Commission demands an adoption of common definitions for key cost variables (European Commission, 2000a, p. 15).[21] WATECO has identified as components of the financial cost: operating costs, maintenance cost, capital cost for new investments, depreciation, opportunity costs for capital costs, administrative costs and other direct costs and has given some rough guidelines for the calculation (WATECO, 2002, Annex IV.I.14ff.), especially for the adequate depreciation of investments.

"Environmental costs" are defined by the Commission (European Commission, 2000a, p. 10) as the costs of damage that water-uses impose on the environment, ecosystems and those who use the environment (e.g., a reduction in the ecological quality of aquatic ecosystems or the salinization and degradation of productive soils) (similar WATECO, 2002, Annex IV.I.18 and Kahlenborn, 1999, p. 18). They may also consist in imputed risk costs. They refer to societal risks related to the provision of a water service. For instance, if the provision of a water service increases the likelihood of floods or industrial accidents, a risk premium reflecting actual or hypothetical insurance costs should be included as a price component.[22]

"Resource costs" are defined as the costs of foregone opportunities which other uses suffer due to the depletion of the resource beyond its natural rate of recharge or recovery (e.g., linked to the over-abstraction of groundwater).[23] They are therefore not simply opportunity costs[24] of the use or consumption of the water service *within* the said limits.[25] If a certain demand cannot be met, there is no compensation necessary.

Environmental and resource costs cover some external or social costs, but not all. Induced costs for the society resulting, e.g., from the effects on employment are not covered by environmental and resource costs (WATECO, 2002, Annex IV.I.14; cf. Roth, 2001, pp. 13ff.), but may be considered in the scope of the derogation clause in Art. 9 (1) 3rd subpara. WFD and not only at the cost-effectiveness analysis for the programmes of measures (Art. 11 WFD) as WATECO suggests (2002, Annex IV.I.14). Taking these costs into account the WFD outstrips economic theory, which usually does not regard these effects as economically relevant. Generally speaking environmental and resource costs above all are relevant if the environmental objectives of the WFD are missed due to economic activities that constitutes water services and in accordance with the relevant exception clauses of the WFD. If the environmental objective "good status" is reached, the economic activities often do not produce environmental and resource costs related to surface or ground waters.

Art. 9 (1) 1st subpara. 2nd indent WFD additionally requires the contribution of water uses that are not water services to the recovery of the cost of water services, if they are responsible for a fraction of these costs.[26] For example, agriculture has to bear the costs of using nutrients that increase the cost of raw water treatment for drinking water purposes. These costs have to be distributed according to the polluter pays principle, but again the creation of "polluter groups" from individual polluters is allowed, but restricted by Art. 9 (1) 1st subpara. WFD.

4. ECONOMIC AND LEGAL CHALLENGES AND OBSTACLES FOR THE IMPLEMENTATION OF COST RECOVERY

Under perfect circumstances full cost-recovery pricing of a water service reflects all financial and all other costs connected with the provision of the service. Below, different areas of problems are discussed from an economic and legal perspective in order to highlight the specific demands stemming from the call for cost-recovery prices as well as the obstacles hampering the consistent transposition of the WFD. Regarding the cost-recovery principle, these problems relate both to the *emergence* of financial, environmental and resource costs and to the *imposition* of these costs on relevant water users and polluters. Many authors claim that the cost-recovery requirement is essentially fulfilled in Germany with regard to financial costs and that environmental and resource costs are already covered by water extraction charges and waste-water charges (Michel, Pejas, & Quadflieg, 2002, p. 11; RP Gießen, 2002, p. 42; Kahlenborn, 1999, p. 36; WATECO, 2002, Annex IV.I.13). Whether this position can be upheld has to be investigated in the following.

4.1. Problems with Financial Costs

Regarding the financial costs, an adequate price of a water service should include all variable and fixed costs. Capital costs can be considered in terms of replacement costs or in terms of the current value, as there is hardly any technical progress in some parts of the provision of water service (WATECO, 2002, Annex IV.I.16). This calculation is taken out under the hypothesis of sustaining the water service in quantity and quality over time and unchanging natural conditions. It has to be adjusted to long-term forecasts of supply and demand for water that are required within the economic analysis according to Annex III of the WFD (cf. Kahlenborn, 1999, p. 12). Subsidies disable price incentives for using resources in a sustainable manner and have, therefore, been disregarded for calculation of the cost-recovery rate (WATECO, 2002, Annex IV.I.36). Subsidies may consist of direct ones, e.g., for investments or for on-going purposes (cf., e.g., for the Middle-Rhine area WATECO, 2002, Annex IV.I.30).[27] As mentioned above, they are still common in Germany and amount to a significant part of the overall costs (ibid.), if the water extraction charges can be regarded as representing the environmental and resource costs, which is questionable (see Section 4.3). Otherwise the charges paid by the utilities exceed the subsidies by far and a full cost recovery is already reached. Indirect subsidies result from, e.g., cross-subsidies between different user groups, regions or within one user group.[28]

As regards waste-water treatment, the situation is different. The amount of public subsidies for investments and ongoing purposes is far above the figures for waste-water discharge fees and the degree of cost-recovery amounts to 63%, if *all* public allocations and subsidies are excluded (e.g., for the case Middle-Rhine see RP Gießen, 2002, p. 44).[29] Public subsidies for investments may be justified (in the sense of Art. 9 (1) 3rd subpara. WFD) by the fact that after German unification, large investments in sewage infrastructure in Eastern Germany were necessary for achieving the West German standards in sewage disposal. High subsidies were paid by the state and in a recent study it has been calculated that only about 70–75% of the operational costs of sewage disposal is reflected in Saxony sewage disposal prices (Lenz, 2003).[30] Cost-recovery merely of financial cost in this field is still a challenge.

Regarding the financial costs, the accounting standards for public as well as for private enterprises in the field of drinking water supply and waste-water treatment have to be harmonized on the one hand, and the relevant differences between public and private enterprises in this respect have to be identified or defined on the other hand. The aggregation of these costs to the river basin level does not provide any special problem in Germany as the water supply system is largely decentralized.

4.2. Monetary Evaluation of Environmental and Resource Costs and their Aggregation

Since many environmental and resource costs generated by water services' provision have not yet been evaluated, it would be a great effort to execute this task of monetary evaluation in the coming years in order to consider the emergence of these costs in the shaping of prices, tariffs and charges. This is not only a challenge due to the large number of local water services and their many environmental effects, but a methodological challenge as well. There are difficulties involved in cost evaluation of effects like receding ground water or structural changes in landscapes and ecosystem functions (having use values as well as non-use values) caused by water services' provision. The reason for this is that final effects are sometimes uncertain or even unknown and sometimes only emerge in the long run, affecting future generations (cf. DG ECO2, 2004, p. 5). The methodologies for evaluating environmental costs that WATECO (2002, Annex IV.I.21) discusses and suggests are:

- market methods,
- cost-based valuation methods,
- revealed preference methods and
- stated preference methods.[31]

These methods are well known from the traditional economic benefit–cost analysis (BCA). Although in recent years much progress has been made in improving the BCA evaluation instrument (cf., e.g., Bateman & Willis, 1999), the different evaluation methods still represent a plethora of weaknesses (cf. Kahlenborn, 1999, pp. 22ff. and Bartolomäus, Beil, Bender, & Karkow, 2004, pp. 232f.). Regarding BCA methods, which try to find a relationship between specific market information or revealed preferences and the marginal value of an environmental good, their largest limitation is due to the fact that they cannot be used for evaluation of economic non-use values – such as the pure existence value or the bequest value of an environmental good (cf., e.g., Randall, 1991, p. 303; Santos, 1998, p. 68). What is more, these indirect methods tend to handle observed market data as equilibrium data (e.g., travel cost approach, hedonic pricing) (Hanley & Spash, 1993, p. 80), and it is often assumed that preference functions of different people in different regions are identical in order to create a demand function (Endres & Holm-Müller, 1998, p. 56). These assumptions are sometimes not valid for real economic life and therefore the results are questionable in this case.

The contingent valuation method (a stated preference method), which has the potential to evaluate both use and non-use values (cf. DG ECO2, 2004, p. 16), tries to ascertain the willingness-to-pay of individuals for environmental goods by means of direct interrogation. However, regarding this method, there are also several critical points to be mentioned. Among others, people are only confronted with a hypothetical and not a real market situation and have therefore the opportunity to manipulate the results with strategic answers. It has also been observed that many individuals do not distinguish between a single environmental good (like for instance a forest) and a superior environmental good which does contain the single good (like for instance a mountain landscape). This is called "embedding effect" and leads to the fact that people over- or underestimate specific environmental goods (Randall & Hoehn, 1996). This effect especially calls for caution, when regional results are aggregated on a river basin district level (cf. Kahlenborn, 1999, p. 32). Finally, in contingent valuation interviews many hindrances may occur leading to a distortion of results, e.g., people may just not understand the ecological complexities involved, the interviewer might ask leading questions or the interviewees might state too high a willingness-to-pay for an environmental good in order to appear as an ethically decent fellow (Hausman, 1993).

If we bear up against these criticisms or even ignore them, there is still the very problematic aspect that the economic evaluation results still vary sharply depending on the specific evaluation methods used and the assumptions made.[32] In addition, there are only individual economic studies for evaluating specific environmental goods or environmental functions, so that the results are often too specific and difficult to transfer to other regions or circumstances. Therefore, if these BCA-related instruments are used as the standard for monetary evaluation of environmental and resource costs, EU-wide rules are needed in order to guarantee a uniform application. In the United States, where BCA is much more prevalent in politics than it is in Europe, guidelines have been published for the evaluation of environmental damage (US Department of Commerce, 1996). By contrast, in Germany work only began on drafting a convention for the application of BCA by the Federal Environmental Agency in 2001 (UBA, 2001), and no such guidelines have yet been drawn up for the EU and especially not by the WATECO (2002) and DG ECO2 (2004). It will be a major task of EU-Common Implementation Strategy, of which WATECO forms a part, to furnish consistent guidelines for the application of these methods. Otherwise, we will frequently have to accept that work is being carried out using differing and partly inconsistent evaluation techniques.

For the estimation of resource costs no well-established methods exist either (WATECO, 2002, Annex IV.I.19). They are seldom integrated in market prices. Therefore, it will be necessary to rely on estimates of foregone demands and economic values. This requires basic data on the economy of water resource use and a forecast on the demand and supply of water services within the river basin district. In general, if there is enough water available to fulfill all current and foreseeable future demands, there are no resource costs (WATECO, 2002, Annex IV.I.19, and similar Kahlenborn, 1999, p. 18).

For the other elements of the external costs (induced costs, e.g., the effects on employment) that are possibly needed for application of the exception clause in Art. 9 (1) 3rd subpara. WFD, there is no generally accepted methodology either and nothing provided by WATECO 2002.

For the application of Art. 9 (1) 3rd subpara. WFD, for the use of derogation clauses in Art. 4 WFD and for development of the most cost-effective programmes of measures, all these data are to be collected by means of economic analysis by 2004 according to Article 5 and Annex III of the WFD (cf. DG ECO2, 2004, p. 8). This gathering and structured presentation of data on the supply and demand of water services entails additional problems. Important economic data on water use are often available at the state level gathered by the state statistical offices. However, the methods and the comprehensiveness of collecting and structuring data sets vary from Member State to Member State within the EU. Therefore, it will be difficult to obtain consistent and complete data sets for a whole river basin from the economic analysis. It is therefore unlikely that the necessary database will have been compiled for all river basin districts by 2004.

4.3. Defining 'Polluter Pays' Pricing

Assuming that the problems of defining and calculating financial costs and of measuring and estimating environmental and resource costs of providing water services are resolvable, then there still remains the challenge of imposing these costs in accordance with the polluter pays principle on water services consumers by means of tariffs and prices for water services and/or environmental taxes and charges. Each water service has to be considered separately.

Drinking water supply. Regarding the supply of drinking water, it has to be stated that from the financial costs' point of view the costs of supply systems due to operation and maintenance of the pipeline and sewage network often enclose a high proportion of fixed costs (70–90%; cf. Kraemer & Piotrowski, 1998; Schönbäck, Oppolzer, Krämer, Hansen, & Herbke, 2003a) that are independent from the quantity of water consumed. The polluter pays principle claims that tariff structures match cost-structures of the water service. Any significant mismatch hence may be regarded as a cross-subsidy. Therefore, the usual structure of tariffs in Germany that have a small fixed-cost component and a vast variable-usage component[33] is incompatible with the polluter pays principle, if applied on the household level. From a financial point of view, which is the only one relevant for the federal states' Municipality Charges Law, this tariff structure does not comply with the polluter pays principle (cf. Gawel, 1995, p. 177). Only if the narrow perspective of financial cost recovery is left and broadened to an economic perspective that includes all external costs of providing water service, an outweighing portion of the variable cost in linear or progressive tariffs is possible (cf. Gawel, 1995, pp. 177f.), depending on the extent of external costs produced by the single water service. In the case of drinking water supply in Germany, the environmental and resource costs are likely to be low,[34] as Germany is essentially a water surplus area (Rothenberger, 2003, p. 32), so that the financial costs remain dominant.[35] The current tariff structures represent, therefore, transfers from the family-households to the single-households[36] (Rothenberger, 2003, p. 42). This effect is mitigated by the fact that, usually for higher supply capacity, the fixed costs are higher and the price per unit of water is lower. However, a strict application of the polluter pays principle could be viewed as contradictory to the aim of providing incentives for efficient use of water, as required by Art. 9 (1) 2nd subpara. 1st ind. WFD. But from an economic point of view, this deviation of the polluter pays principle remains unfounded: if external costs are taken into account, there is no economic argument for an "additional incentive" to use less quantities of the water service (cf. Gawel, 1995, p. 193); this contradiction is grounded in Art. 9 WFD.

In the case of drinking water supply, sometimes there is the problem that blocks of rented flats often only contain a single water meter for the whole building, not for each household.[37] Landlords then allocate the water costs as they see fit (e.g., according to the size of households), instead of according to actual consumption, and are only restricted by the Ordinance on Operational Costs (as part of laws governing tenancy). Hence, when setting the basis for calculation of water prices, the proportion due from each household normally cannot be calculated on the basis of *actual distribution*. Therefore, the incentive intended by the common structure of the tariffs is not accomplished.

In its Communication on pricing policies for enhancing the sustainability of water resources, the European Commission has stated that the overall price of a water service should comprise a fixed-cost component and a variable usage component, if it is to encourage people to save water and reduce pollution (European Commission, 2000a, p. 16). For many years water tariffs in most Member States have consisted of a fixed-cost and a variable cost component. Indeed, some Member States countries have introduced progressive block tariffs. Since the price of water is raised in response to increasing consumption, block tariffs are a strong incentive to reduce water consumption (cf. OECD, 1999, pp. 52ff.). Progressive tariff structures are to be found in household water supply in the hotter EU countries (Italy, Portugal, Spain) (European Commission, 2000b, p. 10). Progressive tariffs are only justified economically and with regard to the polluter pays principle, if the environmental and resource costs are included in the prices and if they increase marginally with increasing consumption.

This may be the case in water shortage areas. The additional incentive, in this case, is indeed only justified if the polluter pays principle is applied on the individual level and not on the household level. Hence this structure is no model for the states with a more moderate climate.

The polluter pays principle is also relevant for the distribution of costs between different user groups. In Germany water supply tariffs often include discounts for mass consumers. It is often argued that their consumption ensures the continuous flushing of the pipeline network and thus contributes significantly to the prevention of sanitary problems. As the mass consumers, on the other hand, allow for economies of scale,[38] these discounts may be justified in some cases and compatible with the polluter-pays-principle.[39] Like Germany, especially many northern EU countries offer such discounts to major industrial water consumers in return for high water consumption, whereas the variable component in their prices is often regarded as too small to have an impact on consumption behavior (cf. OECD, 1999, pp. 63–66).

Water supply for commercial purposes (*self-supply*). As regards water supply for other purposes (cooling, industrial production, irrigation) in Germany, extraction from surface waters or groundwater is substantially done by the users themselves (circa 97%; Schönbäck et al., 2003a, p. 356), amounting to 87% of the overall extraction of water (ibid.). As there are no specific subsidies for the providers of the services – being identical with the users – they bear fully the financial costs of these services themselves. Environmental and resource costs of these services can only be recovered by taxes or charges imposed on the extraction. The legal justification of present extraction charges of the federal states according to the Basic Law does not allow one to regard them as a recovery instrument. Extraction of water is subject to authorization. The charges are meant to "skim off" the special advantage that the authorized individuals gain with their permit, over those who do not use (or, at least, not to the same extent) this public good. This sounds like recovery of opportunity costs, but the charges are imposed regardless of whether there are actually others who want to obtain an extraction permit, and also regardless of whether abstraction exceeds the natural rate of recharge or recovery. This will be the case only in a few selected areas. In addition, the common distinction of rates between different purposes of water extraction (e.g., in Baden-Württemberg acc. Annex to Art. 17a (3) Water Act: surface water extraction: €0.05113 for public water supply, €0.01023 for cooling, €0.00511 for irrigation and €0.02045 for other purposes; groundwater extraction: €0.05113 for all purposes) cannot be justified at first sight with different environmental or resource costs.

Therefore, these charges in their present shape cannot be identified as instrumental in the recovery of environmental and resource costs, as RP Gießen (2002, p. 42) and WATECO (2002, p. 17) claim (similar Görlach, Interwies, Pielen, & Rathje, 2004, p. 11). For that purpose they have to be modified, especially with regard to a regional differentiation according to available quantities. This could be done by a segmentation of the river basin in different areas or building of classes of abstraction sites. Such discrimination would be in accordance with the legal justification of the extraction charge, as the advantage conferred by the permit could be considered to vary corresponding to local and regional scarcity. As the description of the "elements of the charge" has to be in abstract-general terms, the schedule of charges can only depict the diverse regional conditions and their differences coarsely.[40] Another question is also whether a seasonally differentiated schedule of charges may be developed. Systematically, the problem is that if the charges are used to recover environmental and resource costs, the structure of rates should match the development of the marginal environmental and resource costs. This is especially difficult if they increase with growing consumption, as it is often assumed (cf. Kahlenborn, 1999, pp. 15f.). But this is probably not always the case – e.g., for reservoirs: once they are built, their environmental costs do not change very much according to the quantity used (within their capacity). A simple schedule like the abstraction charge covers also the abstraction, where no environmental or resource cost occurs and is, therefore, too high. However, even an accordingly gradual/stepwise tariff would not transport the relevant information to the individual consumer whose additional consumption causes the particular environmental and resource cost (cf. Kahlenborn, 1999, p. 15), if it is possible to identify this particular consumer.

This could only be the case if not the abstraction is charged but the individual consumption via an excise duty, where, if necessary, a certain basic quantity is free of charge, calculated from the maximum (without environmental and resource costs) available quantity of water divided by the number of individuals to be supplied. The questions remain, who defines the available sources for the community and what happens if some of the costumers use less water than granted free of charge, so that others may consume more than the said quantity without causing environmental and resource cost? These considerations would lead to a type of certificate for the individual consumption of water to be introduced. Another problem is whether there should be a hierarchy of consumers in such a way that drinking water supply gains priority over any other abstraction of water. As in Germany, the abstraction of water for drinking water supply amounts only to about 15% of the overall abstraction, there would hardly be any environmental and resource costs stemming only from abstractions for drinking water supply purposes. But apart from the fulfillment of basic needs, no general priority can be given to households over industry or agriculture.

Waste-water collection and treatment. In the case of sewage disposal of, and waste-water treatment for, households, the financial costs are borne by them via municipality charges apart from subsidies for investments.[41] Here again the problem arises that the proportion of fixed cost for running the sewerage and treatment plants are high, while their proportion in the tariffs is rather low. This fact eases the problem that the quantity of waste water or freshwater (as the usual parameters for the calculation of the charges in Germany) is only a rough indicator for the intensity of the utilization of the service, as the quantity does not give any information on the load of pollution. This problem increases with regard to the possible environmental and resource costs, caused by substances not eliminated in the treatment facility and discharged in, e.g., surface water (cf. Gawel, 1995, pp. 178, 195). However, here again it is to be stated that not all material loads of sewage generate environmental costs, as long as the natural self-purification capacity (or natural absorption rate) – to be defined with respect to the environmental objective "good status" – is not exceeded. The natural self-purification capacity is not considered in waste-water charge,[42] but only compliance with the best available treatment technology resulting in a reduced rate – regardless of the compliance with the environmental quality objective and regardless of the regional environmental situation. The charges are calculated on the basis of the value fixed in the discharge permission. Since the discharges are only monitored selectively, they do not represent the actual burden. But as there are heavy fines in case the allowed values are exceeded, the actual burden is likely to be below the allowed load (Bode, 1999, p. 250). All these features of the waste-water charge in Germany do not allow them to be regarded, at present, as an instrument for recovering the environmental and resource costs, as RP Gießen does (2002, p. 42). Also the sometimes-used additional charges for heavily polluted water do not adequately fulfill the task of cost distribution according to the polluter pay principle (Gawel, 1995, p. 184).

Additionally, regarding toxic (or for other reasons problematic) substances, no natural self-purification capacity usually does exist and for many of the 100,000 substances used in European industry, scientific knowledge and appropriate analytical methods for examining their toxicological and chemical effects are lacking if they are released into the environment (Reemtsma & Klinkow, 2001). If they are disposed in private households over their water toilets – which is often forbidden – there is no chance for the environmental authorities to backtrack this pollution path.

4.4. The Adequate Contribution to the Cost Recovery by Water Uses that are not Water Services

Full cost recovery requires identifying water users that are not customers of water services. This is especially difficult when surface water is used as a pollutant sink. Whereas discharges from point sources (e.g., industrial and sewage plants) can be measured by a suitable monitoring system, the sources of discharges from diffuse sources (e.g., agriculture and traffic) are far more difficult to pinpoint. Actually, the current policy path for regulating diffuse emissions in the EU (and as well in the US, see Boyd et al., in this volume) is not correlated to existing pollution load levels in the environment and is often not aimed at environmental-quality objectives. Rather, technology standards prevail which focus on the reduction of emissions of the single source. But due to increasing activities, especially in the traffic area, these standards do not really contribute to an improvement in the ecological and chemical status of water bodies. And in the area of agriculture, current best practice standards are not sufficient for preventing the pollution of ground water with fertilizer and pesticides. The Directive concerning the protection of waters against pollution caused by nitrates from agricultural sources (European Community, 1991) has proved to be of little effect (cf. European Commission, 2002), especially in Germany, as the directive has not been implemented properly (ECJ 14.3.2002 Case C-161/00; ECR 2002: I-02753). However, as the proportion of diffuse emissions, which are not regulated according to the polluter pays principles, becomes larger and larger, the economic misallocation effects due to external costs are rising, too. An additional problem results from the fact that these emissions are only relevant in the context of Art. 9 WFD, if they cause an increase in costs of the provision of water services which is difficult to identify. However, water-service providers often conclude agreements with farmers and pay compensations for reducing fertilizer and/or pesticide use in order to reduce pollution of water, which is meant to be used as drinking water. Similarly, farmers are often compensated for management restrictions beyond the best practice standards (Art. 19 (3) and (4) Federal Water Act). This way, the polluter is compensated for not polluting and the water service costumers have to pay – this is quite the opposite of the polluter pays principle and in contradiction with Art. 9 WFD. Up to now there is no strategy for calculating an adequate contribution to the costs of water services (cf. Görlach et al., 2004, pp. 11, 21). At present, the best feasible solution is the implementation of charges for diffuse polluters in the form of charges on, e.g., fertilizers, and pesticides. However, in Germany this approach causes significant problems with respect to the Basic Law, which cannot be developed further here (cf. for this problem Unnerstall, 2004, pp. 271f.).

5. SUMMARY AND OUTLOOK

Full cost-recovery water prices according to the polluter pays principle have the potential to inform the water users about the total costs connected to their own water service consumption. They can act as an incentive for reducing water consumption and pollution to sustainable levels, and they may contribute to the financing of watershed protection activities. In this respect and in view of the history of economic incentives in European water legislation, the new European WFD takes the historical plunge in order to introduce this innovative water policy instrument. However, it emerges that the imposition of adequate 'polluter pays' prices for water services as called for by the WFD presents a very difficult challenge. It is also apparent that, given the numerous obstacles, a consistent, comprehensive implementation of the WFD requiring full cost recovery in 2010 will be impossible.

The environmental authorities responsible for the implementation of the WFD have recognized these problems. In order to handle it, it was, for example, decided in Germany to start with the introduction of full-cost water prices only for drinking water supply and sewage disposal (LAWA, 2003). Indeed, this way it is possible to avoid the difficulties linked with evaluation of environmental and resource costs of water services like water provision for inland navigation, management of reservoirs and flood protection. Furthermore, management of drinking water supply and sewage disposal already operates on a solid operational basis with easy-to-grasp subsidies from the state. Actually, considering sewage disposal only from an operational point of view, it can be ascertained that full cost prices have not yet been achieved as subsidies of the state for investment costs and subsidies of the municipality for operational expenses amount to a significant portion of the overall financial costs. Therefore, the widely spread opinion among representatives of environmental authorities that full-cost water prices already exist in Germany for the two major water services is questionable and it must be stated that this attitude is very optimistic, even if only drinking water supply and sewage disposal are viewed as water services.

In addition to this argument, it must be realized, despite other opinions, that actual environmental and resource costs of drinking water supply and sewage disposal are not adequately included in German water prices. Despite the fact that charges and fees are levied for groundwater abstraction and waste-water disposal in many German federal states, these charges and fees are neither based on, nor constitutionally justified in, actual economic evaluation results, but are simply established by the Federation and the federal states. Furthermore, as illustrated above, the problems of tariff-schemes and cross-subsidies, which are diametrically opposed to the polluter pays principle, is another obstacle to full-cost recovery water prices. Thus, the economic substantiation of German water prices is still a challenging task to be carried out in the future.

Summing up, even the limited application of full-cost-recovery water prices to only two water services already involves a long list of problems with regard to their implementation. Extending their application to other water services, which have even higher impacts on the environment, would multiply these problems. The concrete implementation of this challenge is an open process, which will probably take place in very different ways in the individual Member States of the EU. Nevertheless, introduction of full cost recovery in accordance with the 'polluter pays' principle has the potential to ease the strain on European surface and ground water bodies, even if their implementation will take more time and effort than initially planned.

The requirement of full cost recovery and the critical awareness of subsidies for water services has also to be seen in the wider context of the strong tendency for liberalization of markets and privatization of public enterprises in the field of services of general interest (e.g., telecommunication) driven by the EU. This trend has not yet reached the water sector in the same intensity as other sectors (cf. Pöcherstorfer, 2003), but the establishment of an internal market in the water sector is on the political agenda (European Commission, 2003, pp. 13, 46). According to the European Commission, the results of its assessment of the water sector will be presented by the end of 2004 (2004, p. 33) after getting an expert opinion on the application of competition rules in the European water sector (WRc & Ecologic, 2002). The next step, already taken, is the recently completely revised Directive on coordinating the procurement procedures of entities operating in the water (…) services sectors (European Community, 2004), defining the conditions for awarding contracts in this area. Whether the requirement of cost recovery that would have to be enforced against, e.g., private water suppliers, will change the suitability of the prices of water to be used for competition purposes is difficult to answer.[43] However, the problems inherent in designing tariffs for drinking water supply according to the polluter pays principle come back in the form of public price control, if it is privatized.

NOTES

1. Cost recovery means income received from the direct sale of services or in any other way directly associated with the operation of a service, or: Extent to which the production or supply costs of a specific good or service is covered by the revenues (DG ECO2, 2004).

2. Water services can be defined, preliminarily, as all services providing abstraction, impoundment, distribution and treatment of surface water or groundwater, including the provision of drinking water and waste water treatment. The scope and content of the relevant definition used in the WFD will be presented and discussed below in Section 3.

3. In favour: Oppermann (1999) Mn. 1157, Schoo in Schwarze (2000) Art. 269 Mn. 19; against: Waldhoff in Callies and Ruffert (1999) Art. 269 Mn. 13 and Kirchhof (2003, p. 1362).

4. Jansen (2003, p. 263) alleges that these competencies can also be used to introduce new taxes.

5. Unclear at this point Jans (2000, 60f.), who regards Art. 175 (2) EC as a basis for an environmental tax introduced by Community and for the Community, but only if the provision is extensively interpreted.

6. OJ (Official Journal) L 076, 23/03/1992, pp. 1–13.

7. OJ L 316, 31/10/1992, pp. 19–20. It contains especially minimum charges for different types of mineral oil and different consumption purposes.

8. OJ L 187, 20/7/1999, p. 42.

9. Directive 2003/87/EC of the European Parliament and of the Council of 13 October 2003 establishing a scheme for greenhouse gas emission allowance trading within the Community and amending Council Directive 96/61/EC OJ L 275, 25/10/2003, pp. 0032-0046.

10. Council Directive on the approximation of the laws of the Member States relating to detergents of 22/11/1973 (73/404/EEC), OJ L 347, 17/12/1973, p. 51; Council Directive on the approximation of the laws of the Member States relating to methods of testing the biodegradability of anionic surfactants of 22/11/1973 (73/405/EEC), OJ No. L 347, 1973, p. 53; Council Directive of 16 June 1975 concerning the quality required of surface water intended for the abstraction of drinking water in the Member States (75/440/EEC), OJ No. L 194, 25/7/1975, p. 26; Council Directive of 4 May 1976 on pollution caused by certain dangerous substances discharged into the aquatic environment of the Community (76/464/EEC), OJ No. L 129, 18/5/1976, p. 23; Council Directive on the protection of groundwater against pollution caused by certain dangerous substances of 17 December 1979 (80/68/EEC), OJ No. L 20, 26/1/1980, p. 43; Council Directive of 21 May 1991 concerning urban waste water treatment (91/271/EEC), OJ L 135, 30/5/1991, p. 40; Council Directive of 12 December 1991 concerning the protection of waters against pollution caused by nitrates from agricultural sources (91/676/EEC), OJ L 375, 31/12/1991, pp. 1ff.; Council

Directive of 3 November 1998 on the quality of water intended for human consumption (98/83/EC), OJ L 330, 5/12/1998, pp. 32ff.

11. § 54 Prussian Water Act of 7/4/1913 (PrGS: 53); Art. 119 Württembergisches Wassergesetz vom 1.12.1900 (RegBl: 921).

12. Art. 73 (1) Bavarian Water Act of 23/3/1907 (GVBl: 157) and § 41 Badisches Wassergesetz vom 26.6.1899 (GVBl: 250).

13. In the German Democratic Republic fees for water use were imposed at least from 1971 onwards, for the history see Sanden (1994, pp. 76ff.).

14. See e.g. Annex to § 17a (3) Water Act Baden-Württemberg.

15. For the history see Ewringmann (2002, pp. 270ff.). For development in the German Democratic Republic, that introduced a wastewater charge in 1971, see Sanden (1994, 89ff.).

16. Critically to the design of the wastewater charge see Ewringmann (2002) and SRU (2004) Mn. 478.

17. Art. 28 Basic Law also guarantees this; cf. Laskowski (2003, pp. 3ff.). It is also a legal obligation; cf. Wolfers (2004, p. 116). To the decentralized structure of the drinking water supply and waste water management sector cf. Petry/Dombrowsky (in this volume).

18. As the application of Art. 3a is restricted to undertakings whose turnover exceeds 40 Mill. EUR (Art. 4(2)(b)), only few of the water supplies are affected by this provision (Ewers et al., 2000, p. 32).

19. Cf. ECJ of 24/7/2003 Case-C280/00 and Baumeister (2003) for field of regional public transport; for the debate in general see European Commission (2004, pp. 16f.).

20. The WATECO has not dealt with this problem: 2002, Annex IV.I.28.

21. Implicitly also the Court of the European Community, when it required for the fixing of the adequate compensation for discharging public service obligations as "basis of an analysis of the costs which a typical undertaking, well run ..., would have incurred in discharging those obligations, taking into account the relevant receipts and a reasonable profit for discharging the obligations", ECJ of 24.7.2003 Case-C280/00 No. 93.

22. Direct damage to individuals caused by the provision of water services are recovered by the system of civil law. They are part of the maintenance costs.

23. Similar WATECO (2002, Annex IV.I.17). According to WATECO overabstraction of groundwater is also to be applied on a yearly basis (2002, Annex IV.I.19). Cf. Dasgupta and Heal (1979, p. 164); Messner (1999, pp. 47ff.).

24. Instructive examples are provided by Kahlenborn (1999, p. 14).

25. Now with explicit reference to WATECO (2002) dissenting DG ECO2 (2004, p. 2); critically Görlach et al. (2004, pp. 9 and 13f.).

26. As to environmental services that are impaired by diffuse pollution: they are beyond the scope of the cost recovery principle.

27. As subsidies may also be regarded covering the operational losses of public or publicly owned utilities, as the public authorities are liable for the deficits.

28. WATECO (2002, Annex IV.I.36) regards income transfers also as a direct subsidy.

29. For unknown reasons allocations and subsidies for on going purposes are not calculated as subsidies, so that the degree of cost recovery is stated as 80.6%.

30. However, it should be commented that rational arguments underpin these sewage subsidies. Since capital investments in sewage infrastructure will have benefits for the next hundred years, it should not only be paid within the next 30 years. Rather, low depreciation rates should reflect the long term character of this capital good or the state should pay subsidies in order to enable the investments, if they would otherwise be omitted, which is likely as there would be no loans from private financial institutions for financing an investment that runs a hundred years.

31. More detailed now DG ECO2 (2004, Annex 2).

32. Because different evaluation methods may measure different things (DG ECO2, 2004, Annex 2; cf. also Messner and Drechsler (2001)).

33. According to Kraemer and Piotrowski (1998, p. 12) 9% of the revenue of the public drinking water supplier stems from the fixed price component and 91% from the variable component.

34. The most important environmental costs are probably those caused by the reservoirs that are built for drinking water supply purposes.

35. The water extraction charges of the some federal states, which some regard as representing the environmental and resource costs, amount to less then 20% of the financial costs in the Case of the Middle-Rhine area (cf. RP Gießen, 2002, p. 42).

36. The reference to households is, strictly speaking, already a deviation from the polluter pays principle, for it has to be applied to the level of individuals.

37. In 1999 only 60% of private households in Germany had their own water meter. The proportion with their own water meter in France was 88%, compared to less than 30% in Italy (OECD, 1999, p. 46).

38. Critically for the case of Italy Antonioli and Filippini (2001).

39. The overall importance of the industrial consumption is rather low in Germany, as less 20% of the publicly offered water supply is consumed by commercial users (Schönbäck et al., 2003a, p. 356).

40. The manifold legal difficulties with respect to the German constitution (the Basic Law) and its provisions on taxes, charges and fees cannot be developed and examined here due to space restrictions.

41. Some of the subsidies or additional ones could stem from reductions in the waste water charge or from the revenue of the waste water charge, but they are not displayed as such (e.g. RP Gießen, 2002, p. 44). However, the total amount of subsidies is a multiple of the charges paid in the case of public sewage treatment system in the Middle-Rhine area (RP Gießen, 2002, p. 43).

42. This capacity is the possible justification for the introduction of free of charge disposal, as Scholl (1996) suggests in order to increase the incentive effect of the charge.

43. Pöcherstorfer (2003, p. 189) expects that the interest of private companies in ongoing privatisations will decline. At least it is questionable whether the vast variety in annual water charges (€350 in Berlin to €50 in Rome), that are emphasized by the Commission for illustrating potential gains of liberalization (European Commission, 2003, p. 14) won't shrink tremendously, if the principle of cost recovery is implemented critically, regarding the aspect of prices comparisons, see also SRU (2002, No. 658), Bode (1999) and Laskowski (2003, p. 5). The possible gain of liberalization in Germany is estimated by Ewers et al. (2000, pp. 23f.) to be around

10–20%. Cf. for some other European states Schönbäck, Oppolzer, Krämer, Hansen, and Herbke (2003b). For the legal-technical difficulties of privatisation cf. Wolfers (2004).

REFERENCES

Antonioli, B., & Filippini, M. (2001). The use of variable cost function in the regulation of the Italian water industry. *Utilities Policy*, *10*, 181–187.

Bartolomäus, C., Beil, Th., Bender, S., & Karkow, K. (2004). Kontingente Bewertung – und was dann? In: R. Döring & M. Rühs (Hrsgs.), *Ökonomische Rationalität und praktische Vernunft – Gerechtigkeit, Ökologische Ökonomie und Naturschutz* (pp. 229–247). Würzburg: Königshausen & Neumann.

Bateman, I. J., & Willis, K. G. (1999). *Valuing environmental preferences – Theory and praxis of the contingent valuation method in the US, EU, and developing countries*. Oxford: Oxford University Press.

Baumeister, H. (2003). Der EuGH und die Finanzierung der kommunalen Daseinsvorsorge – vor allem des öffentlichen Nahverkehrs. *Neue Zeitschrift für Baurecht*, *3*(10), 550–553.

Baumol, W. J., & Oates, W. E. (1971). The use of standards and prices for the protection of the environment. *Swedish Journal of Economics*, *73*(1), 42–54.

Bode, H. (1999). Produktkosten der deutschen Wasserwirtschaft. *gwf Wasser/Abwasser*, *140*, 244–252.

Brackemann, H., Ewens, H.-P., Interwies, E., Kraemer, A. R., & Quadflieg, A. (2002). Die wirtschaftliche Analyse nach der EG-Wasserrahmenrichtlinie (Teil 1). *Wasser und Abfall*, *4*(3), 38–43.

BVerfG (Bundesverfassungsgericht/German Federal Constitutional Court). (1995). Beschluss des Bundesverfassungsgerichts vom 7.11.1995 – 2 BvR 413/68 und 1300/93. *Entscheidungen des Bundesverfassungsgerichts (BVerfGE)*, *93*, 319–352, Tübingen: Mohr Siebeck.

Callies, Ch., & Ruffert, M. (1999). *Kommentar zum EU-Vertrag und EG-Vertrag*. Neuwied: Luchterhand.

Dasgupta, P., & Heal, G. (1979). *Economic theory and exhaustible resources*. London: Cambridge University Press.

DG ECO2. (2004). (Common Implementation Strategy Working Group 2B, Drafting Group ECO2) (2004). Assessment of Environmental and Resource Costs in the Water Framework Directive, June.

Endres, A., & Holm-Müller, K. (1998). *Die Bewertung von Umweltschäden-Theorie und Praxis sozioökonomischer Verfahren*. Stuttgart, Berlin, Köln: W. Kohlhammer.

European Commission. (1992). Proposal for a Council Directive to introduce rational planning techniques in the electricity and gas distribution sectors, COM/95/0369 final – SYN 95/0208. *Official Journal C 001*, 04/01/1996, Brussels, 6.

European Commission. (1993). White Paper on growth, competitiveness, and employment: The challenges and ways forward into the 21st century. COM (93) 700 final, 5 December 1993, Brussels.

European Commission. (1995). Amended proposal for a Council Directive introducing a tax on carbon dioxide emissions and energy. COM/95/172 final, Brussels.

European Commission. (1997a). Commission Communication, Environmental Taxes and Charges in the Single Market. COM (97) 9 final 26.03.1997. *Official Journal C 224*, 23/07/1997, Brussels, 6–19.
European Commission. (1997b). Proposal for a Council Directive establishing a framework for Community action in the field of water policy, COM/97/0049 final – SYN 97/0067. *Official Journal C 184*, 17/06/1997, Brussels, 20–40.
European Commission. (2000). Opinion of the Commission pursuant to Article 251 (2) (c) of the EC Treaty, on the European Parliament's amendments to the Council's common position regarding the Proposal for a European Parliament and Council Directive establishing a framework for Community action in the field of water policy (COM(97)49 final, COM(97)614 final, COM(98)76 final and COM(99)271 final) amending the proposal of the Commission pursuant to Article 250 (2) of the EC Treaty; COM(2000)219 final, 1997/0067 (COD). 05/06/2000, Brussels.
European Commission. (2000a). Communication from the Commission to the Council, the European Parliament and the Economic and Social Committee – Pricing policies for enhancing the sustainability of water resources; COM (2000) 477 final of 26.07.2000, Brussels.
European Commission. (2000b). Water pricing policies in theory and practice – Accompanying document to the Communication by the Commission COM: Pricing policies for enhancing the sustainability of water resources. SEC(2000) 1238, Brussels.
European Commission. (2002). Implementation of Council Directive 91/676/EEC concerning the protection of waters against pollution caused by nitrates from agricultural sources; COM (2002) 407, Brussels.
European Commission. (2003). Communication from the Commission to the Council, the European Parliament, the European Economic and Social Committee and the Committee of the Regions – Internal Market Strategy Priorities 2003–2006; COM(2003) 238 final of 07.05.2003, Brussels.
European Commission. (2004). Communication from the Commission to the European Parliament, the Council, the European Economic and Social Committee and the Committee of the Regions – White Paper on services of general interest, COM(2004) 374 final of 12.05.2004, Brussels.
European Community. (1973). Declaration of the Council of the European Communities and of the representatives of the Governments of the Member States meeting in the Council of 22 November 1973 on the programme of action of the European Communities on the environment. *Official Journal C 112*, 20/12/1973, Brussels, 1–2.
European Community. (1975a). Council Recommendation (75/436/Euratom, ECSC, EEC) of 3 March 1975 regarding cost allocation and action by public authorities on environmental matters. *Official Journal L 194*, 25/07/1975, Brussels, 1–4.
European Community. (1975b). Directive 75/439/EEC on the disposal of waste oils. *Official Journal L 194*, 25/07/1975, Brussels, 23–25.
European Community. (1991). Council Directive 91/676/EEC of 12 December 1991 concerning the protection of waters against pollution caused by nitrates from agricultural sources. *Official Journal L 375*, 31/12/1991, Brussels, 1–8.
European Community. (2000). Directive 2000/60/EC of the European Parliament and of the Council of 23 October 2000 establishing a framework for Community action in the field of water policy. *Official Journal L 327*, 22/12/2000, Brussels, 1–72.
European Community. (2000a). Commission Directive 2000/52/EC of 26 July 2000 amending Directive 80/723/EEC on the transparency of financial relations between

Member States and public undertakings. *Official Journal L 193*, 29/07/2000, Brussels, 75–78.

European Community. (2004). Directive 2004/17/EC of the European Parliament and of the Council of 31 March 2004 coordinating the procurement procedures of entities operating in the water, energy, transport and postal services sectors. *Official Journal L 134*, 30/04/2004, Brussels, 1–113.

European Council. (1999). Common Position (EC) No 41/1999 adopted by the Council on 22 October 1999 with a view to the adoption of a Directive 1999/./EC of the European Parliament and of the Council establishing a framework for Community action in the field of water policy. *Official Journal C 343/01*, 30/11/1999, Brussels, 1–72.

European Parliament. (1999). Proposal and amended proposals for a Council Directive establishing a framework for Community action in the field of water policy (COM(97)0049 – C4-0192/97, COM(97)0614 – C4-0120/98 and COM(98)0076 – C4-0121/98 – 7/0067(SYN)), A4-0261/98. *Official Journal C 150*, 28/05/1999, Brussels, 388–419.

European Parliament. (2000a). Recommendation for the second reading on the common position adopted by the Council with a view to the adoption of a European Parliament and Council directive on establishing a framework for Community action in the field of water policy (9085/3/1999 – C5-0209/1999 – 1997/0067(COD)), A5-0027/2000, PE 231.746. 3 February 2000, Brussels.

European Parliament. (2000b). European Parliament legislative resolution on the common position adopted by the Council with a view to the adoption of a European Parliament and Council directive establishing a framework for Community action in the field of water policy (9085/3/1999 – C5-0209/1999 – 1997/0067(COD)), A5–0027/2000, 16 February 2000. *Official Journal C 339*, 29/11/2000, Brusses, 136–154.

European Parliament. (2000c). Report on the joint text approved by the Conciliation Committee for a European Parliament and Council directive establishing a framework for Community action in the field of water policy (C5-0347/2000 – 1997/0067(COD)), A5–0214/2000. 29 August 2000, Brussels.

Ewers, H.-J., Botzenhart, K., Jekel, M., Salzwedel, J., & Kraemer, A. (2000). *Optionen, Chancen und Rahmenbedingungen einer Marktöffnung für eine nachhaltige Wasserversorgung.* BMWi-Forschungsvorhaben (11/00).

Ewringmann, D. (2002). Die Emanzipation der Abwasserabgabe vom Ordnungsrecht im Rahmen der EG-Wasserrahmenrichtlinie und eines Umweltgesetzbuches. In: E. Bohne (Hrsg.), *Perspektiven für ein Umweltgesetzbuch* (pp. 265–293). Berlin: Duncker und Humblot.

Gawel, E. (1993). Novellierung des Abwasserabgabengesetzes - ein umweltpolistisches Lehrstück. In: *Zeitschrift für Umweltrecht* (Vol. 4, pp. 159–164).

Gawel, E. (1995). *Die kommunale Gebühr*. Berlin: Duncker und Humblot.

Gawel, E. (1999). Anmerkungen zur Definition und zum Verständnis "betriebswirtschaftlicher Grundsätze" im Kommunalabgabenrecht der Länder – dargestellt am Beispiel des § 6 KAG NW. *Verwaltungsarchiv, 90*, 442–465.

Geiger, A., & Freund, A. (2003). Europäische Liberalisierung des Wassermarktes. *Europäische Zeitschrift für Wirtschaftsrecht, 16*, 490–493.

Gießen, R. P. (Regierungspräsidium). (2002). Pilotprojekt, Bewirtschaftungsplan Mittelrhein". 2. Statusbericht, September 2002.

Görlach, B., Interwies, E., Pielen, B., & Rathje, B. (2004). *Assessing Environmental and Resource Costs in the Water Framework Directive: the Case of Germany.* Final Report, (http://www.ecologic.de/download/projekte/1950-1999/1970-01/erc_germany.pdf) (accessed: 14.3.2006).

Hanley, N., & Spash, C. L. (1993). *Cost–benefit analysis and the environment*. Brookfield: Edward Elgar.

Hansjürgens, B. (2001). Das Verusacherprinzip als Effizienzregel. In: E. Gawel (Ed.), *Effizienz im Umweltrecht-Grundsatzfragen writschaftlicher Umweltnutzung aus rechts-, writschafts- und sozialwissenschaftlicher Sicht* (pp. 381–396). Baden-Baden: Nomos.

Hausman, J. (1993). *Contingent valuation: A critical assessment*. Amsterdam: Elsevier Science.

Jans, J. H. (2000). *European environmental law* (2nd rev. ed.). Groningen: Europa Law Publishing.

Jansen, B. (2003). Ökologische Steuern in der Europäischen Gemeinschaft. *Zeitschrift für Umweltrecht, 14*, 257–264.

Kahlenborn, W. (1999). *Umwelt- und Ressourcenkosten von Wassernutzungen vor dem Hintergrund der künftigen Wasserrahmenrichtlinie der Europäischen Gemeinschaft (im Auftrag des Bundesministeriums für Umwelt, Naturschutz und Reaktorsicherheit)*. Endgültiger Endbericht, November 1999.

Kahlenborn, W., Buck, M., & Kraemer, A. (1999). *Kostendeckung bei Wasserpreisen und Abwassergebühren vor dem Hintergrund der künftigen Wasserrahmenrichtlinie der Europäischen Gemeinschaft. Berlin: ecologic* (http://www.ecologic.de).

Kaika, M., & Page, B. (2003). The EU water framework directive: Part 1. European policy-making and the changing topography of lobbying. *European Environment, 13*(6), 314–327.

Kirchhof, F. (2003). § 38 Umweltabgaben – Die Regelungen in den Europäischen Gemeinschaften und ihren Mitgliedsstaaten. In: W. Rengeling (Ed.). *Handbuch zum europäischen und deutschen Umweltrecht* (pp. 1335–1390). Köln: Heymann.

Kloess, A. (1908). Das Deutsche Wasserrecht und das Wasserrecht der Bundesstaaten des Deutschen Reiches. Halle a.S.

Kraemer, A., & Piotrowski, R. (1998). Wasserpreise im europäischen Vergleich (Kurzfassung), Berlin: ecologic (http://www.ecologic.de).

Laskowski, S. R. (2003). Die deutsche Wasserwirtschaft im Kontext von Privatisierung und Liberalisierung. *Zeitschrift für Umweltrecht, 14*, 1–10.

LAWA (Länderarbeitsgemeinschaft Wasser/German Working Group of the Federal States on water issues). (2002). Arbeitshilfe zur Umsetzung der EG-Wasserrahmenrichtlinie: Themenbezogenes Arbeitspapier Nr. 5 (Stand 27.2.2002).

LAWA (Länderarbeitsgemeinschaft Wasser/German Working Group of the Federal States on water issues). (2003). *German Guidance Document for the implementation of the EC Water Framework Directive (Updated Version of 30.4.2003)*.

Lenz, P. (2003). *Untersuchung zu Kosten, zur Kostendeckung von Wasserdienstleistungen (Wasserversorgung und Abwasserbeseitigung) im Freistaat Sachsen/Regierungsbezirk Leipzig in Bezug auf die Wasserrahmenrichtlinie Artikel 9, Sächsisches Staatsministerium für Umwelt und Landwirtschaft*. Unpublished study, Dresden, October 2003.

Messner, F. (1999). *Nachhaltiges Wirtschaften mit nicht-erneuerbaren Ressourcen – Die vier Ebenen einer nachhaltigen Materialnutzung am Beispiel von Kupfer und seinen Substituten*. Frankfurt/M.: Peter Lang Verlag.

Messner, F., & Drechsler, M. (2001). Monetäre versus multikriterielle Bewertung. In: H. Horsch, I. Ring & F. Herzog (Hrsg.), *Nachhaltige Wasserbewirtschaftung und Landnutzung – Methoden und Instrumente der Entscheidungsfindung und – umsetzung* (pp. 225–234). Marburg: Metropolis Verlag.

Michel, B., Pejas, W., & Quadflieg, A. (2002). Die Wirtschaftliche Analyse nach EG-Wasserrahmenrichtlinie (Teil II). *Wasser und Abfall*, *4*(6), 8–12.
OECD (1999). *The price of water. Trends in OECD countries*. Paris: OECD.
Oppermann, Th. (1999). *Europarecht*. München: Verlag C. H. Beck.
Pöcherstorfer, W. (2003). Daseinsvorsorge und Marktöffnung durch Gemeinschaftsrecht – auch in der Wasserwirtschaft? *Zeitschrift für Umweltrecht*, *14*, 184–190.
Randall, A. (1991). Total and nonuse values. In: J. Braden & C. Kolstad (Eds.), *Measuring the Demand of Environmental Quality* (pp. 303–321). Amsterdam: North Holland.
Randall, A., & Hoehn, J. (1996). Embedding in market demand systems. *Journal of Environmental Economics and Management*, *30*, 369–380.
Reemtsma, T., & Klinkow, N. (2001). Untersuchungsstrategie für gefährliche Stoffe in Abwassereinleitungen der Industrie. Umweltbundesamt, Texte 07/01, Berlin.
Roth, E. (2001). Water Pricing in the EU. EEB (European Environmental Bureau), Publication 2001/002.
Rothenberger, D. (2003). *Report zur Entwicklung des Versorgungssektors Wasser*, (http://www.mikrosysteme.org/documents/Report_Wasser.pdf (accessed: 14.3.2006).
Sanden, J. (1994). *Wasserrecht im Wandel: Die wasserrechtlichen Abgaben der ehemaligen DDR und ihre Relevanz für das bundesdeutsche Umweltrecht*. Baden-Baden: NOMOS.
Santos, J. M. L. (1998). *The economic evaluation of landscape change – theory and policies for land use and conservation*. Cheltenham, UK/Northhampton, US: Edward Elgar.
Scholl, R. (1996). Zur Novellierung der Abwasserabgabe: Einführung einer Messlösung. In: J. Köhn & M. J. Welfens (Eds), *Neue Ansätze in der Umweltökonomie* (pp. 355–366). Marburg: Metropolis.
Schönbäck, W., Oppolzer, G., Krämer, R. A., Hansen, W., & Herbke, N. (2003a). Internationaler Vergleich der Siedlungswasserwirtschaft, Band 1: Länderstudien, (http://www.arbeiterkammer.at/pictures/d22/153_Band1-5.pdf) (accessed: 21.3.2006).
Schönbäck, W., Oppolzer, G., Krämer, R. A., Hansen, W., & Herbke, N. (2003b). Internationaler Vergleich der Siedlungswasserwirtschaft, Band 2: Systemvergleich vor europapolitischem und ökonomischem Hintergrund (http://www.arbeiterkammer.at/pictures/d22/153_Band1-5.pdf) (accessed: 21.3.2006).
Schwarze, J. (Hrsg.) (2000). *EU-Kommentar*. Baden-Baden: NOMOS.
SRU (The German Advisory Council on the Environment) (2002). *Umweltgutachten*. Baden-Baden: NOMOS.
SRU (The German Advisory Council on the Environment) (2004). *Umweltgutachten*. Baden-Baden: NOMOS.
UBA (Umweltbundesamt, German Federal Environmental Agency). (2001). *Kosten und Nutzen des Umweltschutzes. Methodenkonvention zur Schätzung externer Kosten*. Unpublished draft, Berlin, September 2001.
Unnerstall, H. (2004). Rechtliche Grundlagen gegenwärtiger und zukünftiger ökonomischer Instrumente im Naturschutz und in der Flächenhaushaltspolitik. In: R. Döring & M. Rühs (Hrsg.), *Ökonomische Rationalität und praktische Vernunft – Gerechtigkeit, Ökologische Ökonomie und Naturschutz* (pp. 249–274). Würzburg: Königshausen & Neumann.
Unnerstall, H., & Köck, W. (2004). The implementation of the EU water framework direktive into federal and regional law in Germany. *Journal for European Environmental and Planning Law*, *1*(3), 207–217.
US-Department of Commerce. (1996). National Oceanic and Atmospheric Administration, 61 Federal Register 14261, March 25. US Government Printing Office, Washington, D.C.

WATECO (CIS Working Group 2.6 on Water and Economics). (2002). Economics and the Environment – The Implementation Challenge of the Water Framework Directive. Guidance Document and Annexes. http://forum.europa.eu.int/Public/irc/env/wfd/library?l = /framework_directive/guidance_documents/economic_wateco&vm = detailed&sb = Title (accessed: 12-03-2003).

Wolfers, B. (2004). Privatization of the Berlin water works – a blueprint for balancing public and private interests? *Journal for European Environmental and Planning Law*, *1*(2), 116–124.

WRc and Ecologic. (2002). Study on the Application of the Competition Rules to the Water Sector in the European Community. Study contract No. COMP/2002/E3/S12.334052 for the European Commission – Competition Directorate General.

Wüsthoff, A. (1962). *Einführung in das deutsche Wasserrecht*. Berlin: E. Schmidt.

TRADING AS A U.S. WATER QUALITY MANAGEMENT TOOL: PROSPECTS FOR A MARKET ALTERNATIVE

James W. Boyd, Leonard A. Shabman and Kurt Stephenson

ABSTRACT

The paper reviews current experience with water quality trading programs and evaluates trading's potential as a future water quality management tool. The relative virtues of cap and trade (CAT) versus regulatory offset programs are discussed, as are administrative and technical barriers to trading. Several existing trade programs are discussed in detail. The article places particular emphasis on the relationship between water quality trading and watershed-based regulatory initiatives such as the total maximum daily load program.

1. INTRODUCTION

Market-based approaches to water quality regulation have long been advocated by economists as an alternative to command-and-control

Ecological Economics of Sustainable Watershed Management
Advances in the Economics of Environmental Resources, Volume 7, 385–407

ISSN: 1569-3740/doi:10.1016/S1569-3740(07)07016-2

regulation.[1] In theory, market-based policies such as effluent fees and discharge allowance trading can both reduce the costs of meeting an environmental goal and promote innovations in pollution prevention. In practice, market-based approaches have had only limited application, in part because of concerns over the monitoring and enforcement of discharge limits under such systems.[2] Still, in recent years, "trading" has received renewed attention in the United States. Early in 2003, the EPA issued a "Water Quality Trading Policy" to signal its support for trading and to provide guidelines for program design (U.S. EPA, 2003a). At the same time the agency has been funding pilot trading programs to foster experimentation with trading concepts (U.S. EPA, 2003b).

There are several explanations for EPA's support for trading. First, trading is associated with concepts like government reinvention and regulatory flexibility. This has political value and is consistent with regulatory reform positions advocated by both the former Clinton and current Bush administration (U.S. EPA, 1996a, 2003a). Second, as will be described in more detail below, trading has become a mechanism to motivate and finance pollution reductions by sources that are not regulated under U.S. environmental statutes.[3] Third, trading is seen as a way to implement watershed load limits called for by the evolving total maximum daily load (TMDL) program (Stephenson, Shabman, & Boyd, 2005). Trading is not synonymous with TMDLs. However, TMDLs do set total pollutant load caps and make initial allocations of allowable discharges to sources – allocations that sources could, in principle, buy and sell.

A market-based approach to water quality regulation conjures images of multiple sources with discretion to determine the best way to control pollutant discharges making exchanges of discharge allowances over a wide area. An initial point to make is that no existing water quality management trading program conforms to this image of a market, though some programs do include significant discharger decision-making discretion. More common are programs, labeled as trading, that focus not on discharger flexibility but on flexibility for regulators in their role as permitting authority.

This distinction – whether decision-making authority resides with the discharger or the regulator – can be used to differentiate between two fairly dissimilar forms of water quality trading program in the U.S.: cap and trade (CAT) and offset programs. CAT programs are more "market-like," featuring control flexibility by sources. Offset programs also provide flexibility, but in a much more limited way that is tied closely to existing permit regulations.

We argue that trading of either form can be a worthy modification to current regulatory practice. We first describe the current approach to U.S.

water quality management and features that help explain the interest in, and practice of, both forms of trading. We then describe the basic features of CAT and offset programs, as applied to water quality. After illustrating these systems by describing four existing programs, we conclude with an assessment of the future of trading programs for water quality management in the U.S. The assessment will highlight limitations and challenges inherent in any water quality management approach based on achieving ambient water quality goals by securing pollutant load limits at diverse sources within a watershed. We will argue that the U.S. commitment to organizing water quality management around the watershed approach called for under the TMDL program can help advance CAT programs as a particularly desirable form of trading.

2. THE MOTIVATIONS FOR TRADING

2.1. Current Water Quality Management

The centerpiece of U.S. water quality regulation is the national pollutant discharge elimination system (NPDES) permit program. The NPDES program requires individual sources of pollutants to obtain permits, or licenses, that specify pollutant amounts that can be legally discharged.[4] The discharge limits are set based on an EPA regulatory determination of the "best conventional" control technology or "best economically achievable" control technology and depend on the type of pollutant and the discharging industry (or other source). Note that the identified technology need not be at the so-called limits of technology (LOT) if EPA deems the costs of such technology to be prohibitively high. Also, note that the control technologies on which limits are based may be changed over time.

The control technology standard is used by the regulatory authority (typically a state) to set numerical concentration and aggregate volume limits on the discharge of specified pollutants from each regulated source. It is these numerical limits, called effluent standards, that are binding. While the pollutant control approach used to meet the effluent standard is in principle up to the discharger, in reality sources usually adopt the waste treatment technology used to set the effluent standard (Davies, 2001; Environmental Law Institute, 1998). In fact, the EPA identified technology itself is often specified as a requirement in the NPDES permit. Thus, one essential characteristic of this system is that numerical, quantity-based limits tend to be uniformly applied across relatively broad classes of facilities. Also, each permitted source must meet the imposed limit, even if another, different type of source in the watershed could achieve the reductions at lower cost.

NPDES permits are only required of so-called point sources. Point sources tend to be larger industrial and commercial facilities and public treatment facilities. Also, the current NPDES system can impose more stringent controls on point sources if the waterbody to which discharges occur is in violation of state water quality standards. These *ambient* water quality standards can therefore also lead to tighter effluent discharge standards. The tightening of permit conditions only works, however, if point sources are the source of the water quality violation.

Isolated runoff – nonpoint pollution – from farms, roads, and lawns is typically unregulated.[5] Nonpoint sources are a significant water policy issue in the U.S. since most of the U.S.'s remaining water quality problems are due to nonpoint pollution (U.S. EPA, 1996b). For a given waterbody, if ambient impairments are caused by nonpoint sources, a regulatory mechanism other than the NPDES system must be called into play. This is the rationale for the so-called TMDL program. A long-neglected aspect of the Clean Water Act (CWA), TMDL provisions require states to identify waters that are not in compliance with water quality standards, establish priorities, and implement improvements – including improvements that rely on nonpoint source reductions (Houck, 1999; Boyd, 2000; National Research Council, 2001). The focus placed on nonpoint contributions to water quality impairment is a virtue of the TMDL approach.

But it is important to emphasize again that nonpoint pollution is not directly regulated under the CWA and the TMDL program offers no new implementation authority over nonpoint sources even in the event of a nonpoint source-related impairment. Accordingly, current programs to address nonpoint sources bear little resemblance to the command-and-control oriented NPDES program. Nonpoint sources are managed by encouraging changes in land-use practices, either through landowner education programs or government subsidy payments for prescribed best management practices (BMPs). Payments tend to be financed with general tax revenues and are distributed under a wide variety of programs. Trading is advocated either as an alternative or adjunct to the NPDES and subsidy programs described above.

2.2. *Trading Programs Defined*

It is important to distinguish between the two basic types of trading: CAT systems and permit offset systems. Because CAT systems feature true control flexibility and decentralized decision making they can be thought of as "real trading" (Shabman, Stephenson, & Shobe, 2002). It needs to be emphasized, however, that water quality trading is rarely of this form. Offset programs, described later, are more common.

CAT Programs. CAT programs have several basic characteristics. First, a cap on total releases from a set of sources is established. Second, the regulator allocates initial levels of control responsibility to the sources. Third, the regulated sources themselves determine the best way to limit discharges to stay within their allowances and determine whether they are better off buying or selling allowances. However, the sum of allowances is always equal to the cap, even as transfers are made. Fourth, there is a continuous financial incentive to seek out and implement pollution prevention because improved controls means allowances no longer need to be purchased or can be sold or rented at a profit.

Note that a CAT scheme differs from conventional regulation in several significant respects. Conventional regulation does not impose an aggregate cap on emissions to a waterbody, but rather imposes limitations on individual sources. From an environmental standpoint the "cap" in CAT is a virtue, since the enforcement and evaluation of a trading program can be focused on the environmental issue of greatest concern: is the cap on releases to a given waterbody being met? The current point source, rather than ambient, focus of existing regulation can distract from that basic question (Swift, 2001).

From the standpoint of economic efficiency, a well recognized virtue of allowance trading lies in the ability to reallocate control activity, and the ability to do so in a decentralized manner. This is particularly true when there is a significant disparity in pollutant control costs across sources. When control costs vary across sources, uniform control requirements are inefficient. The ability to reallocate, and thereby move away from uniform control, means that sources with lower control costs can be given a correspondingly large responsibility for reductions. They accept this responsibility because they are paid to do so by high-cost sources. Accordingly, with a trading mechanism, high-control-cost firms abate less and low-control-cost firms abate more. When high-control-cost firms abate less than low-control-cost firms, a given level of pollution reductions is achieved at the least cost.

The decentralization of decision-making implies that sources can control releases in whatever manner they see fit. This promotes innovation in control technology and practices. Centralized allocations imply government-imposed rigidity in control practices. If compliance is geared toward a particular technology standard there is little incentive to discover and implement alternatives.

At best, new technology would be subject to regulatory review and eventually become a new standard. Under conventional permitting, the reward for innovation is a new, tighter standard. In contrast, under a trading scheme, performance is what is measured, not the means by which performance is achieved. Accordingly, innovations that generate reductions can be translated into units of control that can be sold. This financial incentive is not present under conventional command-and-control regulation.

In principle, a centralized regulator could reallocate control activity to achieve the efficient outcome. For example, with source-specific knowledge of pollution control costs, a regulator can simply require firms to engage in the pattern of control activity that is efficient. In other words, the efficient, non-uniform control requirements could be directly mandated via permits specifying targeted, non-uniform treatment levels. But a regulator's ability to generate the efficient outcome breaks down in the face of real-world constraints. The most important of these constraints is that regulators have limited information regarding firm-specific control costs (Speir, Stephenson, & Shabman, 2000). They have even less information about potential innovations' effect on costs. Decentralized markets achieve the efficient allocation without a centralized evaluator.

Offsets. In practice, the legal flexibility, administrative capability, and technical sophistication necessary to implement water quality CAT schemes are significant barriers to their broad implementation. Accordingly, most current trading systems take a different, more limited form that is better described as an "offset" approach. These offset programs grant limited decision-making flexibility to regulators, not to dischargers.

Offset programs allow the purchase of control activity (the "offset") by a regulated source from another party. Offsets are superficially similar to trades made under a CAT system. After all, there is cash payment from one source to encourage another to undertake reductions they would otherwise not make. But in other important respects, offsets do not resemble CAT programs.

Offset programs require the ex ante participation and approval of a centralized regulator, since they are authorized as an NPDES permit condition – the permit being a precondition for facility operation (Shabman et al., 2002). The regulator is also involved on a case-by-case basis. In contrast, regulators in a CAT program focus on the ex post performance of reallocation agreements. With CAT, regulators do not veto reallocations ex ante, but rather focus on penalizing non-performance of reallocation obligations or reallocations that fail to comply with the aggregate cap.[6] Also, many offset programs do not allow a regulated source to decrease its control level in exchange for offsetting increases in control by another source. This is something a CAT program would allow. Finally, offset trades, like the conventional regulatory system to which they are an adjunct, are not associated with a watershed-level load cap. This severs the linkage between ambient water quality attainment and aggregate control activity that is the centerpiece of a TMDL-based CAT program.[7]

Given these important differences, and the relatively limited nature of the flexibility granted, what is to be gained by offset programs? The answer to this question requires thinking of offsets, not as an ideal form of trading, but as a pragmatic response to constraints on regulatory discretion under the CWA. Offset programs are an understandable response to problems created by the relatively high cost or technical infeasibility of additional controls on sources regulated under the NPDES permitting program, and the need to motivate controls at unregulated nonpoint sources.

Consider a case where regulated industrial sites or municipal waste treatment plants are discharging to a river that is not achieving desired water quality goals – a fact attributable to unregulated nonpoint discharges. The permitted sources may be faced with demographic changes (e.g., population growth) or facility expansion opportunities that result in more waste being generated. However, if a facility is located in a water quality impaired area it may be called on to make reductions beyond those originally deemed "economically attainable" or be prohibited from expanding production. And even if LOT treatment is employed the added control still may not result in zero discharge.

The regulator is now faced with an unpalatable choice. They can require the limit of technology and allow the source to make a discharge, thus further degrading water quality. Or they can deny the discharge and prohibit important new economic activity in the watershed. Offsets are a way to avoid this choice. They allow the regulated source – typically after having gone to the LOT – to purchase controls from unregulated sources or to invest in off-site pollutant reduction projects (e.g., constructed wetlands). These purchased controls offset the additional loading to the water body. As a result of the offset, economic activity can proceed while water quality degradation is avoided, or even improved.[8] It deserves emphasis, however, that these offset actions are pre-selected by the regulator. They are not the result of independent, bilateral agreements between sources.

With the above example in mind note that offsets can also be viewed as a financing mechanism for nonpoint controls. Offsets are not a particularly fair way to raise such money, since they require already-regulated facilities to bear remediation costs arising from water quality problems that are in some cases quantifiably generated by nonpoint sources. However, offsets are a pragmatic regulatory response to the CWA's failure to mandate nonpoint source pollution controls.[9] Point source NPDES requirements exert significant leverage over regulated facilities, particularly in water quality limited areas. Offsets exploit this leverage to generate financing for nonpoint projects. It is widely recognized that a more effective, efficient, and fair approach would be to regulate nonpoint sources directly.[10] Unfortunately, meaningful regulation of nonpoint sources has been politically unpalatable at both the federal and state level.

3. EXISTING TRADING PROGRAMS

The above description of CAT and offset programs oversimplifies the range of compliance flexibilities that are labeled as trading. A 1999 summary of trading projects lists 37 such programs at various stages of development (Environomics, 1999). In 2001, a second EPA report, *The United States Experience with Economic Incentives for Protecting the Environment*, provided a similar inventory of trading programs. The raw number of projects should be interpreted carefully, however. First, they include a very wide range of activities, including facility-specific agreements and source-water protection programs. Second, most of the listed projects are still in the development phase. Third, these inventories are somewhat outdated and include programs that were not in fact implemented and omit programs that have since been approved.[11]

In this section, we describe a set of established programs typically identified as trading programs. The first two programs we describe bear the closest resemblance to CAT systems. We then describe two programs that are best described as offset systems. As noted earlier, however, offset programs are much more common.

3.1. Tar-Pamlico Point Source Program

North Carolina's Tar-Pamlico river basin program offers an example of how a CAT system can be implemented in the water quality context. The program is made possible, in a legal sense, because it is geared toward a reduction in nitrogen, which is not subject to the same NPDES permitting requirements as other pollutants.

In the late 1980s, North Carolina's Tar-Pamlico Sound faced a number of water quality problems associated with excessive nutrients. In 1992, the state imposed a cap on industrial and municipal dischargers equivalent to a 30% aggregate nutrient load reduction from 1990 levels (Hall & Howett, 1995; NCDENR, 2001). An association of 13 point source dischargers (primarily wastewater treatment plants) was formed and the cap was applied to them collectively. The association was assigned a fixed number of nutrient allowances and the state established enforceable financial penalties for failure to meet the allowance cap. The association itself was responsible for allocating allowances among its members consistent with the cap. Once members received their allowances, they were able to reallocate allowances freely among themselves under the association's internal rules.

The program is often thought of as a point–nonpoint trading program, but this is not in fact an appropriate description.[12] Rather, the program required point sources to purchase nonpoint reductions *if* they failed to meet the load cap. The point sources – via reallocations among themselves – have been able to meet the annually declining cap every year of the program. Accordingly, the success of Tar-Pamlico is as a point–point trading arrangement.

North Carolina grants the association considerable discretion to determine how discharges will be controlled and provides a reasonably stable setting for investment in aggressive pollution prevention activities. Individual dischargers are not required to use specific control practices, nor are their operational choices constrained by technology-oriented permit requirements. Because the contractual arrangement between the state and the association focuses on an aggregate cap rather than technology requirements, association members are assured that significant reductions in discharges will not be penalized by more stringent individual permit requirements. Moreover, the state has granted broad power to the association to reallocate allowances among its members without each member having to enter into a formal regulatory approval process with the government.

The role of the association deserves particular attention, particularly as it relates to the process of reallocating control activities. Unlike an atomistic market, reallocation in the Tar-Pamlico program occurs within the internal confines of a private organization – the association. This is made possible by the relatively small number of participating sources and perhaps by their homogeneity. But the association achieves the same end as a more open market – a low cost mechanism for dischargers to reallocate a fixed number of allowances. In practice, the point sources optimized treatment practices and installed biological nutrient removal technologies at a couple of facilities sufficient to meet the aggregate cap. The result was non-uniform, but aggregate, point source compliance with a cap at a cost far lower than had been anticipated when the program was first put in place (NCDENR, 2001).

3.2. Grassland Drainage Area

The Grassland Drainage area is another example of a CAT program. The program covers a pollutant associated with agricultural production. However, even though the pollutant is from agriculture it does not enter the waters of concern as a diffuse flow, but instead is discharged through irrigation drainage canals. As a result, the monitoring and negotiation over control reallocations occurs at the level of irrigation and drainage districts, a form of farmer cooperative (Woodward, Kaiser, & Wicks, 2002).

The Grassland Drainage area is an irrigated, 97,000-acre agricultural area in California's San Joaquin valley. The soil is laden with selenium, which in high concentrations can be toxic to wildlife. In the late 1980s and early 1990s, irrigation led to elevated selenium levels in drainage that poisoned wildlife in and around receiving wetlands and reservoirs. Because the source of the pollution was from agriculture, it was deemed "nonpoint" and no NPDES-based permitting program could be used to limit selenium discharges. However, in 1996 the area's six irrigation and drainage districts applied for permission to release drainage water to a large conveyance canal controlled by the U.S. Bureau of Reclamation. This request was granted subject to the six irrigation districts, agreement to a declining cap on total selenium discharges to the canal. Failure to meet the cap triggers fines and possible revocation of permission to discharge to the canal.

The contract resulted in the formation of a Grasslands Association, comprising the six drainage districts that release to the canal. The association has the authority to flexibly allocate control activities among and within the drainage districts, subject to aggregate compliance with the cap. An important characteristic of the program is that the aggregate load can be measured by monitoring concentrations in the Bureau's drainage canal. This focused, rather than diffuse, outflow allows for precise monitoring of the cap. District-specific monitoring is also possible, by measuring releases at pumps prior to release to the canal. This provides credibility to district-to-district load allocation agreements.

As a CAT program, the focus is on aggregate releases, not source-specific mandates and the decisions on how and where to control rest with the dischargers. As an EPA report (U.S. EPA, 2001) on the program suggests, "the theory is that the region will meet its selenium load target at the lowest possible cost because reduction measures will be taken where they are cheapest to achieve. In addition, the program should spur innovation by bringing selenium reduction decisions to a more localized level."

A number of district-to-district trades have occurred (Austin, 2001). But to date the most significant impact of the program has been in stimulating innovation at the district level. Some districts have instituted water pricing to promote re-use and conservation of irrigation water – and thereby reduce drainage. Water recycling and planting practices have also been instituted in order to reduce drainage. As a result, drainage levels and selenium, salt, and boron loads have fallen significantly since 1996 (Austin, 2001). Like Tar-Pamlico, an association, composed of a small number of administrative entities responsible for meeting load limits, is a key component of the program's administration. The program is a somewhat special case in that it involves a water conveyance and drainage network amenable to monitoring. As a general rule, nonpoint runoff and caps are more difficult to monitor. The legal trigger – the need to discharge to a Bureau of Reclamation drain – that gave rise to the program is also somewhat unusual.

3.3. Cherry Creek

The Cherry Creek offset program is managed by an independent "quasi-municipal" authority that was established by the state of Colorado in 1985 to raise revenues, allocate loads, and oversee watershed planning.[13] The authority finalized the implementation of a trading program in 1997. The driver for the program was the authority's determination that municipal treatment facilities would be unable to make additional phosphorous load reductions as the local population grew. Further reductions were impractical because the existing facilities were already near the limits of control technology.[14] Construction of a plant dedicated to loads from new development would be both extremely expensive and legally questionable given water quality impairment in the basin. Lacking plant treatment options, development in the region could be severely constrained due to the violation of water quality standards. The alternative to plant treatment is construction of offset mitigation projects in the watershed to sequester, filter, or degrade pollutants in the water.

The offsets in the Cherry Creek program are created at government-operated, off-site projects that mitigate the effects of land runoff from existing and new urban development (Paulson, Vlier, Fowler, Sandquist, & Bacon, 2000). Examples include shore stabilization projects and constructed retention ponds. Property taxes, development charges, and recreational fees initially raised funds for the mitigation projects. The authority plans to recover mitigation project costs by a charge on customers of the sewage treatment facilities.

Cherry Creek is often cited as a good example of a trading program. It is also a good example of why the term "trading" can be deceptive. Clearly, the offset purchase requirements give regulators an alternative to much less palatable choices – development prohibitions, costly off-system septic treatment, or degraded water quality. Note, however, that the trading program does not actually result in the reallocation of control activities across existing sources. Phosphorus, particularly associated with residential and commercial development-related discharges, is not directly regulated under the CWA. Accordingly, mitigation is not offsetting NPDES permit requirements, but instead is paying for reductions in TMDL-based load allocations.[15] In fact the sewage treatment facilities are expected to be at, and stay at, the highest levels of technological control available. In addition, customers of the sewage treatment facilities, and not those who are the cause of the urban runoff, are expected to pay for the mitigation projects. This suggests that the Cherry Creek "trading" program might be more accurately understood as a program to finance nonpoint mitigation projects.

Decision making is centered on the regulatory authority that constructed and operates the projects. The authority defines the rules by which credits are generated and awarded. Credits are only made available to treatment facilities facing a demonstrable need to increase wastewater flows arising from population and employment growth in the area. The way in which credits are priced is also at odds with what one might normally think of as a market. Prices are not the product of exchange between buyers and sellers, but rather are calculated from cost-recovery formulas employed by the trading authority. Accordingly, prices are defined not by market forces, but by the authority's desire to share costs equitably.[16]

3.4. Rahr Malting

The Rahr Malting permit issued by the Minnesota Pollution Control Agency is often identified as one of the first and most innovative trading projects in the nation (Minnesota Pollution Control Agency, 1997a; Grumbles, 2002). Rahr, like Cherry Creek, is an example of an offset program. Nonpoint source controls are paid for by a private firm and enforced via their inclusion in Rahr's NPDES permit. The Rahr program can be thought of as trading in that off-site mitigation is allowed to substitute for certain extreme (e.g., extremely costly) on-site control activities. The reallocation is tightly controlled and administered by the Pollution Control Agency, however. The program is not decentralized nor does it represent a shift to performance-based goals.

The Rahr plant formerly paid fees to discharge to a public wastewater treatment facility. In order to save money, the plant wished to construct and operate its own dedicated treatment facility. Doing so, however, required an NPDES permit, since the on-site treatment would be a point source requiring permitting. However, the stretch of the Minnesota river where the facility is located is of poor quality and loads are fully allocated under a TMDL to existing point sources. Because of the water quality and TMDL allocation constraints the firm faced a "zero discharge" permit requirement. When the treatment plant's highly advanced on-site treatment technologies failed to achieve zero discharge, the Pollution Control agency allowed (or, rather, required) Rahr to pay for agricultural nonpoint source reductions upstream of its treatment facility to offset the load increase resulting from the plant's operation (Minnesota Pollution Control Agency, 1997b).

Specific BMPs were identified and analyzed by the regulatory agency and significant attention was paid to the question of nonpoint load equivalency.[17] Rahr established a trust fund to finance BMPs that are nominated by farmers, municipalities, and landowners. A range of offset projects have been completed to date, including bank and river channel stabilization, rotational grazing, and floodplain restoration actions (Studders, 2000). These nonpoint source controls became conditions in the Rahr point source permit. Thus, the installation and maintenance of a BMP on an upstream and unregulated farm, for example, became the legal responsibility of Rahr.

The Rahr permit does not offer the plant any additional authority over how to manage its own discharges. Rahr is still required to install and maintain limits-of-technology controls. Administratively, the state environmental agency must approve each BMP submitted for approval and the BMPs come from a regulator-identified menu of acceptable nonpoint source controls. Negotiating such deals through the permit process is costly to both the discharger and the regulatory agency. Rahr's modified permit, for instance, took over 2 years to negotiate.

The Rahr permit is a good example of the way in which stringent permit conditions are leveraged into financing for unregulated, nonpoint source control activities. The situation on the Minnesota river – in particular the zero discharge requirement – gave the state an opportunity to experiment with this form of permit requirement. The permit is innovative in the context of NPDES permitting by its inclusion of nonpoint sources. But it deserves emphasis that all of this was made possible by Rahr's lack of alternatives given the zero discharge constraint. Certainly many other point sources have and will face similar constraints. But financing extracted from point sources that wish to expand on water quality limited waters will not be adequate to deal with the nation's widespread nonpoint source and ambient water quality issues.

4. IS THERE A FUTURE FOR CAP AND TRADE?

The relative abundance of offsets relative to CAT programs is due to their legal and administrative consistency with the existing NPDES regulatory approach.[18] CAT programs demand more significant administrative, monitoring, and enforcement innovations. However, the cost minimization, pollution prevention innovation, and ambient water quality goal achievements realized in the Tar-Pamlico and Grasslands programs highlight the potential of CAT programs. Moreover, the TMDL program and its emerging role as a centerpiece for the U.S. approach to water quality management creates an opportunity to advance CAT programs.

4.1. CAT as a Complement to the TMDL Process

A CAT program design begins with a desired water quality standard that is translated into an aggregate pollutant loading limit for a defined watershed area. CAT programs and TMDL-based regulation are complementary because the definition of the load limit is built directly into the TMDL process. A TMDL "specifies the amount of a particular pollutant that may be present in a waterbody, allocates allowable pollutant loads among sources, and provides the basis for attaining or maintaining water quality standards."[19]

Also a TMDL plan specifies an initial allocation of load limits for all point and nonpoint sources (note that aggregate and source-specific load limits were a feature of the Tar-Pamlico and Grasslands programs). More generally, any CAT program design requires that the load cap be divided as a load limit among sources and that individual discharges not violate the limit. Source-specific initial load limits (termed allowances) are a specific amount of allowable discharge (for example in kilograms, pounds, or tons) during a specified time-period. It is allowances derived from the TMDL plan that would be bought and sold under the CAT.

Trading and TMDLs are complementary for another important reason: they both demand sophisticated assessment of the fate and transport of pollutants released at different locations in a watershed. Trading redistributes the location of control activity across the landscape and within a watershed. This creates a challenge for trading programs because the environmental equivalence of individual trades must be assessed.[20] In order to avoid "hotspot" concentrations of pollutants and reallocations that lead to a degradation of water quality due simply to their hydrological location, the environmental impact of trades must be evaluated spatially. Trades can be adjusted, if they are not initially equivalent, via the use of ratios designed to achieve environmental equivalence (Randall & Taylor, 2000). Technically such "equivalence" analysis is accomplished through watershed-level fate and transport modeling.

Because fate and transport models are highly specific to local conditions they are technically complex and not easily generalized.[21] This creates an administrative barrier to CAT programs because location-specific modeling is costly and demands significant technical expertise. Moreover, the data and model uncertainties associated with such assessments can themselves create a barrier to trade. If approved trades must include margins for error in the determination of appropriate trading ratios, the margins implied by model uncertainty can significantly erode the gains from trade.

Existing command and control regulations avoid this challenge because they are not directly concerned with the attainment of water quality goals. However, as U.S. water quality regulation evolves toward a more water quality driven approach technical resources, including monitoring and modeling tools, will increasingly be available to assess the consequences of trades.

Note, though, that any regulatory approach that seeks to measure the ambient impact of diverse sources on a waterbody demands the development of such modeling and assessment techniques. The TMDL approach, because it is focused on water quality, is already generating the development and application of such models around the country. Over the next decades, our ability to monitor and model impacts is likely to improve significantly.

4.2. The Special Challenge of Diffuse Sources

In a fully capped CAT program, all sources, point and nonpoint, must obtain allowances. When trades involve nonpoint sources an additional "equivalency" concern arises. Because discharges from nonpoint sources are not as readily monitored or enforced, promised reductions by nonpoint sources are not equivalent – from an enforcement perspective – to reductions by point sources.[22] In practical terms, this creates reluctance to accept promises of nonpoint reduction in return for an increase in point source discharges (Bartfeld, 1993). In fact, it is this concern that leads regulators to require LOT controls before nonpoint source offsets can be purchased (as illustrated in the Cherry Creek and Rahr programs). It is also one of the reasons that trades involving nonpoint sources typically feature trading ratios that adjust for this difference in enforceability. Diffuse sources present a challenge for CAT systems, but these challenges are not insurmountable. Recent experience, including the Grasslands program, shows that creative opportunities for monitoring and enforcement can be employed even for agricultural runoff (Austin, 2001; Goldstein & Ritter, 1995).

Because the largest gains from trade are likely to come from reallocation of control activity toward nonpoint sources, lack of confidence in nonpoint source monitoring and enforcement is a significant impediment to realizing the gains promised by a CAT program. Of course the nonpoint source monitoring problem must be addressed, irrespective of whether trading programs are used. For example, TMDL-driven command-and-control style nonpoint regulations (though currently rare) will require some form of monitoring and enforcement. Even if we do not regulate nonpoint sources, but instead rely on subsidies alone, it is important to know what reductions those subsidies purchase. Any regulatory programs – not just trading programs – that seek meaningful future nonpoint controls must deal with the nonpoint monitoring issue.

In the meantime, the most promising approach to CAT program introduction will be to apply CAT only to sources where compliance with an allowance holding can be reliably monitored and measured. This is the reason that the CAT program for the Tar-Pamlico river only covered point sources – it was a partial cap program – and left nonpoint source loads to be addressed in other ways. A partial cap approach can be designed to expand the coverage of sources over time. For example, a cooperative of municipal sources might agree to extend their service areas to serve households with failing septic tanks (a nonpoint source) in return for an increase in the cap allocated to the cooperative (Woodward, 2003). Program expansions could in principle be extended to other nonpoint sources as the ability to monitor such sources improves.

4.3. CAT Administrative Requirements

Decentralized decision-making by sources with an incentive to meet performance, rather than technology-based requirements, is the core of CAT program design. Decentralized decision-making gives dischargers discretion to determine for themselves what effluent control strategies and technologies they will use to limit discharges, the number of allowances they will hold, and who they will trade with. Innovations in permitting and monitoring are needed to provide this kind of decision-making discretion.

First, when load control and discharge reallocation decision making is decentralized, regulatory authorities must direct their attention to monitoring and enforcing pollutant load limits at individual sources to determine whether the loads are consistent with the allowances held. At the watershed level, monitoring must establish whether the cap is being violated.

Second, the program's rules must be legally and administratively clear and stable in order to motivate innovation and investment in new pollution prevention actions. A predictable set of CAT rules allows dischargers to have well-informed expectations upon which to make long-term investments in pollutant reductions. In particular, once created, the total number of allowances must be fixed and equal to the discharge cap for some predictable period of time. If the number of allowances can be changed by unpredictable regulatory fiat, then the market value of the allowances is undermined and the incentive to invest in pollution prevention in order to realize the return for selling or renting allowances is reduced.

One program characteristic that should be avoided is the retirement of a source's allowances whenever it undertakes significant control activity. This kind of regulatory response will significantly discourage long-term pollution prevention investment. In order to have allowances to sell, sources must in effect over-control relative to their total allowances. When they do so, however, they face a regulatory risk under the CWA. Over-compliance signals to the regulator that greater levels of control can be achieved. If the logic of the NPDES process, a process focused on maximizing controls at every source, is followed this can lead to a recall of allowances that formed the basis for an initial allocation. Regulators might be compelled to act in such a way to secure continuous pollutant reduction toward zero discharge (Steinzor, 2002). In such situations, the incentive to over-comply through innovation is thereby undermined (Stephenson, Shabman, & Geyer, 1999).

Unfortunately, given fundamental uncertainties about biological and physical stressors and processes, standards and caps may need periodic revision (National Research Council, 2001). How these revisions are managed will be a fundamental challenge to be faced by CAT programs and must be addressed in the design of such programs.

Finally issuing permits to groups of dischargers, rather than individual sources, can itself be a source of flexibility.[23] The group permit also fosters new organizational arrangements among groups of dischargers that can help deal with a potential weakness of CAT markets: namely, their "thinness." The larger the number of trading partners, the more reliable are price signals. When markets are illiquid (thin) prices may fail to reliably transmit signals of the real value of control activity. Institutional arrangements – such as associations, cooperatives, or group compliance permitting – can foster coordination and information exchange between participants that trade infrequently or that are few in number (Shabman et al., 2002).

Group permitting also addresses another important design consideration – minimizing transactions costs among allowance traders. If allowance trades require complex individualized legal contracts and case-specific reviews, as is the case in offset programs, the CAT program will be severely hampered. The group permitting process allows individual trades without trade-specific approval.

5. CONCLUSION

Much of what is referred to as trading does not in fact involve decentralized, flexible reallocations of allowances among dischargers. Rather, the dominant form of U.S. water quality trading today is offset programs. Offset programs grant limited decision-making flexibility to regulators setting permit conditions. We have described some of the advantages of offset programs in this paper. However, real trading, in the form of CAT programs, promises a much greater up-side, though CAT programs will require significant innovation relative to current point source-focused permitting.

CAT programs address what we consider to be the ultimate goal: a system of regulation driven by the attainment of ambient water quality standards by means of an aggregate cap on releases throughout a watershed. U.S. water quality regulation is fitfully moving in this direction anyway, the TMDL approach being the best and most important example. Many of the demands of this kind of regulation complement the design of a real CAT approach to water quality trading.

CAT programs can reduce costs by allowing reallocation of control activities. Perhaps more important, however, is that CAT programs create an incentive to innovate. Pollution prevention innovations generate discharge reductions that can be sold or used to meet a facility's trading allowance. Innovation thus has real value to sources under a CAT scheme.

From an environmental perspective CAT programs raise a frightening spectre in many minds: namely, a shell game wherein regulated sources increase their well-monitored controls by buying reductions from sources whose behavior is relatively difficult to monitor.[24] Accordingly, a CAT program requires watershed assessment, monitoring, and enforcement tools that can simultaneously provide safeguards to ensure environmental quality and preserve the flexibility necessary for potential participants to engage in trade. A tall order, to be sure. Such a commitment will require significant investment by both the public and private sectors. However, and as we have argued, many of these monitoring challenges are associated with any watershed management program. Even under programs without trading there is a need to effectively assess the impact of spatially differentiated sources on water quality and to monitor and credibly deter nonpoint releases. Nonetheless, CAT programs should be expanded slowly and include only sources whose waste discharge behavior can be readily monitored.

Programs that focus on total watershed loadings and that seek to cap releases on a wider range of sources may be the future of water quality regulation. CAT programs are a natural complement to such programs and share many of the same challenges. Moreover, only CAT programs promise the decision-making flexibility necessary to stimulate significant innovations in pollution control activity. For this reason, CAT water regulation must be seriously explored. Without the innovations it promises, environmental goals will take longer to achieve and come at greater economic cost.

NOTES

1. Command-and-control is a general term for regulations where specific control activities, rather than performance goals, are mandated by regulation. Specific control activities can include mandatory technology installations or management and operations practices. See Section 2.1.
2. See Boyd (2003) for a description of the use of effluent fees as a water quality regulatory tool.
3. So-called nonpoint sources, which include most agricultural, commercial, residential, and construction-related sources are not regulated under the Clean Water Act.
4. EPA is required to devise technology-based performance standards for conventional pollutants (defined as biological oxygen demanding substances, pH, total suspended solids, and fecal coliform) and toxic pollutants. Nutrients, such as nitrogen and phosphorus, are not defined as either conventional or toxic pollutants (Stephenson et al., 1999).
5. In common parlance, pollution from nonpoint sources is "runoff caused primarily by rainfall around activities that employ or cause pollutants." *United States v. Earth Sciences, Inc.*, 599 F.2d 368, 373 (10th Cir. 1979). Over the years, sources that

at one time were considered nonpoint sources and therefore not regulated under the NPDES program have been brought under the NPDES or some other regulatory umbrella. For example, urban storm water and now some large agricultural operations, are now considered point sources.

6. This is an over-simplification. Most cap and trade programs include guidelines that limit certain types of reallocations, such as trades that result in undesirable geographic concentrations of emissions (e.g., "hot spots").

7. Trading schemes associated with "group permits" partially resolve this issue. Group permits are discussed in more detail below.

8. The offset is usually not one-to-one. Trading ratios are applied for a variety of reasons as discussed in more detail in Section 4.

9. This overstates the case somewhat. The CWA does not legally limit the EPA's authority to bring sources under the NPDES program, or another technology based control requirement. As noted earlier, pollutant sources such as storm water were initially not governed by NPDES permits, but now are.

10. According to one commentator on the movement toward trading, "we've gone through a tremendous convoluted thought process to try to figure out how to get point sources to pay for farmers' problems. We need to keep at the front of the stage that it is a farmer's problem, and we don't want to lose sight of the direct approach." (Fox-Wolf Basin 2000, 2000).

11. For example, Virginia is cited as having a trading program in place. While a form of trading program was considered at one point no such program is now operating.

12. Point sources finance nonpoint source controls if they exceed the cap, which has not happened. Even if it did, however, it would not trigger real point–nonpoint trading. Instead, financial penalties imposed for failure to meet the cap would simply be allocated to the North Carolina agricultural nonpoint source cost share program.

13. The authority is composed of two counties, four cities, and seven special water and sanitation districts.

14. "Due to the high level of phosphorus treatment already being achieved at Cherry Creek publicly owned treatment works and the incredibly high costs that would be associated with achieving additional reductions, the authority determined that no viable non-trading phosphorus reduction options existed for these facilities ..." Paulson et al. (2000, pp. 3–4).

15. In the mid-1980s phosphorus loads were allocated across a range of sources to satisfy state water quality standards as part of a TMDL plan.

16. To quote one analysis of the program's characteristics "In Cherry Creek, the decision regarding pricing and cost recovery was decided largely on equity issues... The option of pricing credits at $ 0 came up in some discussions. Equity issues arise, however, if some authority members need and draw/purchase credits disproportionately to their financial contribution to the authority." Paulson et al. (2000, pp. 4–7).

17. BMPs were analyzed in order to ensure that they were verifiable and at least offset, with conservativism, the load increase from the plant.

18. It deserves emphasis that offset programs can entail large transaction costs associated with the need to modify individual permits on a case-by-case basis.

19. Proposed Revisions to the Water Quality Planning and Management Regulation, 64 Fed. Reg. 64 46,012 (to be codified at 40 C.F.R. pt. 130), proposed Aug. 23 1999, at 46,013.

20. The transport, diffusion, and cross-pollutant interactions are idiosyncratic to location so that similar releases in different geographic and hydrologic settings in the same watershed can have different environmental consequences. This has long been recognized. For example, Kneese and Schultze (1975, p. 88) state "the impact on water quality from the wastes of any firm depends on the firm's location along the river basin and the hydrology of the stream, …."

21. Schnoor (1996, p. 2) notes that "regardless of how much monitoring data are available, it will always be desirable to have an estimate of chemical concentrations under different conditions, results for a future waste loading scenario, or estimates at an alternate site where field data do not exist. For all these reasons we need chemical fate and transport models, and we need models that are increasingly sophisticated in their chemistry, as we move toward site-specific water quality standards…."

22. Pollutants deposited on land, such as pesticides and fertilizers, are carried into surface waters by runoff. Runoff is determined by climatic conditions, in particular rainfall, and by the geology of the region, including soil types and elevation gradients. Some of the pollution will degrade naturally, some will be deposited in soils and groundwater, and some will impair surface waters.

23. Note that group permits are a feature of the Tar-Pamlico and Grasslands programs.

24. These fears are not unfounded. Offset programs under the Clean Air Act, so-called Open Market Programs, have been roundly criticized for a lack of enforcement and oversight activity on the part of regulators (U.S. EPA, 2002).

REFERENCES

Austin, S. (2001). Designing a selenium load trading program to reduce the water quality impacts of discharge from irrigated agriculture. *Harvard Environmental Law Review, 25*(2), 337–403.

Boyd, J. (2000). The new face of the clean water act: A critical review of EPA's Proposed TMDL rules. *The Duke Environmental Law and Policy Forum, 11*(1), 39–87.

Boyd, J. (2003). Water pollution taxes: A good idea doomed to failure? *Public Finance and Management, 3*, 34–66.

Bartfeld, E. (1993). Point–nonpoint source trading: Looking beyond potential cost savings. *Environmental Law, 23*, 43–106.

Davies, T. (2001). *Reforming Permitting*. Resources for the Future Report, Washington DC.

Environomics. (1999). A Summary of U.S. Effluent Trading and Offset Projects. Draft consultant report, Bethesda Maryland.

Environmental Law Institute. (1998). Barriers to Environmental Technology Innovation and Use. ELI Project #960800, Washington.

Fox-Wolf Basin 2000. (2000). Issues forum panel probes tensions, challenges that trading faces. *Basin Link, 2/2000*, 9.

Goldstein, A. L., & Ritter, G. L. (1995). A performance-based nonpoint source regulatory program for phosphorus control in Florida. In: K. Steele (Ed.), *Animal Waste and the Land-Water Interface* (pp. 427–440). Boca Raton: Lewis Publishers.

Grumbles, B. (2002). Statement before the Subcommittee on Water Resources and Environment. Committee on Transportation and Infrastructure, U.S. House of Representatives, June 13.

Hall, J. C., & Howett, C. M. (1995). Guide to establishing a point-nonpoint source pollution reduction trading system or basin-wide water quality management: The Tar Pamlico River Basin Experience. North Carolina Department of Environment and Natural Resources, July 1995.

Houck, O. A. (1999). *The clean water act TMDL program: Law, policy, and implementation.* Washington: Environmental Law Institute.

Kneese, A., & Schultze, C. (1975). *Pollution, prices, and public policy.* Washington, DC: Brookings.

Minnesota Pollution Control Agency. (1997a). Rahr Malting Company Trading Permit Fact Sheet. March 1997.

Minnesota Pollution Control Agency. (1997b). National Pollutant Discharge Elimination System and State Disposal System Permit MN 0031917. Rahr Malting Company, Minnesota.

National Research Council (2001). *Assessing the TMDL approach to water quality management.* Washington, DC: National Academy Press.

North Carolina Department of Environment and Natural Resources (NCDENR). (2001). Frequently Asked Questions about the Tar-Pamlico Nutrient Trading Program. Accessed online at: http://h2o.enr.state.nc.us/nps/FAQs9-01prn.pdf

Paulson, C., Vlier, J., Fowler, A., Sandquist, R., & Bacon, E. (2000). Phosphorus credit trading in the Cherry creek basin: An innovative approach to achieving water quality benefits. Water Environment Research Foundation, Project 97-IRM-5A.

Randall, A., & Taylor, M. A. (2000). Incentive-based solutions to agricultural environmental problems: Recent developments in theory and practice. *Journal of Agricultural and Applied Economics*, *32*, 221–234.

Schnoor, J. (1996). *Environmental modeling: Fate and transport of pollutants in water, air, and soil.* New York, NY: Wiley.

Shabman, L., Stephenson, K., & Shobe, W. (2002). Trading programs for environmental management: Reflections on the air and water experiences. *Environmental Review*, *4*, 153–162.

Speir, C., Stephenson, K., & Shabman, L. (2000). Command-and-Control and Effluent Allowance Markets: Roles of Economic Analysts. Paper presented at the American Agricultural Economics Association meetings, Tampa, Florida. http://agecon.lib.umn.edu/cgi-bin/pdf_view.pl?paperid = 2080&ftype = .pdf

Steinzor, R. (2002). Toward better bubbles and future lives: A progressive response to the conservative agenda for reforming environmental law. *Environmental Law Reporter*, *32*, 11421.

Stephenson, K., Shabman, L., & Boyd, J. (2005). Taxonomy of trading programs: Concepts and applications to TMDLs. In: T. Younos (Ed.), *TMDLs: Approaches and Challenges.* PennPress.

Stephenson, K., Shabman, L., & Geyer, L. L. (1999). Watershed-based effluent allowance trading: Identifying the statutory and regulatory barriers to implementation. *The Environmental Lawyer*, *5*, 775–815.

Studders, K. (2000). Statement to the U.S. House Committee on Government Reform. Subcommittee on National Economic Growth, Natural Resources and Regulatory Affairs, September 13.

Swift, B. (2001). The forum: Need economic-environment alignment. *The Environmental Forum*, *18*, 50–51.

U.S. EPA. (1996a). Draft Framework for Watershed-Based Trading. Office of Water, EPA 800-R-96-001.

U.S. EPA. (1996b). The Quality of Our Nation's Water. Overview of States §305(B) Water Quality Reporting For The Year 1996.

U.S. EPA. (2001). Grassland Bypass Project: Economic Incentives Program Helps to Improve Water Quality. Accessed online at http://www.epa.gov/owo/nps/Section319III/CA.htm

U.S. EPA. (2002). Open Market Trading Program for Air Emissions Needs Strengthening. Office of Inspector General, 2002-P-00019, 2002.

U.S. EPA. (2003a). Water Quality Trading Policy. Office of Water, Washington, DC, released January 13.

U.S. EPA. (2003b). EPA Releases Innovative Approach to Cleaner Water: 11 Pilots Receiving More than $ 800,000 in Funding. Office of Water, Washington, DC, accessed online at http://www.epa.gov/epahome/headline_011303.htm, released January 13.

Woodward, R. (2003). Lessons about effluent trading from a single trade. *Review of Agricultural Economics*, *25*, 235–245.

Woodward, R., Kaiser, R., & Wicks, A.-M. (2002). The structure and practice of water quality trading markets. *Journal of the American Water Resources Association*, *38*(4), 967–980.

www.ingramcontent.com/pod-product-compliance
Ingram Content Group UK Ltd.
Pitfield, Milton Keynes, MK11 3LW, UK
UKHW020846290926
15470UKWH00004B/1575

9 780762 314485